ENCYCLOPEDIA OF MAMMALS

Maurice and Robert Burton

Introduction by

Dr L. Harrison Matthews
FRS

in association with
Phoebus

Introduction

Mammals are generally regarded as the highest animals in the scale of evolution, perhaps partly because man is himself a mammal. But we may not be right in thinking of animals as 'high' or 'low', because all seem to be equally well equipped for their various ways of living. It may be better not to try making an exact definition of the relative 'status' of the different animals because, if, as is now generally agreed, all living things have arisen from a common source of life, all are collateral descendants. They are the topmost twigs of the same family tree, all are level, so that none is higher than its neighbours. We do, however, illogically say that we are the top-dogs in the world of living things, we are undoubtedly mammals, and therefore the mammals are the highest animals.

The mammals are said to be more intelligent than other animals, so that they are less dependent upon fixed patterns of instinctive or 'innate' behaviour. They certainly have relatively large brains, though none of them can think in the way that we do because they do not possess language. But many of them can alter their behaviour to match changing circumstances in a way that animals governed almost entirely by instinct cannot. Although much of the behaviour even of mammals is innate, we can agree that on the whole the mammals show greater intelligence than the others.

Although there are only about 4 000 different kinds of mammals now living – there are some 8 500 birds and 23 000 fishes – they form a diverse group of animals. In addition to the familiar land mammals they include the bats and the whales, which are so different from the others that simple people, even today, think they are birds and fishes. There are very few characters common to all mammals by which the class can be defined.

Mammals take their name from their most obvious character, which they share with no other animals. The females suckle their young so that they are fed for some time after birth on the milk secreted from the mammary glands of the mother, *mamma* being the Latin for a breast – and one of the first words uttered by the human suckling.

The other obvious character of mammals is the covering of hair or fur on the surface of the body. Most mammals are nearly completely clothed in hair, and even the almost naked whales have a few hairs on the snout and chin at least while they are young. No other animals have hair of the same type; the hair-like filaments that you see on the body, of a plucked chicken, for example, are not true hairs but modified feathers. The pterodactyls, the flying reptiles of long ago, also had a covering of fur or hair of some kind, for they were warm-blooded like birds and mammals, and had to keep their heat in.

All the other features that distinguish mammals from the rest of the animal kingdom are found in the details of their internal anatomy, such as the structure of the jaw and middle ear, the presence of a midriff, the arrangement of the main artery leaving the heart and the large size of the fore-brain

First published 1975 by
Octopus Books Limited
59 Grosvenor Street, London W1

Reprinted 1979

ISBN 0 7064 0392 4

Produced by Mandarin Publishers Limited
22A Westlands Road
Quarry Bay, Hong Kong

Printed in Hong Kong

This book is adapted from 'Purnell's Encyclopedia of Animal Life', published in the United States under the title of 'International Wild Life'.
It has been produced by Phoebus Publishing Company in cooperation with Octopus Books Limited.

and the complexity of the nerve-paths in it.

Most of the mammals form a group called the Eutheria or placental mammals, in which the young are carried in the body of the mother until sufficiently developed to be born. The Metatheria, the marsupials or pouched mammals, are a smaller group found only in Australasia and South America, in which the young are little more than embryos when they leave the mother's body so that they have to be put back into it, in the pouch, to finish their development. A third, small group, the Prototheria, confined to Australasia, contains the duck-bill or platypus and the spiny anteaters or echidnas, the only mammals that lay eggs. Although the three groups are descended from common ancestors, the placentals, the 'highest' mammals, are not thought to be descended from either of the other groups.

Mammals of one kind or another are able to make a living in practically every part of the world. Land mammals inhabit all the continents except the Antarctic, roaming the surface, burrowing underground, flying in the air, or swimming in the rivers and lakes; and marine mammals, the seals, whales and dolphins, live in all the oceans from the tropics to the polar seas. They have made good use of every possible habitat, and exploited every kind of food.

Most of the mammals are specialists in their diet, their teeth and digestions being adapted for dealing with one kind of food only. But some are omnivorous and can eat practically anything. Man himself is the most versatile of the omnivorous mammals, so that there is hardly anything not actually poisonous that he has not at some time, somewhere, included in his food. His wide appetite has led to the invention of agriculture for the production of plant products, and of animal husbandry to give him meat, milk, butter and cheese for his table, and many products for other uses.

Primitive man, ancient and modern, has always been a hunter of animals and a gatherer of plant foods. The hunter naturally prefers a large to a small prey, and consequently the larger mammals have been specially attacked.

When human populations were small the hunting of wild animals for food did no harm to the mammal populations because there is always a 'sustainable yield' that can be taken without damaging the stocks. During the last few hundred years, however, the human population of the world has been increasing with ever greater speed so that the numbers of many kinds of mammals, and even their very existence, are threatened.

This has come about in two ways. In the first man has over-cropped his prey, taking more than the sustainable yield, so that stocks dwindled until they got so low that it no longer paid to hunt them. We have all heard how the great Antarctic whaling industry ruined itself by killing too many whales, so that some kinds that were once abundant are now rare.

The other way in which man has damaged the fauna of wild mammals throughout the world is due to his population 'explosion'–he is taking over the wild places of the earth for his own occupation and leaving no room for wild animals. This is even more destructive than direct hunting because it alters the environment so much that the animals can no longer make a living–the 'ecosystem' becomes unbalanced or destroyed.

In Europe such changes happened so long ago that comparatively few of us regret the near or final extermination of such creatures as the wild ox, bison, wolves or bears. In Asia, much of which has been settled by man since earliest times, most of the mammals have been able to exist alongside man. Even in densely populated countries like India tigers were until recently plentiful–and ate over a thousand men every year. But with the political changes of the last 30 years, the growing human numbers, land development and the use of modern firearms, the tiger, like many others, is rapidly dwindling in numbers. It was in Asia that an enclosed park first preserved a large mammal from extinction: the Imperial hunting park near Peking held the only living herd of the peculiar Père David deer, an animal extinct in the wild for centuries.

In the lands settled by Europeans in colonial times the position is very different. The early settlers in Africa, America, and Australia found huge populations of indigenous mammals which they exploited for food and other useful products. Small populations of men exploiting large populations of mammals at first seem unable to reduce the stocks seriously, but when the human population grows the damage soon becomes obvious. In addition the frontiersmen are always pressing on to bring more land into cultivation, and kill off wild mammals to make way for crops and domestic animals.

In North America the destruction of the fauna was carried out from earliest times until towards the end of the last century when public opinion began to regret the loss of the spectacular herds of bison and other mammals. The desire to halt the destruction led to what is now called 'conservation', with the founding of the Yellowstone National Park in 1872. Thanks to conservation the bison and the pronghorn 'antelope' are now numerous, though the great herds of former days cannot be restored because they have lost so much of their territory to man. Over most of settled North America, too, the brown bear and its huge variety, the grizzly, have been exterminated because large bears and human settlement just cannot exist together.

Although the fur-bearers have been trapped for centuries none of them except the giant sea mink have become extinct; the sea otter came near it until hunting was stopped, after which it increased greatly. Wolves have been killed off in the developed parts of the country, but the coyote remains plentiful in spite of shooting, poisoning, and trapping–it seems to have come to terms with man and to be able to co-exist as does the fox in Europe.

In Africa the enormous numbers of mammals that once swarmed nearly everywhere dwindled away under the influence of man. The destruction started in northern Africa in ancient times, and in southern Africa in modern ones. The last stronghold of the great game mammals is in eastern Africa where they are increasingly being conserved in national parks.

Australia was the last continent to be colonised, and although the time has been comparatively short the fauna has suffered as much or even more than those elsewhere. The wonderful array of different kinds of marsupial or pouched mammals was looked upon as fair game by the early settlers, to be hunted to extermination for meat and furs, or to make way for farming. The destruction went on until well into the 20th century, and to some degree continues, as in the killing of kangaroos to make canned pet food. Another thing that devastated the native mammals was the introduction of placental mammals.

Everyone knows of the vast economic damage done by the introduced European rabbit, but introduced foxes, dogs, cats, horses and camels have all 'gone wild' to the great damage of the environment and the native mammals. The placentals have a more elaborate brain structure than the marsupials, and when they come into competition with them the marsupials lose every time. A few have been able to adapt and co-exist with man, notably the brush-tailed possum.

The Australians for long seemed to be uninterested in their marvellous fauna of marsupial mammals and to care nothing that it was often being wantonly exterminated, but a great change in public opinion has taken place in the last 30 years. They seem suddenly to have woken up to the value, interest and beauty of their mammals. Their scientists are studying them, and they have made laws to protect them and to stop them being exported. The future for the Australian marsupials now looks brighter than it has for over a century.

This book illustrates and describes a selection of the world's mammals, showing their diversity in form and behaviour, and where they are to be found. Apart from the scientific and general natural-history interest, most of the mammals have a strong appeal to our appreciation of beauty in the animal kingdom, so that we enjoy the pleasure of seeing them in their natural setting. They have for centuries been exploited and harrassed by man, but the modern wish to conserve rather than to destroy may save them yet. Long may they flourish!

L. Harrison Matthews
M.A., Sc.D., F.R.S.

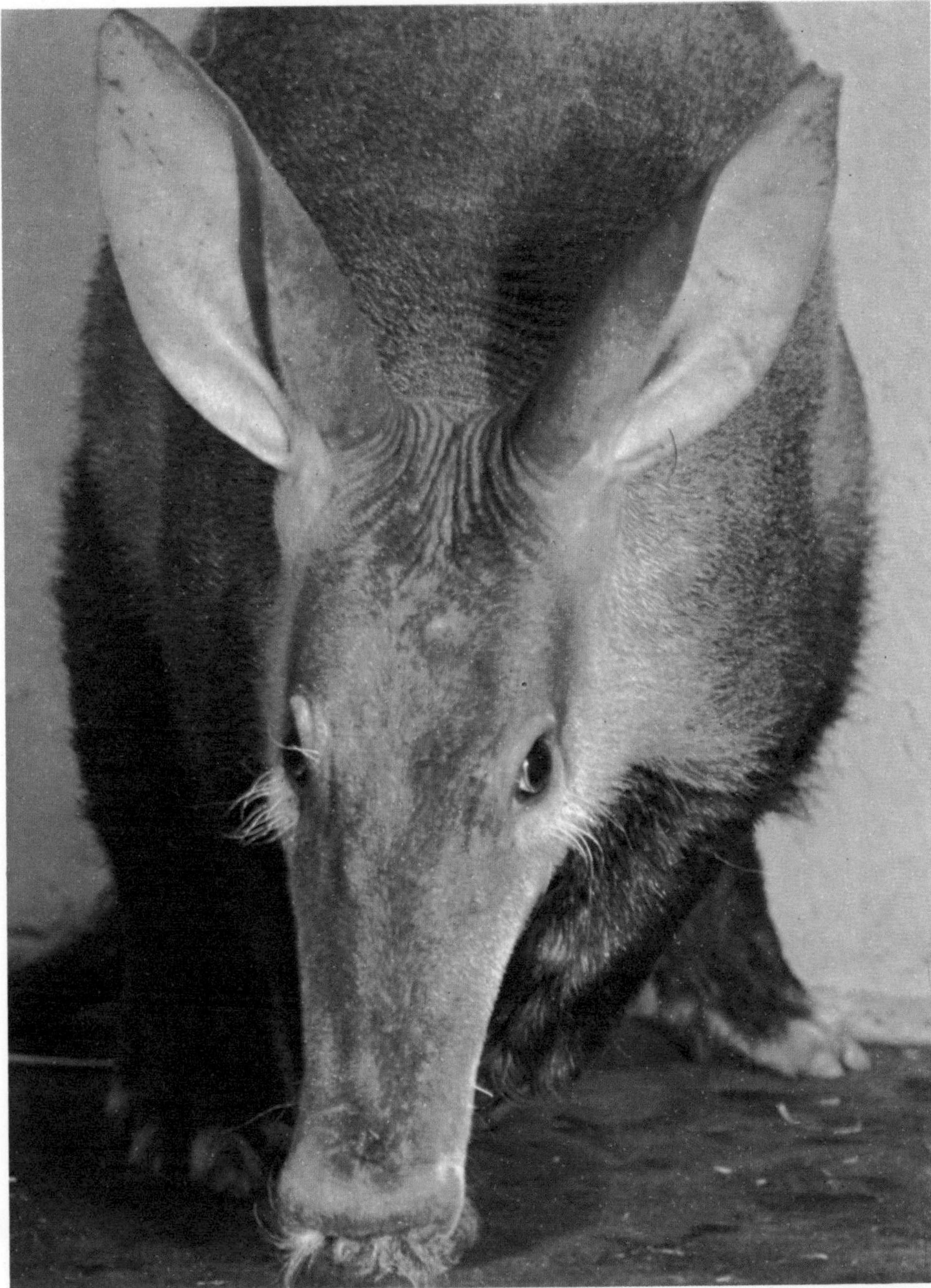

△ *The aardvark's nose is guarded by a fringe of bristles and it can also close its nostrils, as a protection against termites.*

▽ *Aardvark at home in African scrub close to a termites' nest where it has been feeding on these soft-bodied insects.*

Aardvark

African mammal with a bulky body, 6 ft long including a 2 ft tail, and standing 2 ft high at the shoulder. Its tough grey skin is so sparsely covered with hair that it often appears naked except for areas on the legs and hind quarters. The head is long and narrow, the ears donkey-like; the snout bears a round pig-like muzzle and a small mouth. The tail tapers from a broad root. The feet have very strong claws— four on the front feet and five on the hind feet. The name is the Afrikaans for 'earth-pig'.

Distribution and habits

The aardvark has powerful limbs and sharp claws so it can burrow into earth at high speed. This it does if disturbed away from its accustomed burrow. There are records of it digging faster than a team of men with spades. When digging, an aardvark rests on its hind legs and tail and pushes the soil back under its body with its powerful fore feet, dispersing it with the hind legs.

The normal burrow, usually occupied by a lone aardvark, is 3–4 yd long, with a sleeping chamber at the end, big enough to allow the animal to turn round. Each animal has several burrows, some of them miles apart. Abandoned ones may be taken over by warthogs and other creatures.

Years can be spent in Africa without seeing an aardvark, although it is found throughout Africa south of the Sahara, except in dense forest. Little is known of its habits as it is nocturnal and secretive, though it may go long distances for food, unlike other burrowing animals.

Termite feeder

The aardvark's principal food is termites. With its powerful claws it can rip through the wall of termite nests that are difficult for a man to break down even with a pick.

Its method is to tear a small hole in the wall with its claws; at this disturbance the termites swarm, and the aardvark then inserts its slender 18 in. tongue into the hole and picks the insects out. It is protected from their attacks by very tough skin and the ability to close its nostrils–further guarded by a palisade of stiff bristles.

As well as tearing open nests, the aardvark will seek out termites in rotten wood or while they are on the march. It also eats other soft-bodied insects and some fruit, but–unlike the somewhat similar pangolin (see page 168), which has a muscular, gizzard-like stomach filled with grit for crushing hard-bodied insects–it cannot deal with true ants.

Breeding cycle

The single young (twins happen occasionally) is born in midsummer in its mother's burrow, emerging after two weeks to accompany her on feeding trips. For the next few months it moves with her from burrow to burrow, and at six months is able to dig its own.

Digs to escape enemies

The aardvark's main enemies are man, hunting dogs, pythons, lions, cheetahs and leopards, and also the honey badger or ratel, while warthogs will eat the young. When suspicious it sits up kangaroo-like on its hind quarters, supported by its tail, the better to detect danger. If the danger is imminent it runs to its burrow or digs a new one; if cornered, it fights back by striking with the tail or feet, even rolling on its back to strike with all four feet together.

On one occasion, when an aardvark had been killed by a lion, the ground was torn up in all directions, suggesting that the termite-eater had given the carnivore a tough struggle for its meal. However, flight and—above all—superb digging ability are the aardvark's first lines of defence for, as with other animals with acute senses like moles and shrews, even a moderate blow on the head is fatal.

The last of its line

One of the most remarkable things about the aardvark is the difficulty zoologists have had in finding it a place in the scientific classification of animals. At first it was placed in the order Edentata (the toothless ones) along with the armadillos and sloths, simply because of its lack of front teeth (incisors and canines). Now it is placed by itself in the order Tubulidentata (the tube-toothed) so called because of the fine tubes radiating through each tooth. These teeth are in themselves very remarkable, for they have no roots or enamel.

So the aardvark is out on an evolutionary limb, a species all on its own with no close living relatives. Or perhaps we should say rather that it is on an evolutionary dead stump, the last of its line.

What is more, although fossil aardvarks have been found—but very few of them—in North America, Asia, Europe and Africa, they give us no real clue to the aardvark's ancestry or its connections with other animals.

class	**Mammalia**
order	**Tubulidentata** *sole representative*
family	**Orycteropidae**
genus & species	***Orycteropus afer***

◁ *The claws of the aardvark are so powerful that it can easily rip through the wall of a termite nest which is so hard it is difficult for a man to break down even with a pick-axe.*

The termites are so disturbed by having their nest opened that they swarm about and the aardvark then puts its pig-like muzzle into the nest to eat them.

It has an 18 in. long, slender, sticky tongue with which it captures and eats the swarming termites that make up the main food of aardvarks.

▷ *A day-old aardvark. It depends on its mother for six months until it can dig its own burrow. The aardvark's snout and round, pig-like muzzle earn it the Afrikaans name for 'earth-pig'.*

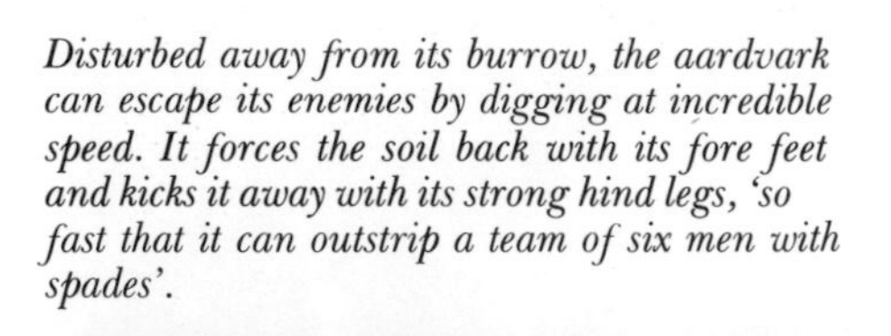

Disturbed away from its burrow, the aardvark can escape its enemies by digging at incredible speed. It forces the soil back with its fore feet and kicks it away with its strong hind legs, 'so fast that it can outstrip a team of six men with spades'.

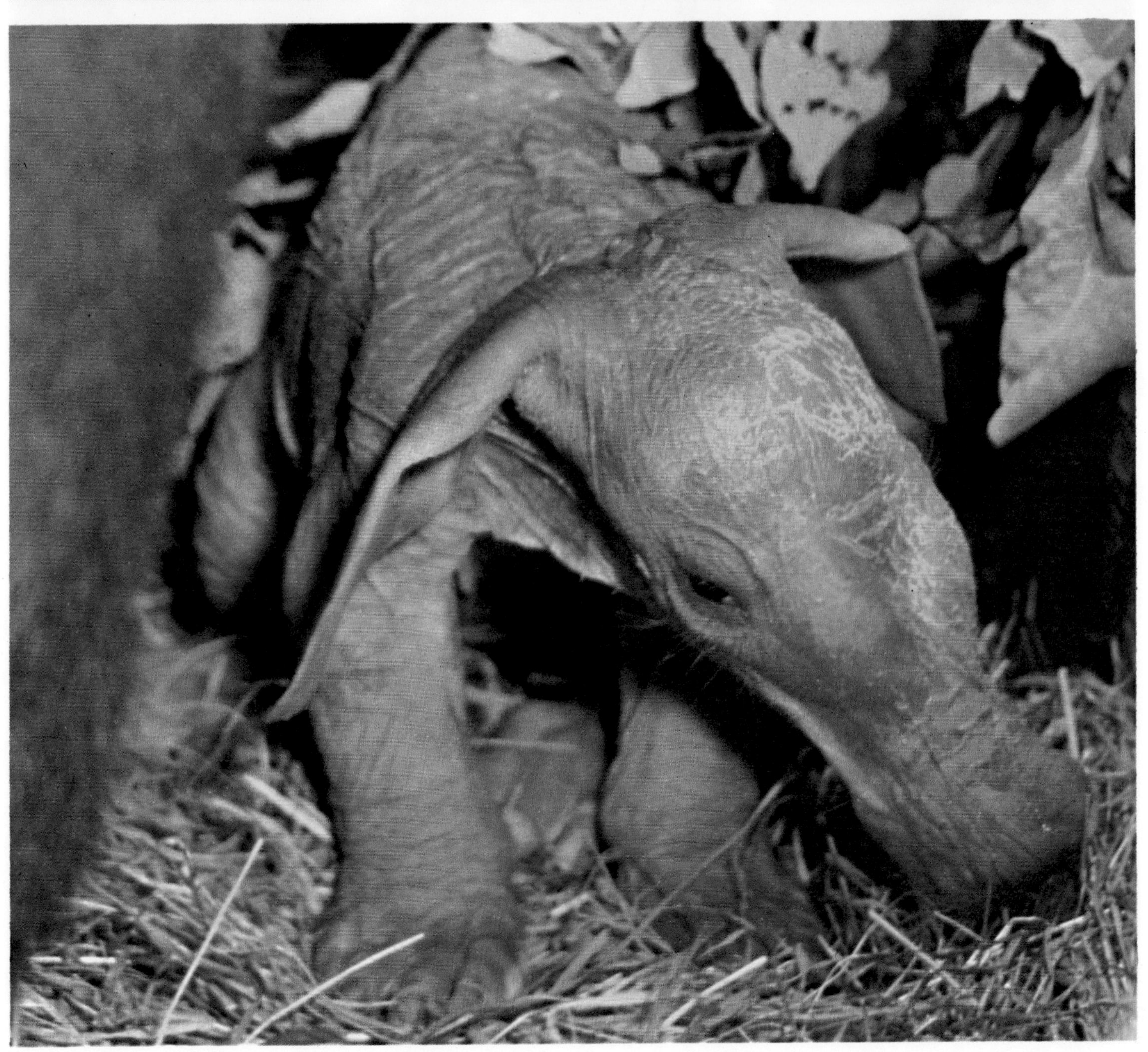

Badger

The one European species of badger ranges right across Europe and Asia, from Ireland to China, yet it is one of the most elusive of mammals. It is nocturnal and generally so wary that it is rarely seen, even by many who set out with the intention of watching it.

The badger is a bear-like animal with a stocky 3ft body, short tail and short but powerful legs armed with strong claws on the front feet. It also walks on the soles of its feet like a bear, but the resemblance ends there. Both bear and badger are members of the Carnivora, or flesh-eaters, but they are in different families. The badger is placed in the family Mustelidae, along with the otter, stoat and weasel. A character of the mustelids is that their footprints show five toes. In this way the footprints of a badger can be distinguished from those of a dog, which show only four toes.

At a distance, the badger's coat looks grey but the individual hairs are black and white. Most animals are lighter in colour on the underside of the body, but the badger has black on its belly and legs. The most striking part of the badger, however, is its head. This is white with two broad, black stripes running from behind the ears almost to the tip of the muzzle. The small eyes are placed in the black stripes and so are inconspicuous.

The American badger is widespread in North America from south-western Canada, south to central Mexico. It is a stocky animal with a 30 in. flat and wide body, short legs, and a stump of a tail. Its weight may be up to 25 lb. It is very similar in appearance to the European badger except in colouring. It has a dark-brown face, with white cheeks, a black spot in front of each ear, and a narrow white stripe running from its nose along the back of its neck. During the winter the American badger lives in a state of semihibernation in its sett, which is lined with dry leaves and grass. It comes out from hibernation from time to time in order to find food.

In China there is another animal resembling the badger. This is the hog badger which has very similar habits, but can easily be distinguished from the European badger by the naked pig-like snout, from which it got its name, and a much longer tail.

Nocturnal sett dwellers

The badger must have been well established in European folklore, as it has given its name, or local variants of it, to many towns and villages. For example, the old English name was brock, meaning particoloured, and there are in 'badger country' such places as Brockenhurst, Brockhampton, and Brocklesby, and in Germany, where the name for badger is 'dachs', we have such place names as Dachsbach, Dachsberg and Dachsfelden.

Badgers can be found throughout Britain, but are rarest in flat areas such as East Anglia. They often live surprisingly near to the centre of large cities and are found in practically all European and Asian countries, from just south of the Arctic Circle to the Mediterranean and the Himalayas. In the northern parts of their range they will hibernate, but as far south as Britain they are active all winter.

Badgers are so rarely seen because of their nocturnal habits. Many supposedly nocturnal animals are often active during the day as well, but it is extremely rare for a badger to be seen about during daylight hours. Badgers emerge from their setts during the long, dark nights of autumn and winter, regularly one hour after sunset. The short nights of the summer months result in an earlier emergence.

If there is any disturbance, suspicious

◁ *American badger has a dark head with a white line on forehead and nape. It is known to US foresters as a silver tip after its fur. This usually inoffensive animal can put up a very tough fight as its defensive posture with hair on end and wicked-looking teeth implies.*

▷ *Badgers are known to follow the same well-trodden tracks on their nocturnal food-searching rambles. Because of this habit small gates can be built in wire fences, constructed to keep out rabbits and other pests from farmlands and forests, but allowing the beneficial badgers to come and go as they wish.*

▽ *The fantastically strong claws on the front feet are used for digging the badger's home or sett.*

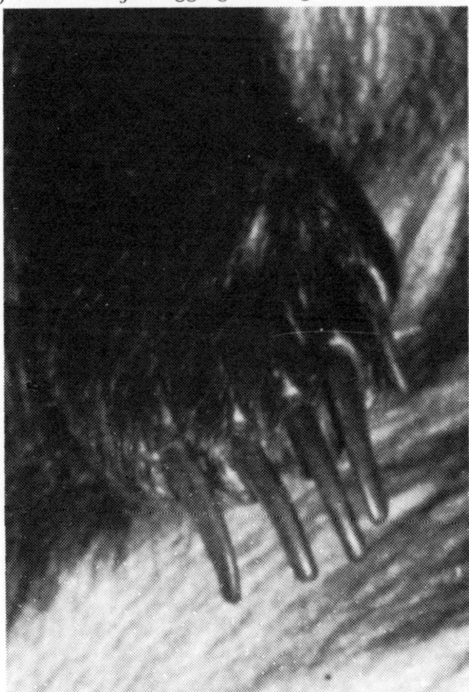

sound or scent, the badger may remain underground for the night, and they will often fail to come out on bright moonlit nights.

The setts, as badgerholes are called, are easily distinguished from the dwellings of foxes or rabbits by their large size and the mass of earth and stones that lies at the entrance. Confirmation that there are badgers in residence must, however, be made because foxes and rabbits will take up residence in them, sometimes when the badgers are still in occupation.

Signs of the badgers' presence are unmistakable. It has regular, well-trodden paths leading from the sett which may be followed for some distance, often running to a stream or pond where the badgers habitually drink. Where they pass under barbed wire or brambles, wisps of distinctive black and white hairs can be found stuck in the barbs or thorns. Near the entrance to the sett there are more definite signs of activity. Scratching posts indicate where a badger has stood on its hind legs and scratched the trunk of a tree with its forepaws. Around the mouth of the hole and along the paths leading to it fresh vegetation will be strewn—bracken, bluebells and other plants that have been collected for bedding. The badger gathers these in its fore legs and shuffles backwards leaving a trail of plants.

Badgers have a reputation for being especially clean animals, largely because they frequently change their bedding and also dig latrines, shallow pits that can be found within 20 yd or so of the sett.

The favourite sites for setts, which may be vast underground systems with many entrances, are in woodland, preferably those bordered by pastures. Sandy soil seems to be preferred.

Cubs born underground

The male, or boar, and the female, or sow, badgers probably pair for life. Mating takes place in July or August but the embryos do not begin to develop until December or January. This is another example of delayed implantation. That is, the fertilised egg undergoes its first few divisions, after which growth is arrested for some time. The young, one to five in number, are born during February or March. At birth they measure no more than 5 in., of which $1\frac{1}{2}$ in. is tail.

For the first 6–8 weeks of their life they stay underground, then make their first, tentative visits to the outside world. The routine for leaving the sett is that the sow appears first, sniffing the air and behaving even more cautiously than usual. Then she turns back to the entrance and coaxes her cubs to come out behind her. At first they stay out for only a very short time, scampering back to safety at the slightest alarm. After a week or so they become bolder and start exploring the neighbourhood and playing rough and tumble games. Later they are taken out to learn to feed themselves and they eventually leave their parents in October.

Main food is earthworms

Badgers belong to the Carnivora or flesh-eaters, and an examination of a badger's skull suggests an animal equipped for attacking and consuming large prey. The teeth are strong and there are long ridges around the hinges of the jaws that prevent them from being dislocated. Yet the badger lives on a wide variety of soft food. Earthworms are a major item, as were young rabbits before the epidemic rabbit disease, myxomatosis, made them rare. Mice, voles, moles, frogs, snails, beetles and even hedgehogs and wasps make up the animal content of the diet, while windfall apples, bulbs, acorns, blackberries and grass are also eaten. Fungi are sampled, and cereal crops suffer when badgers flatten large areas to get the ears. Poultry killing occurs occasionally but is not typical; the majority never touch them. The reason is usually scarcity of normal food or a single animal that has acquired the habit. Badgers have sometimes been found in hen houses, with none of the inhabitants disturbed. The

American badger can burrow rapidly, digging out the rabbits and rodents on which it feeds.

The proportions of the different items taken varies with season and weather. On wet nights, for instance, badgers will go to pastureland to feed on earthworms, and suckling sows normally eat little else but earthworms.

Enemies

Badgers have little to fear, except from man, who in past times has trapped them to provide sport by badger-baiting with dogs, or, at all times, because they have killed his poultry. Today they probably suffer most in Britain from rabbit clearance societies, who regularly gas rabbit warrens. Although gassing badgers is illegal, rabbits in a sett are sufficient excuse.

Recognition marks

It is often said that the grey body and the conspicuous black and white stripes on the head serve as camouflage, simulating shafts of moonlight coming through the trees. Yet badgers avoid moonlight and when they do show themselves they are surprisingly conspicuous. White objects will stand out if there is any light at all.

Another theory is that the stripes are a warning coloration, a warning to other animals to keep away from the badger's bite. Certainly there is a tendency for animals with a powerful form of defence to have a striking colour pattern. For example, the skunk, with its powerfully deterrent odour, has a conspicuous band of white along the head and back and a bushy white tail. Similarly, bees and wasps have warning colours of black and yellow. Predatory animals learn to associate these with an unpleasant experience, and they are aware that they should leave well alone.

There is a drawback to this idea with badgers. Against whom is this warning to operate? It is difficult to find a serious enemy of the badger even by examining the animals that lived in Europe a thousand years ago. Bears normally do not prey on animals of this sort, wolves are mainly active by day and lions go for a different type of prey. A further difficulty is that other badgers, the American badger and the hog badger, for instance, do not have these facial markings, yet survive. The probability is that the white markings help badgers to recognise each other in the dark.

The badger's reaction to danger is quite spectacular. If startled, it emits a violent snorting, enough in itself to scare anything not expecting it. Then it literally bristles, the hair of the body standing on end so that it looks twice the normal size. Anyone who suddenly met a badger when it has 'blown up' will be only too aware of the shock-value it has.

△ Two young badgers at the mouth of their sett. This is a series of tunnels which may be quite deep and very ancient.

▽ Badger caught by photographer changing bedding. When bringing in bedding it shuffles backwards without looking where it is going.

class	**Mammalia**
order	**Carnivora**
family	**Mustelidae**
genera & species	***Meles meles*** *European badger* ***Arctonyx collaris*** *hog badger* ***Taxidea taxus*** *American badger*

Beaver

The beaver is the second largest rodent, exceeded in size only by the capybara. Stout-bodied, with a dark brown fur, it is up to $3\frac{1}{2}$ ft long including 1 ft of broad scaly tail, and it may weigh between 30 and 75 lb. Its muzzle is blunt, ears small, and it has five toes on each foot. Those on the front feet are strongly clawed, used for digging, manipulating food and carrying. The hind feet are webbed, with two split claws for grooming the fur and spreading waterproofing oil. The body oil, as well as the dense underfur and the heavy outer coat of guard hairs, not only act as waterproofing but also as insulation against the cold. When a beaver submerges, its nostrils and ears are closed by valves, and it can remain underwater for 15 minutes. The tail is used for steering and sometimes for propulsion through the water. It also forms a tripod with the hind legs when the beaver stands up to gnaw trees or when carrying, with the fore feet, mud or stones for building.

There are two species of beaver, both so alike in appearance and habits that we are fully justified in speaking merely of the beaver. The first, the European beaver, must at one time have been very abundant throughout Europe, even in England, where its bones may still be found. On the Continent of Europe it is still present in small numbers in Scandinavia, along rivers in European Russia, in the Elbe and Rhône valleys, and, where given protection, it shows signs of increasing numbers.

The Canadian beaver formerly enjoyed a wide range across the North American continent, from northern Canada south to beyond the US – Mexico border. Today, in severely depleted numbers, its range extends from Canada into some parts of the northern US.

Habits

Beavers live in loose colonies, each made up of a family unit of up to 12, including the parents, which mate for life. Their home may be in a burrow in a bank, with an underwater entrance or in a lodge in a 'beaver pond', a pool made by damming a river until it overflows. The lodge is built of sticks and mud, often against a clump of young trees, with underwater entrances, a central chamber which is above water level and a ventilating chimney connecting the chamber with the top of the lodge. Secondary dams are built upstream of the lodge, with usually one secondary dam downstream of the main dam. Young trees are felled, cut up and carried to the site, and, if necessary, canals are dug to float logs to the pond.

Intelligence of beavers

Many people are convinced that beavers are unusually intelligent, largely because their dams are such fine examples of engineering works. The structure of the

△ Beaver in the Canadian autumn. Its tail makes a tripod with the back legs so it can sit up and gnaw tree tunks.

▽ A beaver lodge. It is made of sticks and mud, in a pond, which is made by the beavers damming a stream so it overflows.

Beaver swimming. It can dive instantly if danger threatens, simply by depressing its rudder-like tail, here stretched out behind. When submerged its nostrils and ears are closed by valves and it can stay underwater for 15 minutes.

beaver brain, however, gives no indication of any greater mental capacity than is found in other rodents. Moreover, some of a beaver's actions which appear to be the result of a high order of reasoning can be shown to be due to instinct, the result of an inborn pattern of behaviour.

The lodge is a conical pile of branches and sticks 2–6 ft long, compacted with mud and stones, the upper half of which projects above the surface of the water. From an engineering standpoint it could hardly be improved. It has a central chamber just above water level, one or more escape tunnels leading from the chamber to below-water exits, well-insulated walls and a vertical chimney or ventilating shaft for regulating the temperature inside and to give air-conditioning. The evidence gained from dissecting a lodge suggests that it is built by laying sticks more or less horizontally to construct a pile, with an admixture of mud which stops short a foot or so from the top of the pile. Then the beavers chew their way in, to make the entrance tunnels and the central chamber. The absence of mud packing from the top of the pile means that spaces between the sticks serve for ventilation. In other words, there is no more intelligence required than any other rodent uses to dig in the ground.

The dams are text-book examples of engineering, and beaver dams give way no more frequently than do man-made dams. One reason for this is that a beaver dam is resilient, subject to immediate repair, is under constant surveillance and is supported by subsidiary dams. All the actions used in its construction can, however, be shown to be the result of a succession of instinctive actions, just as are those that result in the building of a bird's nest.

It is often said that beavers not only show skill in felling trees but use intelligence, by dropping the trees so that they fall towards the nearest water. This is not the case. Moreover, beavers are not uncommonly killed by the trees they fell falling on them.

Beavers often do other stupid things. The classic example is of a small lake in New York that was created by an artificial barrier of stones and cement litter. This was occupied by a family of beavers who were seen to 'repair' the dam with branches and mud although it was fully effective without these. Moreover, although the level of the pond was, so far as could be seen, satisfactory for their needs, the beavers built a subsidiary dam upstream of the pond, the only result of which was to flood the adjacent land to no purpose.

Everything considered, the achievements of beavers are more a tribute to the effectiveness of evolution in developing inherited behaviour patterns than a sign of unusual intelligence. There is, however, one qualification to be made. Young beavers stay with the parents for 2 years. During that time they must learn a great deal by following the example of the adults. It may be justifiable, therefore, to speak in terms of a cultural inheritance, and this alone would give an appearance of a greater intelligence.

Further evidence for the theory of innate behaviour can be added. Until a few years ago, beavers in the Rhone valley were hunted and they had long taken to burrowing in the river banks. Then they were protected by law, and shortly afterwards they began once more to build lodges and dams. A beaver may live for 20 years, but here they were reverting to a former pattern of behaviour after centuries of suppression due to persecution, using techniques they could not have learned. Only great intelligence, or a very strong instinct, could have brought about this behaviour.

Aspen and willow as food

Beavers eat bark, mainly of aspen and willow, from the smaller branches cut when building. Twigs and branches are stored around the base of the lodge. These have always been regarded as for the winter use of all members of the colony. Recent research has shown that the bulk of these are eaten by youngsters; older beavers live on their fat and eat little during winter.

Life history

Beavers, which are monogamous, mate in January to February. Gestation is 65–128 days and in April, May or early June two to eight kits, sometimes more, are born, with a coat of soft fur and eyes open. At birth each weighs about 1 lb and is 15 in. long including $3\frac{1}{2}$ in. of tail. At one month, each will find and eat solid food, but weaning is not complete until 6 weeks old. Young remain with parents for 2 years, becoming sexually mature at 2–3 years.

Enemies

As with all rodents, beavers are preyed upon by any carnivores of approximately their own weight or more. In this instance the enemies include wolverine, lynx, coyote, wolf, bobcat, puma and bear. A beaver's alarm signal when a predator is in sight is to bring the tail over the back then smack it down with such force on the water that the sound can be heard up to $\frac{1}{2}$ mile away.

Decline of the European beaver

Beavers were once common in Switzerland, as shown by the place names Biberach, Bibersee, Biberstein and Bibermukle (biber is German for beaver). Their extermination was due partly to their valuable fur, but more particularly to slaughter for their glandular secretion used to mark their territories. This, known as castoreum, enjoyed a vogue as a cure-all in the 16th and 17th centuries, with a resulting insensate slaughter of the luckless animals. Analysis has shown castoreum to contain salicylic acid, one of the ingredients of aspirin.

The former presence of beavers in the British Isles, too, is commemorated in many place-names, such as Beverley, Beverege, Bevercotes, Beverstone and Beversbrook in England, and Losleathan in Scotland and Llostlydan in Wales, both meaning broadtail. The animal seems to have still been plentiful in Britain up to the mid-16th century. The value of its fur can be gauged from prices fixed by the Welsh prince, Howel Dha, in the 10th century, at 120 pence a skin as compared with 24 pence for a marten pelt and 18 pence for otter, wolf and fox. This undoubtedly led to the animal's extermination, and in France to its almost complete elimination except in the Rhône Valley. In the 16th century, Henry IV of France, impressed by the demand for beaver pelts for hats, trimmings, fur linings and leather for shoes, sought to increase the economic strength of his country by sending men to Nova Scotia and Newfoundland. In due course the British gained this resource, largely through the Hudson Bay Company, and it was the search for more and more furs, particularly beaver, that led to Canada being opened up.

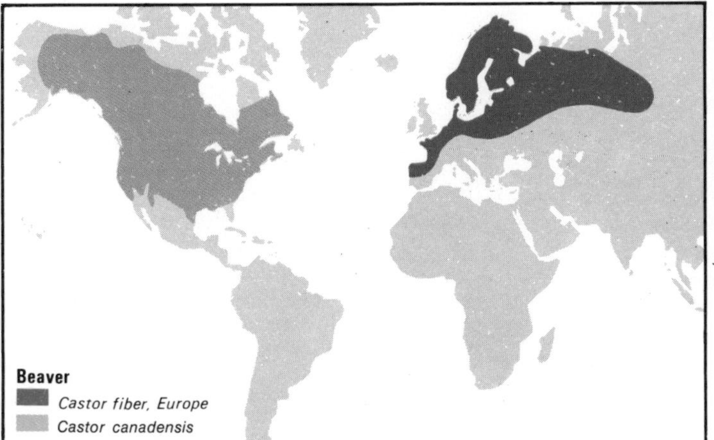

The beaver was on the way to extinction in Europe and America in the 19th century, but has since been rigorously protected and the populations are increasing. They have been so successful in some places that control is necessary.

Profit versus protection

In America, with the arrival of the early settlers, the beaver was recognised as a valuable source of both meat and fur. Trade was soon established with the Indians, who wisely killed only mature animals, so that their hunting could have done little to impair the number of beavers. The brisk fur trade with England that sprang up roused the old spirit of avarice, and before long white trappers joined their efforts with those of the Indians—and killed indiscriminately.

In about 150 years the beaver had been exterminated in the coastal regions of the Eastern States, and seriously reduced elsewhere. The story is, however, patchy, for in some spots their numbers remained relatively unimpaired. Also, beavers in deep rivers were less easy to catch than those in, say, mountain streams. As the North American continent was more and more opened up, so the trade continued unabated, with similar results to those seen in the Eastern States, but on a wider scale.

The Hudson Bay Trading Company was formed in 1670 and such was the growth of its beaver trade that between 1853 and 1877 it marketed nearly 3 million beaver pelts. This steady drain brought about a serious depletion which has been rectified to some extent by official conservation measures.

Beavers were not always killed for profit. At times the animal became a nuisance, either through its inroads on timber, in settled areas, or when it took a liking to the stalks of corn. In places, too, it became a menace to river banks. Nevertheless, it was early recognised that only harm could result from its total elimination. As early as 1866, it became protected by law in the State of Maine, with the result that by the early years of the present century it had increased so much in numbers that some control had to be imposed to protect plantations.

Since that time, both in the USA and in Canada, with increasing speed to the present day, there have been many efforts made at conservation, either by individual landowners, by public bodies, or by Government action, State or Federal. In some cases the reason behind it has been no more than a desire to preserve an interesting animal. In others it is due to a realisation that the work of beavers contributes to the conservation of water in the land and to the preservation of trout streams. It has been found that it is possible, by the intelligent use of closed seasons, limiting the numbers of pelts taken and having them taken only under licence, not only to bring about increased beaver populations locally, but to derive revenue from the surplus. Consequently, there has been considerable reintroductions to re-stock areas where they had been exterminated.

The conservation of water may be summed up in the following quotation from an American water company's report: 'On almost all the mountain streams they (the beaver) should be protected and encouraged. A series of beaver ponds and dams along the headwaters of a mountain stream would hold back large quantities of mountain water during the dangerous flood season and equalise the flow of the streams so that during the driest seasons the water supply would be greatly increased in the valleys. Beaver-ponds not only hold water but distribute it through the surrounding soil for long distances, acting as enormous sponges as well as reservoirs. A series of ponds also increases the fishing capacity and furnishes a safe retreat for the smaller trout and protection from their enemies.'

Fear and flight having failed to rescue this beaver, caught away from the relative safety of its pool, it must now turn and fight for its life against the hungry hunting coyote.

class	**Mammalia**
order	**Rodentia**
family	**Castoridae**
genus & species	***Castor fiber*** *European beaver* ***C. canadensis*** *Canadian beaver*

Bighorn sheep

There are three species of wild sheep ***Ovis canadensis*** *is the bighorn of western North America and north-eastern Siberia;* ***O. dalli*** *is the Dall sheep of north-western Canada and Alaska. Through the mountains of central and south-western Asia from Tibet to Asia Minor and on the Mediterranean islands are found many races of the wild sheep,* ***Ovis ammon*** *from which domestic sheep arose. These range from the gigantic Marco Polo sheep in the Pamir Mountains to the small mouflon of Corsica and Sardinia.*

None of the wild sheep has woolly coats comparable with the domestic sheep. Rather, they have coats of coarse hair ranging in colour from a creamy-white to brown. The coat of the wild sheep is, therefore, not unlike that of a goat. Sheep have, however, characteristically narrow noses with concave foreheads, compared with the convex foreheads of goats. Further differences are that sheep do not have the scent glands at the base of the tail nor a beard, although some male sheep have a fringe of long hair down the front of their necks. There are differences also in the horns and male bighorns have massive horns curving around the back of the head past the neck, and in old animals coming forward to the level of the eye. The females have only small, slightly curving horns a few inches long. The horns are divided into concentric rings. Less obvious growth rings subdivide these into irregular segments. Each segment represents a year's growth, making it easy to tell a bighorn's age.

Habits and distribution

Bighorn are found in many parts of the Rocky Mountains, in the states of New Mexico, Colorado, Nebraska and east to North Dakota. In Canada they occur in British Columbia, Alberta and Saskatchewan. They live in dry upland country and mountains above the treeline, reaching the most precipitous and inaccessible parts.

Outside the breeding season the sheep live in small flocks, of either rams or ewes.

Grass and bud eaters

The preferred food is sedges and grasses, with new aspen and spruce buds being taken in the spring, when bighorn will also dig up leguminous roots. In the autumn and winter, herbs, fungi, lichens and berries are eaten.

Bighorns feed most during the day, especially in winter, but in the summer they will lie up during the heat of the day, finding resting places that give a full view of the area around.

Formidable prey

Full-grown and healthy bighorn will have little to fear from anything except man, but lambs and sick sheep are liable to fall a prey to coyotes, pumas and eagles.

Jousting rams

During the summer the sexes segregate into separate flocks of up to 75, but more usually about a dozen individuals. The groups of ewes, with their young, are usually larger than the ram groups. From October onwards throughout the winter, mating takes place, and there is vigorous rivalry between the rams. During this season, mixed flocks are found, where the young rams have sought refuge among the ewes from the aggressive intolerance of the older rams.

Rams begin breeding when 4 years old. Their necks become thicker and during the mating season they lose much of their fear of man. During November and early December, jousting takes place, not only between rams but between ewes. If an encounter comes to blows the two may fight with their forelegs, kicking and pawing at each other, before indulging in the spectacular, bone-jarring tournaments. The two contestants turn away from each other and, having retreated some distance, turn back and charge. The last few yards may be covered on hindlegs alone, then just before the collision, heads are lowered and the full force of two 8/900lb bodies is taken on the foreheads. The impact is terrific and the contestants may be dazed, standing nose to nose, eyes glazed for half a minute. Then they carefully back away and repeat the course. These combats may go on for hours with no decisive result; the contestants merely wandering away from each other. Sometimes there are 'battles-royal', free-for-all fights with a dozen or so rams taking part. These may go on all day, with individuals retiring for a time while they recuperate.

These conflicts seem pointless, especially the 'battles-royal', and if fighting is an expression of competition between males, as is usually assumed, it is strange that ewes should also batter each other. The fights between male bighorns cannot be related directly to breeding because they do not form harems. There is no competition for females, who are promiscuous, mating with any male. It seems likely that the fighting is mainly a manifestation of aggressive behaviour and a general excitement and emotional stress that builds up during the breeding season.

The peak of mating behaviour is in the second half of November. The rams move from one flock of ewes to another, singling out ewes and chasing them. If a ewe is receptive she will only run a few yards before stopping and allowing a ram to mount her, otherwise a headlong chase takes place, with the ewe being chased around the hills by up to a dozen rams. Eventually she may find a crevice or rocky overhang in which to shelter and rest, but as soon as she comes out the pursuit starts again.

When a ewe is receptive, servicing takes only a few seconds and may be repeated by the same ram several times. Later the ewe will mate with other rams.

Ewes become mature at $2\frac{1}{2}$ years, bearing their first lamb when 3 years old, after a gestation of $6\frac{1}{2}$ months. Just before the lambs are due, in May, the ewes retreat to the remotest parts of the mountains. Each ewe finds her own place among rocky outcroppings, preferably with a warm southerly exposure, to have her lamb. Unlike the

▽ *Sheep crop short plants with the teeth in their lower jaw biting against the gum of the top jaw, which has no teeth at the front.*

△ Old rams have massive horns which curve round and come forward to the level of the eyes. Ewes have short, only slightly curving horns.

▽ Canadian bighorns live in the Rocky Mountains high above the treeline where it is often snowy. They can live off the scant mountainous vegetation, and can reach the most precipitous and inaccessible parts.

fawns of deer, the new-born lambs are not left hidden while the mothers feed, but follow their mothers all the time. When they are strong enough, they are led down to the grassy slopes where the ewes gather in flocks of 60 or more. When in these groups, the lambs will sometimes be looked after by an 'aunt' who seems to keep an eye on them.

Horns show rank

What are the horns of a sheep, the antlers of a deer, or the tusks of an elephant for? The obvious answer is that they are used as weapons, but a critical examination of only their shape often shows that if they are weapons for stabbing or battering opponents, their design is extremely poor. The long horns of an addax or oryx may be an exception, but consider the horns of a bighorn sheep. They sprout from the top of the head then curve away round the neck, well out of the way of any impact. Moreover, the larger they grow, the less efficient they will be as weapons because they are more likely to get snapped off.

In fact, horns are not used for fighting. In serious fights against predators the sheep kick with the hard hoofs, while the ritual fighting of courtship consists of head-to-head collisions, the bones of the forehead being specially built to reduce the effects of the concussion.

Recent observations have suggested that horns act as a badge of rank, like the stripes or pips on military uniforms. When two rams meet they will tilt their heads slightly, so presenting the other with a good view of the horns. If there is a significant difference in size of the horns, the ram with the smaller ones, hence the younger and junior, will retreat, acknowledging the superiority of the other. According to these observations, fighting broke out only when two rams with equal horns met. This, however, cannot be the sole explanation of fighting, because, as we have seen, there may be fights between several individuals at once. It may be that there is a small amount of fighting to establish rank in the flock throughout the year, which flares up into a general melée in the excitement of the breeding season.

On the basis of some observations on one species it is not possible to come to any final conclusion on the function of horns, antlers or tusks. Detailed studies need to be made on as many species as possible. When postulating any theory of horns as signs of dominance or success, it is well to remember that the hornless freaks, or hummels, sometimes found among deer seem to be consistently successful at mating and in rivalry with other males.

class	**Mammalia**
order	**Artiodactyla**
family	**Bovidae**
genus & species	***Ovis canadensis***

This flock of wild sheep lives in dry upland country. Wild sheep have short hair like goats but can be told from them by their characteristically narrow noses and concave rather than convex foreheads. The sexes generally live in separate flocks but a pair of young rams has sought refuge with these ewes from the aggressiveness of the older males.

American bison battling in the winter snows. The males in a herd fight each other but there is no 'harem master' who monopolises sexual activities. These massive ox-like creatures weigh up to $1\frac{1}{4}$ tons, which is over twice the weight of a small car.

Bison

Bison are massive and ox-like, weighing up to 3 000 lb. The largest bison stand 6 ft at the shoulder, which is raised in a distinct hump giving a hunch-backed appearance. The hair on head, neck, shoulders and forelegs is long and shaggy. The forehead is broad and is flanked by two short, curving horns that are carried by both sexes.

The two species of bison alive today are the European bison, or wisent, and the American bison, often called incorrectly the buffalo. The latter name strictly belongs to the African cape buffalo.

The American bison has a longer coat on the neck, shoulders and forelimbs than the European wisent. The horns are smaller and less curved and the hindquarters are smaller. There are two distinct varieties of American bison. The plains bison of the United States is smaller and lighter in colour than the wood bison of Canada but has a heavier head and hump. The wood bison is more like the wisent.

Senseless slaughter

Some 50 million bison once roamed North America but in 1889 there were 540 left. The massacre of the North American bison is matched only by the extinction of the American passenger pigeon. Both appeared to exist in limitless numbers but both succumbed to organised slaughter backed by modern techniques.

When Europeans first settled in North America, bison could be found from northern Canada as far south as the border of Mexico and across the continent from the east of the Rocky Mountains. They were apparently increasing in numbers and it is thought that they would have spread through the passes of the Rockies and onto the plains of the Pacific coast.

As it happened they did not get the chance to spread, for, with the coming of Europeans, they were hunted so relentlessly that they nearly became extinct.

Bison had always played an important part in the economy of the American Indian, but he accounted for relatively few. Then his hunting became more efficient after the introduction of horses by the Spanish conquistadors. Later, European settlers spread across the plains killing bison for their meat and hides and large-scale hunts were organised to get meat for the men building railroads. This was the era of 'Buffalo Bill' Cody, a professional hunter. The railroads then opened up a new market in the east for hides and tongues, and the bones were ground to make fertiliser for the corn-growing prairies. Where there was no market the bison were killed for sport, the aim being to see how many could be killed in one day. Bags of 50 or 60 seem fairly commonplace and totals of over 100 bison shot by one man in one day have been recorded. The carcases were usually left to rot, not a pound of meat or a single hide being collected. The senselessness of this slaughter is not mitigated by the fact that the spread of agriculture would in any case eventually have doomed the bison herds.

The European bison became reduced in numbers over a period of centuries, mainly because of man destroying the forests in which it roamed. The last truly wild individuals lingered on in the Bialowieza forest of Poland but they were exterminated in the first world war and the upheavals that followed. About 30 remained in zoos and by careful management they were built up to 360 in 1959, including a herd in a reserve set up in Bialowieza. In 1965 the overall figure of European bison was said to be as high as 790 – 800, including those in the Bialowieza forest and those in zoos. The future of the North American species is also more secure. The plains bison now numbers several thousand and is hunted under licence and the Canadians are protecting the remnants of the wood bison in a reserve in Alberta.

Distribution and habits

Bisons probably once roamed across Europe, Asia and North America. This is supported by the similarity between the wood bison and the wisent, which both live in woodland. The wisent invaded Asia and Europe from North America at the end of the Ice Age, coming across what is now the Bering Strait but was once a

land bridge. This is the present day view and is contrary to what has often been supposed, that the North American bison is a descendant of the wisent.

Bison live in herds, in the past numbering thousands of individuals, but they are now much smaller. The smallest group is a family of a bull, a cow and her offspring. The cow is the leader of the family group.

Bison are most active during morning and evening. They are fond of wallowing in mud and rubbing themselves against trees and boulders.

Seasonal migrations

American bison feed mainly on grass, making seasonal migrations of hundreds of miles to find the best feeding areas, which vary with the season.

The wisent prefer to eat bark. Their favourite trees are sallow, poplar and aspen. The sprouts of young evergreens are sometimes browsed and in the autumn, acorns are their favourite food.

Breeding

Mating takes place from July to September. The males in a herd fight each other but there is no 'harem-master' who monopolises sexual activities.

The solitary bulls who were once thought to be outcasts from the herd, driven out by a dominant bull, are now known to take part in mating, along with the other bulls.

Gestation lasts about nine months, the calves being born from April to June. The cow leaves the herd to drop her calf and returns when it is able to walk. The whole herd assists in the defence of the calves, whose only enemy apart from man is the wolf. The calves are nursed for a year and stay with their mothers until they are sexually mature at three years.

Bison and the Indians

One of the grievances that the Red Indians had against the white men was that they had driven away or exterminated the bison, which, for many tribes, formed their mainstay. The bison provided meat, both fresh and dried for later use, hides for clothing, bedding, tents and canoes, dung for fuel, bones and hide and sinews for weapons, tools and utensils. Not surprisingly, the bison appeared in the Indians' religion as a powerful figure to be worshipped.

There were also rituals to ensure that there was a plentiful supply of bison, for if the bison did not turn up on their annual migration the Indians would face a lean time. They believed that the buffalo sprang up every year, that it was necessary to lure them to the Indians' hunting ground, and that rites had to be performed to ensure there would be plenty of them. Skulls of bison were considered to be extremely useful for this purpose. They were piled up or displayed in a prominent place because it was thought that the bisons would seek out their 'white-faced companions'.

In other rites, oracles were consulted to find out where the buffalo were. The medicine men who performed these services acquired great merit because they continued their ceremonies and incantations until the bison arrived – as they were bound to do if the Indians were waiting on a regular migration route.

Unfortunately there came a time when the buffalo did fail to arrive. The white man had contrived to kill them all and the Indians starved. To the Indians, wanton killing of bison was the equivalent of cattle-rustling for the white man, and also punishable by death. Wars broke out as the Indians saw both their territory and their livelihood disappearing. Some enlightened men sought to safeguard both, but many others claimed the extermination of bison was the only way of 'civilising' the Indians (always a good excuse for land-grabbing).

Without the bison, the Indian culture died. For centuries their economy had been based on the bison but as has often happened, technological advance has failed to bring happiness. The ability of the Indians to get supplies of bison without horses, firearms and other refinements is shown by a remarkable discovery in Colorado during 1957. Wind erosion uncovered a small gulch or ravine some 6 or 7 ft deep. Within the gulch were the remains of nearly two hundred bison and most of them had been carefully butchered, the limbs dismembered and the bones piled up. With a knowledge of the hunting methods of present-day Indians it was possible to work out in detail what had happened 8 000 years before. A party of Indians had crept up on a herd of bison from downwind. Not being able to smell them, the bison had not realised that they were there until the Indians were on them. The bisons panicked and stampeded towards the gulch and toppled in. The first ones were crushed to death by the others and any that were only wounded were despatched by the waiting Indians.

The carcases were then hauled out and rolled onto their bellies. An incision was made down the back and the skin stripped

off and laid either side of the body to receive the meat as it was cut off. The Indians probably ate as they worked, each man eating 10 or 20 lb of fresh meat a day. The archaeologists who excavated these bones calculated that this one kill provided some 65 000 lb of food and that it probably fed a party of 150 Indians for a month. The bulk would have been eaten at once and some would have been dried. If they were always so lucky in their hunting, these Indians would have lived an easy life, and we can well understand their descendants' dismay at the disappearance of the bison.

class	**Mammalia**
order	**Artiodactyla**
family	**Bovidae**
genus & species	***Bison bison*** *American bison* ***Bison bonasus*** *wisent*

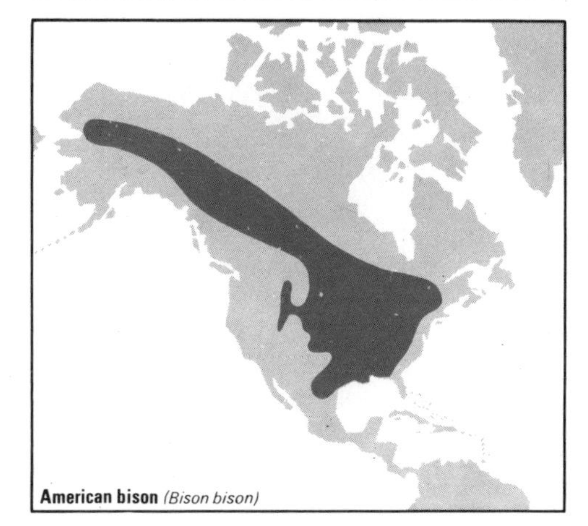

◁△ *A tightly packed bison herd. Any sudden noise will cause the bison to panic and they may stampede for miles before they slow down.*

◁ *Bison rooting for tufts of grass buried beneath the snow in Yellowstone Park, Wyoming.*

△ *A newly born baby calf is tenderly licked clean by mother, who will nurse it for a year.*

▽ *A bison herd grazing on an American plain. Seasonal migrations are made to find the best feeding areas, which vary round the year.*

Black bear

There are five species of black bear, each placed in a separate genus. All are smaller than the brown bears, and all are sufficiently similar for one of them, the one most studied, to serve as a type for the other four. This is the American black bear which originally inhabited practically all the wooded areas of North America from Central Mexico northwards. Its numbers are much reduced now and it has been eliminated from much of its former range, but in national parks its numbers are increasing and elsewhere it survives close to human settlement. Up to 5 ft long with a 4½ in. tail, and weighing 200–500 lb, it has shorter fur, shorter claws and shorter hind feet than the brown bears. The species also shows a number of colour phases: black, chocolate brown, cinnamon brown, blue-black and white with buff on the head and in the middle of the back. This last is most common in British Columbia, where it has been known as Kermode's bear. These different colour phases may occur in the same litter.

△ *Twin cubs being escorted by their mother. They stay with her until at least 6 months old.*

Friendly habits

Black bears are good tree climbers, powerful, quick to react, harmless to people except when provoked, cornered or injured–or through sheer friendliness. In national parks, where they are familiar with human beings and come begging food, visitors to the parks must keep to the protection of cars to avoid inadvertent injury from the bears' claws. Black bears are solitary except during the breeding season, the two partners separating after mating, to wander far in search of food. The American black bear sleeps through the winter–not hibernation in the usual sense–after laying in fat by heavy autumn feeding. It does not feed during the winter although it may leave its den, a hollow tree or similar shelter, for brief excursions during mild spells. When startled, the adult gives a 'woof', otherwise it is silent. The cubs, when distressed, utter shrill howls.

Mixed diet

Insects, berries and fruits, eggs and young of ground-nesting birds, rodents and carrion form its main foods, but young of deer and pronghorn are killed and eaten. Porcupines are killed, the bear flipping them over with its paw and attacking the soft under-belly, often to its own detriment from the quills. Black bears have been found dead with quills embedded in the mouth. Sometimes a black bear may turn cattle-killer.

Enemies

Old or sickly adults are occasionally killed by pumas and wintering bears may be attacked by wolves.

Life history

The breeding month is June and the gestation period is 100 to 210 days. Usually there are two or three cubs in a litter, exceptionally four, rarely five, born in January and February. At birth the 8 in., 9–12 oz cubs are blind, toothless and naked except for scanty dark hair. The mother continues to sleep for two months after the birth, having roused herself sufficiently to bite through the umbilical cords. The cubs alternately suck and sleep during these two months. They stay with the mother for at least six months, and she mates only every other year.

The original Teddy Bear

In 1902, Theodore (Teddy) Roosevelt, who was a keen naturalist as well as President of the United States, captured a black bear cub on a hunting trip, which he adopted as a pet. Morris Michton, a Brooklyn doll manufacturer, used this bear as model for the first Teddy Bear, so named with the President's permission. The popularity of the Teddy Bear as a toy was immediate and world-wide. The black bear, as already stated, is such a favourite in American national parks that they take liberties with visitors. In European zoos a prime favourite with visitors is the Himalayan black bear. Such general favouritism owes much to the human-like qualities of the bears.

We tend to favour in animals, qualities which reflect our own, as with birds that talk or animals that stand erect such as penguins, owls and bears. In man the bipedal stance is habitual; in bears it is but occasional, the usual way of walking being on all-fours. That does not invalidate the comparison, and the effect of the bears' ability to stand erect at times is reinforced by the way they will sit upright, as if on a chair, and also by the characteristic way a bear will wave a fore-paw (or hand) when soliciting food. Another trait which enables us to see ourselves in bears is their way of lying prone, on their backs.

Bears also appear to be intelligent. Whether they are more intelligent than their near relatives, the cats and dogs, has never been adequately tested. At least we know that the cubs stay with the mother for six months, often longer, and some may stay with her until her next cubs are born. A long period of parental care allows for learning by example and a longer period for experience with security. And if bears are by nature solitary they can, if circumstances compel them, as in bear-pits in zoos, live together with little discord, showing they are, like us, fundamentally friendly.

Yet in spite of the comparisons that can be drawn between bears and ourselves, and in spite of our fondness for Teddy Bears, the fact remains that the American black bear, like all other bears, has long been a target for the hunter's gun–and not only the hunter's. In 1953, 700 black bears were killed in British Columbia to provide bearskins, the ceremonial headwear, for the Brigade of Guards, for the coronation of Queen Elizabeth II. An American writer drily remarked: 'Fortunately for the black bear Great Britain's coronations are infrequent . . .'

class	**Mammalia**
order	**Carnivora**
family	**Ursidae**
genus & species	***Euarctos americanus*** *American black bear*

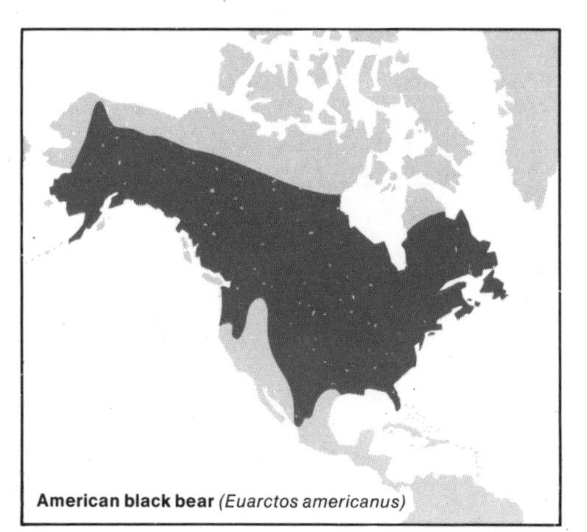

American black bear *(Euarctos americanus)*

Bobcat

The wildcat of America. Although there are other cats, like the puma (page 187) and lynx (page 142), in the United States, the bobcat is much smaller and is comparable with the European wildcat. On average, bobcats weigh 15—20 lb, but a record is 39 lb. The total length of the body is around 2½—3 ft, with the tail accounting for about 6 in. The closely related lynx is readily distinguished as its longer tail has a black tip while that of the bobcat has a black bar on the upper side fringed with white hairs.

The colour of the body varies considerably between different races and can often be linked with the habitat. In general, the colour is a shade of brown spotted with grey or white, but buff bobcats are common in desert country, whereas those from the forests are darker.

The ears are tipped with pointed tufts of hair, less prominent than those of the lynx. Experiments suggest that the tufts improve the efficiency of the ear in collecting sounds and that captive bobcats with clipped ear tufts do not respond so readily to sounds.

The name bobcat is linked with the short tail, and, probably, with a lolloping gait, reminiscent of a rabbit.

Wide ranging hunter

Bobcats are solitary animals and much of their hunting is done by night, so they are not often seen, although a good number of bobcats may be quartering the same area. Their home range varies with the abundance of food. They may roam over an area as much as 50 miles in diameter, or as little as 5 miles. A hunter in Wyoming once caught 39 bobcats in one spot over a period of 13 weeks.

The trails used by the bobcats can be traced, not only by footprints but also by scratches on tree trunks where the bobcats have stretched and sharpened their claws just like a domestic cat. They also have favourite spots for defaecating and urinating. Faeces and urine are covered by scratching a mound of earth over them.

Bobcats are found in all parts of the United States except the midwestern corn belt, throughout Mexico and in the southernmost part of Canada. Because they are small and can easily hide and feed on a variety of prey, they have survived the spread of agriculture much better than their cousin the lynx. Sometimes agriculture benefits the bobcats by providing them with a ready supply of food in the form of calves and lambs.

The bobcat is a good tree climber and will often seek refuge in a tree when hunted by a man with a pack of dogs. The trunk of this tree shows scratch marks where the bobcat has climbed up and down, a typical sign of bobcat country.

Fearless carnivore

Along with its retiring habits, the wide range of food taken has been cited as a reason for the continued abundance of bobcats in areas where the face of the countryside has been greatly changed. Rabbits and rodents such as deermice, wood rats and squirrels form the bulk of their diet. Birds and domestic animals are also eaten.

Other items are eaten occasionally. These include snakes, skunks, opossums, grasshoppers, bats and, very rarely, fruit. Porcupines are often attacked when there is a shortage of other food and porcupine quills have been found in bobcat's faeces, but usually the bobcat comes out second best in such an encounter. Bobcats have often been found with quills in their paws and mouths. In such cases the fight with the porcupine would probably have condemned the bobcat to die of starvation, as a mouthful of quills makes eating impossible.

Bobcats are very strong for their size and will attack and kill adult pronghorns and deer, as well as domestic livestock. The bobcat stalks its prey then throws itself onto the unsuspecting animal's back, biting at the base of its skull and tearing with its claws until the prey drops.

Mother fiercely defends kits

Kits may be born during any month of the year, but usually in late February or March. Occasionally a bobcat will have two litters a year. Gestation takes 50–60 days and the kits are blind for their first week. They are born in a den, in caves, under logs, or even under barns and sheds. There are usually two kits in a litter but three or four are not uncommon. Their mother defends them vigorously and the father is kept well away from the den, until the kits are weaned, when he helps the female collect food.

Hunted by men with dogs

Bobcats have been persecuted by man for their soft fur or merely for sport, and because the cats kill livestock and game birds. In some states bounties are given for bobcats and they are taken by traps or shooting, usually with hounds to flush them. When hunted with packs of dogs, bobcats often run in a wide circle, finally retreating up a tree near the start of the chase. They will often take to water or to swamps where they can outpace the dogs either by swimming or by bounding over shallow water.

Adult bobcats are killed by pumas and the young by foxes and horned owls.

Portrayed on a bank note

The term 'wildcat' has been used in conjunction with various activities, but not with the obvious meaning of ferocity. Wildcat strikes may appear to be acts of savagery to employers but it is difficult to see how the name came to be applied even with active pickets enforcing the strike. Wildcat oil-drilling is even more unlikely.

In these terms 'wildcat' is used to denote a risk or uncertain chance, where the course of events is not settled. Wildcat is applied to oil-drilling taking place in unproven oil-fields where there is no certainty of striking oil and the company can only bore holes and hope that oil will come up. The origin of this term appears to spring from the picture of a wildcat or bobcat that appeared on the notes issued by a mid-western bank during the early part of the last century. These notes were issued with virtually no financial backing and were therefore risky things to deal with. Other banks did similar business and became known as wildcat banks. Gradually the term spread to cover any unsound commercial enterprise involving a risk, and to other fields.

class	**Mammalia**
order	**Carnivora**
family	**Felidae**
genus & species	***Lynx rufus***

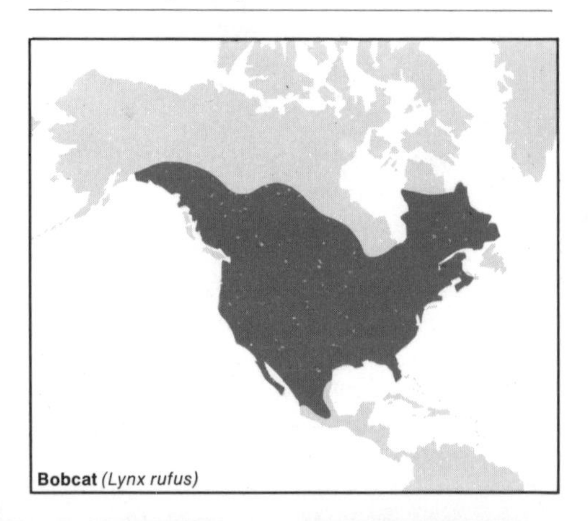

Bobcat *(Lynx rufus)*

◁ The bobcat probably gets its name from the short tail and its lolloping gait, which is rather like that of the rabbit. Its brown coat, spotted with grey, makes a good camouflage. Its range may extend over as much as 50 miles.

△ The ear tufts seem to help in sound detection, and the whiskers feel the way in the dark. This photo-portrait of a bobcat clearly shows the spot on the back of the ears which is present in all the cat family and is used in signalling.

▽ Bobcat on elk kill. The bobcat is very strong for its size and will even kill farm livestock. After stalking its prey it will jump onto its back, biting and clawing at the base of the skull and neck until the animal drops.

Bottlenose dolphin

Also often known as the common porpoise, this is the animal that in the last 20 years has become a star performer in the seaquaria of the United States. It is up to 12 ft long, weighs as much as 440 lb and is black above and white underneath, with a bulbous head and a marked snout. The forehead of the male is more protruding than that of the female. The moderate-sized flippers taper to a point and the fin in the middle of the back has a sharply-pointed apex directed backwards, making the hinder margin concave. It has 20–22 conical teeth in each half of both upper and lower jaw. Although as well suited to life in the water as any fish it is in fact a mammal like whales or, for that matter, man, giving birth to fully-developed young which are suckled on milk. The bottlenose is the commonest cetacean (family name of the whales) off the Atlantic coast of North America, from Florida to Maine. It occurs in the Bay of Biscay and Mediterranean also. It occasionally ranges to Britain, and is also found off West Africa, south to Dakar.

Cooperative schools

Bottlenose dolphins live in schools containing individuals of both sexes and all ages. Apparently there is no leader, but males in the school observe a 'peck order' based on size. When food is plentiful the schools may be large, breaking into smaller schools when it is scarce. The dolphins pack together at times of danger. They also assist an injured member of the school by one ranging either side of it and, pushing their heads under its flippers, raising it to the surface to breathe. In schools they keep in touch by sounds.

They sleep by night and are active by day, although each feeding session is followed by an hour's doze. Females sleep at the surface with only the blowhole exposed and this periodically opens and closes, as it does in stranded dolphins, by reflex action. The males sleep a foot below the surface, periodically rising to breathe.

The main swimming action is in the tail, with its horizontal flukes. This, the flexible part of the animal, is used with an up-and-down movement in swimming, quite unlike that of fishes, with only an occasional sideways movement. The flippers help in steering and balance. The dorsal fin also aids stability, but it is the lungs placed high up in the body that are chiefly responsible for keeping a dolphin balanced.

The depths to which bottle-nosed dolphins can dive has to be deduced from the remains of fishes in their stomachs. These show they go down for food to at least 70 ft, and they can stay submerged for up to 15 minutes. Their lung capacity is half as much again as that of a land animal and in addition they fill their lungs to capacity. Land animals, including ourselves, use only about half the lung capacity and change only 10–15% of the air in the lungs with each breath. A dolphin changes up to 90%.

Tame dolphin leaping some 30 ft into the air to take fish accurately, which proves that its small eyes are still quite useful out of water.

Well equipped for marine life

Since a dolphin's lungs are compressed when diving, air would be squeezed into the bronchial tubes, where no gaseous exchange would take place, unless this were prevented by valves. There are 25–40 of these in the bronchial tubes of the bottlenose dolphin and they act as a series of taps controlling the pressure in the lungs according to whether the animal is diving, swimming on the level or rising to the surface.

At the surface the pulse of the bottlenose dolphin is 110 a minute. When submerged it drops to 50 a minute and starts to increase as the animal nears the surface. The drop is related to the way the blood circulation is shut off so that the oxygen supply goes mainly to essential organs, notably the heart and brain. This extends the time of submergence by reducing the frequency with which visits need be made to the surface to breathe.

Whales and dolphins have an insulating layer of blubber, but they have no sweat glands and they cannot pant, so other means are needed to lose excess body heat. The tail flukes and the flippers are always warmer to the touch than the rest of the body and their temperature is not only higher than that of other parts of the body but varies through a greater range. They also have a much thinner layer of blubber. It is assumed therefore that these parts lose heat to the surrounding water. In brief, whales and dolphins keep cool through their flukes and flippers.

A dolphin's eyesight is not particularly good. Yet the animal can move its eyelids, shut its eyes, even wink. At one time it was thought the eyes were of little value and were quite useless out of water. This last seems proved wrong by the way dolphins in seaquaria will leap out of water and accurately snatch fish from the attendant's hand. Moreover, the visual fields of left and right eyes overlap, so presumably they have partially stereoscopic vision. The sense of smell is, however, either non-existent or almost wholly so.

Hearing is the main sense, apart from taste and touch. This is acute and is especially sensitive to high tones. It is probably second only to the hearing of bats. A dolphin is sensitive to the pulses of an echo-sounder or asdic and will respond to frequencies as high as 120 kilocycles or beyond, whereas we can hear 30 kilocycles at the most. At sea it has been noticed that bottlenose dolphins will avoid a boat that has been used for hunting them but will not be disturbed by other boats. The assumption is that they can recognize individual boats by the sounds they make.

Feeding

Fish form one of the main items in the diet but a fair amount of cuttlefish is eaten, the dolphin spitting out the chalky cuttlebone and swallowing only the soft parts. Shrimps also are eaten. In captivity a bottlenose will eat 22 lb of fish a day, yielding 237 calories per pound of its body weight, compared with the 116 calories/lb taken by man.

Life history

Bottlenose dolphins become sexually mature

△△ *Bottlenose dolphin jaws are well armed with teeth that help it catch its mainly cuttlefish food.*
△ *Baby dolphin with its mother. As with most mammals it stays with its mother for some time and is suckled on milk, being weaned after 5 months or more. Note the blowhole or nostril on top of the head.*

at 5–6 years. The breeding season extends from spring to summer. The gestation period is 11–12 months, births taking place mainly from March to May. The baby is born tail-first and as soon as free it rises to the surface to take a breath, often assisted by the mother usually using her snout to lift it gently up. Just prior to the birth the cow slows down and at the moment of birth she is accompanied by two other cows. These swim one either side of her, their role being protective, especially against sharks, who may be attracted to the spot by the smell of blood lost during the birth process. Weaning may take place between 6–18 months; reports vary considerably.

For the first 2 weeks the calf stays close beside the mother, being able to swim rapidly soon after birth. Then it begins to move away, even to chase fish, although quite ineffectively. However, it readily dashes back to its mother's side or to its 'aunt', the latter being another female that attaches herself to the mother and shares in the care of the calf. The aunt is the only one the mother allows near her offspring.

The calf is born with the teeth still embedded in the gums. These begin to erupt in the first weeks of life but the calf makes little attempt to chew until 5 months old and some take much longer before they attempt to swallow solid food. Even then there may be difficulties and in captivity a calf at about this time has been seen to bring up its first meal, the mother then massaging its belly with her snout.

Suckling is under water. The mother's nipples are small and each lies in a groove on the abdomen. The mother slows down to feed her calf which comes in behind her and lies slightly to one side, taking a nipple between its tongue and the palate. The mother then, by muscular pressure on the mammary glands, squirts the milk into its mouth. Should the calf let go the nipple the milk continues to squirt out. The baby bottlenose must come to the surface to breathe every half minute, so suckling must be rapid. In this species it consists of one to nine sucks, each lasting a few seconds. For the first 2 weeks the calf is suckled about twice an hour, night and day, but by 6 months it is down to six feeds a day.

Can dolphins talk?

It is not all that time ago that it was generally believed that whales, porpoises and dolphins were more or less mute, although the whalers themselves held very definite views to the contrary. It was not until after World War II, when bottlenose dolphins were being first kept in captivity in the large seaquaria, first in Florida and later in California and elsewhere, that it began to be realized fully that they have a wide vocabulary of sounds. Then, a few years ago, came the startling suggestion that these cetaceans might be capable of imitating human speech, even perhaps of being able to talk to people, in a sort of Donald Duck language in which words are 'gabbled' in a very high pitch. These high hopes do not seem to have been realised, but apart from this much has been learned about the noises they make.

One thing that has long been known is that air can be released from the blowhole while the animal is still submerged. This can be seen, by direct observation, emerging as a stream of bubbles. It can be used to produce sounds, and part of the mechanism for it is the many small pouches around the exit from the blowhole which act as safety valves, preventing any inrush of water.

It has been known for some time that some cetaceans are attracted over long distances by the cries of their fellows in distress. Conversely, people have been calling the animals to them by using whistles emitting sounds similar to their calls. Pliny, the Roman naturalist of the first century AD, knew of this, and in modern times the people on the Black Sea coasts have continued to do this. Sir Arthur Grimble also left us an account of what he called porpoise calling in South Pacific Islands. He described how local peoples in this area would call the porpoises from a distance to the shore. These items indicate an acute sense of hearing in dolphins and porpoises and a potentiality for communication by sounds on their part.

Underwater microphones as well as more direct observations in the various seaquaria have established that these cetaceans use a wide range of sounds. These have been variously described as whistles, squawks, clicks, creaks, quacks and blats, singing notes and wailings. It has been found that two dolphins which have been companions will, if separated, call to each other, and that a calf separated from its mother will call to her. Dolphins trained to leap out of water for food have been heard to make sounds at their attendants.

These are, however, only the sounds audible to our ears, which can deal only with the lower frequencies. Much of dolphin language is in the ultrasonic range, and if they are able to understand what we are saying, as one investigator has somewhat unconvincingly suggested, they could be using their own vocalizations to call us by rude names without our knowing it!

It has often been said that if whales cried out in pain we might be less ready to slaughter them. Scattered reports suggest that in fact they do precisely this. Freshly captured bottlenose dolphins placed in the tanks in Florida's Marineland have been heard through the thick plate glass windows to cry with shrill notes of discomfort and alarm. At sea similar distress calls have been heard from injured or wounded whales, porpoises and dolphins.

class	**Mammalia**
order	**Cetacea**
family	**Delphinidae**
genus & species	***Tursiops truncatus***

Bottlenose dolphins leaping out of the water in formation. This demonstrates the powerful swimming action of the tail with its horizontal fluke unlike a fish's tailfin which is vertical. It also shows their sociability.

Dolphins try to find each other by echo-locating 'clicks'. Then they talk in 'whistles' with a few 'grunts' and 'cracks'. Their sounds were analysed in 1965 when TG Lang and HAP Smith of the US Naval Ordnance Test Station put two newly captured bottlenose dolphins, Doris and Dash, in separate tanks linked by a two-way hydrophone system so the experimenters could tap their conversations. The first 4 of a total of 16 two-minute periods shown here indicate how the animals conversed when they were linked by phone and how sporadic noises were made when the line was 'dead'.

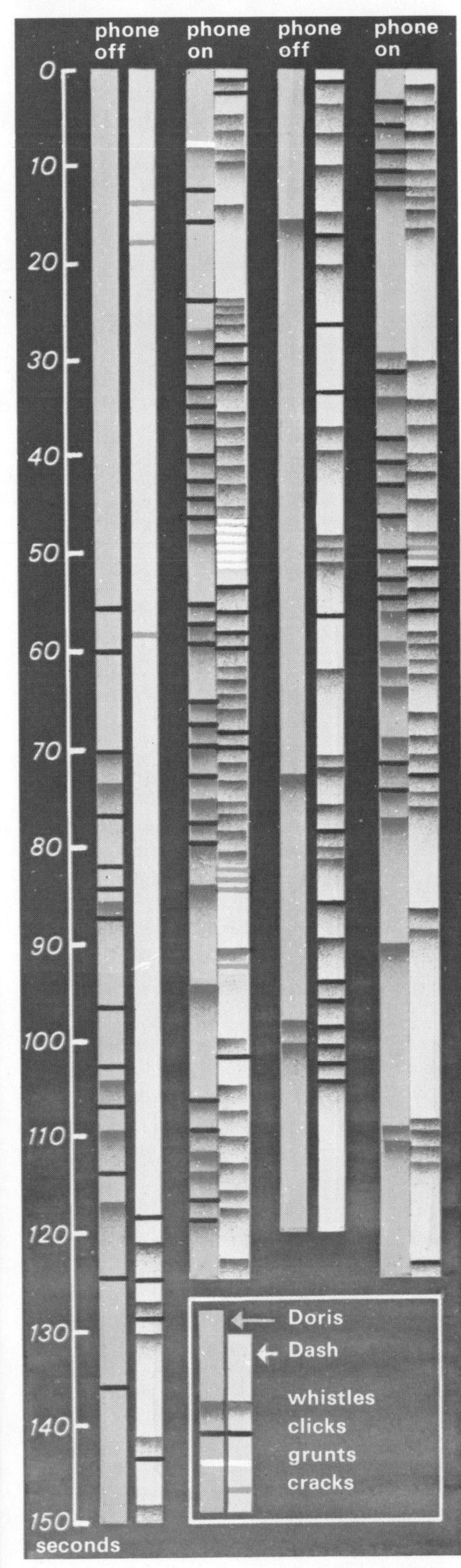

Mother with her cub. When young, the bear can climb, but as it grows older this ability is lost. The cub is born in January or February while the mother is in her winter sleep. It weighs under 1½ lb at birth but by the time it is a year old it weighs about 50 lb.

Brown bear

It is a matter of opinion, almost a matter of taste, whether there are many species of brown bears distributed over Europe, Asia and North America, or whether there is only one.

Authoritative writers are now tending more to the view that the grizzlies and other brown bears of North America, such as the giant Kodiak and Kenai bears, belong to the same species as the European, Russian and Syrian bears, together with the isabelline bear of Central Asia. If we take this view, then the brown bear can be described as a heavily built member of the Carnivora, practically tailless, with broad flat feet, each foot bearing five toes armed with non-retractile claws. Its sense of sight is poor, but smell and hearing are acute. It may be up to 9 ft or more long and weigh up to 1 650 lb. Its coat is usually a shade of brown, but that of the isabelline bear is reddish.

The blue or snow bear, of western China and Tibet, has blackish-brown hairs frosted with slate-grey or silver.

Grizzly's great strength

Brown bears live in wild mountainous country, as well as forests, wandering about singly or in family parties. Their home range averages over a 20 mile radius, and individuals often stray beyond this. The wandering has been shown by finding bear tracks in the snow up one side of a mountain in Canada with a track down the other side where a bear had tobogganed down. They normally walk on all-fours but will at times stand erect and shuffle for a few paces. The dancing bears of former times, led through the streets on a chain with a ring through the nose, were trained to do this to order. Young brown bears climb trees well, though slowly and deliberately, but adults rarely do so. Brown bears are not normally aggressive to man except under provocation or when injured, or when a person comes between a she-bear and her cub. The strength of a bear can be shown by the grizzly's ability to fell a bison larger than itself with one blow of its paw, breaking its neck, and then dragging away the carcase. Bears feed during summer and store fat in their bodies then den up to sleep most of the winter in natural shelters among rocks or in a large hollow tree, even in dens dug in hillsides.

Omnivorous carnivore

Considering that bears are classified in the order Carnivora (flesh-eaters), it is perhaps surprising to find the diet varies greatly with the individual. Some bears are wholly vegetarian, others wholly flesh-eating, but most eat a mixture of plant and animal food. This is made up of berries and fruits, insects and other small animals, honey and the grubs of wild bees. Fish may be taken, flipped from the shallows onto the bank with the forepaws, or seized in the water, as in the salmon runs of western North America. Sometimes the young of deer are killed and eaten, and occasionally a rogue individual takes to killing farmstock.

Life history

This is similar to that of the black bear (see page 26). Mating is in June, the gestation period varies from 180–250 days, and the cubs, normally two, are born in January or February, while the mother is in her winter sleep. Each cub is 1–1½ lb at birth and 8 in. long, almost hairless, blind and toothless. The small size compared with the bulk of the mother is striking and an ancient belief was that they were born shapeless and that the mother 'licked them into shape', hence the well-known saying. The reference here is to the licking the mother gives each cub after birth to clean them, as in most true mammals, of the birth fluids. The mother rouses herself from her dozing sleep to do so. The brown bear cub weighs 50 lb when a year old, and stays with the mother until at least that age. Females may breed at three years. They live for up to 34 years.

Wide difference in size

Half-a-dozen species of brown bears, as well as many subspecies, have been recognized in the past. Over 50 species of grizzly alone have been named, as well as a number of subspecies. In Europe and Asia the position has been only a little better. Clearly this is absurd, especially as all have the same habitat, habits and life history as well as skeletons and general anatomy that are not significantly different. The truth is that this enormous list of species and subspecies was based mainly on different colours of the coat, a most variable feature in any mammal, and especially on differences in size. It is true that there is a big gap between the Syrian bear at 150 lb and the Kenai and Kodiak bears at 1 650 lb. The Syrian bear is the one usually seen in British zoos so the mammoth proportions of the Kenai and Kodiak bears can be imagined. Even the grizzly, scientifically named *Ursus horribilis* (horrible bear), only reaches 880 lb.

It is more likely that the truth about these many species arises from two circumstances: persecution and the absence of reliable records. In the Old World, brown bears once ranged in considerable numbers from Britain to Japan, and as far south as the Mediterranean and the Himalayas. By the 11th century the last had been killed in Britain and today, in Europe, the survivors are largely confined to inaccessible forests in the Pyrenees, Swiss Alps, Carpathians, Balkans, Norway, Sweden and Finland. They are more numerous in parts of the Soviet Union but even there numbers have dropped. It is self-evident that persecution, especially hunting, brings down radically the maximum sizes reached in a species, particularly when the hunters' ambition is to collect record trophies. Bears were killed for their flesh, and their fat, and because their tempers were unpredictable. Above all, their shaggy coats were coveted prizes, and this alone gave a large bear little chance.

In 1904 JG Millais, naturalist and author, wrote: 'No terrestrial mammal varies so greatly, both in size and pelage, as this animal. Between brown bears killed in eastern Norway and those of western Sweden there is a perceptible difference in colour, whilst in the bears of Russia, especially those of the eastern districts, there is a further and much greater difference in size. I have lately seen two enormous bears belonging to the Russian Embassy in London, which measure nearly 9 ft in length (the size of Kenai and Kodiak bears) and are almost black in colour.'

In Germany there are many so-called dragons' caves, including the Drachenfels, in the Sieben Gebirge, where Siegfried is supposed to have killed his dragon. The basis for these legends was the abundance of skulls of cave bears, larger than those of brown bears and characterised by a steeply sloping forehead. Nevertheless, the two bears were often confused until the years 1920-3 when Austria was in such a calamitous state economically. The loss of her cattle during the First World War caused an acute need for fertilizers. These were eventually found in caves, in the thick deposits of guano laid down by bats over the years. In the course of digging out the guano they found thousands of bones of bears, enabling scientists to build up a complete picture of the cave bear and of the modern brown bear that replaced it. The cave bear was larger, almost completely vegetarian, and it went into caves only for its winter sleep. This was clear from the large proportion of baby skeletons found under the bat guano. The study of the bones also revealed that many of the cave drawings until then believed to be of cave bears were, in reality, drawings of the living brown bear. The stratification of the bones in the guano showed that cave bears had died out before most cave drawings had been made.

class	**Mammalia**
order	**Carnivora**
family	**Ursidae**
genus & species	***Ursus arctos*** *brown bear*

The brown bear catches salmon and other fish flipped from the shallows on to the banks with the forepaws (above) or seized in the water (below).

△ *A good sense of smell is very important to the bear which therefore has a well-developed snout with a wet nose, or rhinarium, to increase its sense of smell.*

Brush-tail opossum

The most common and most widely distributed of all species of Australian marsupials. It is found in all parts of Australia and Tasmania, and it flourishes in New Zealand, where it was introduced in 1858. The brush-tail opossum, also known as the vulpine or fox-like opossum, is about 2 ft long, the size of the red fox, has a fox-like head with large ears and a pointed snout. Its eye-catching feature, however, is its tail, which is prehensile at the tip where there is a naked patch on the underside. Its fur is thick, woolly and variable in colour, from silver-grey to dark brown or black.

Adaptable Australian

The brush-tail prefers trees but it will also live in the low bush or in treeless areas, where it will take over rabbit burrows. It has also taken to living in the roof-spaces of houses, even in the suburbs of large towns. Because it is nocturnal, the noise of the brush-tail moving in roof-spaces is unwelcome. So are the stains on the ceiling immediately under the roof. It also damages garden blossoms.

The brush-tail is indifferent to human beings. A wild one visiting the vicinity of houses has allowed itself to be stroked, although it objected to being picked up. It then attempted to bite and scratch although uttering no sound. This contrasts with the noise made in quarrels with its fellows, when it hisses and grunts, and follows with a loud cry which ends in a raucous screech.

Taste for mistletoe

It is largely vegetarian, feeding mainly on buds, and a preferred food is the Australian mistletoe. It is reputed to take eggs and nestling birds, and to eat carrion.

The lizard menace

A brush-tail abroad by day is likely to fall prey to eagles. So also is one with inadequate sleeping shelter, which means that the numbers of brush-tails in an area will be largely controlled by the availability of hollow trees or burrows. Dingoes try to hunt them, and will tear bark from a tree base to reach one. The main enemy is the monitor lizard known as the goanna. As this climbs a tree the brush-tail cries out in fear. The Aborigines, who cook and eat the brush-tail, imitate the goanna's scratching on the bark of a tree to find out if one is inside.

Born climber

Mating takes place in May–June, earlier in southern parts. Gestation is 17–18 days. There is a single young (possibly sometimes two) and this leaves the pouch in July–September, becoming independent of the mother by September–November. At birth it is $\frac{1}{2}$ in. long and weighs less than 1/15 oz, compared with the mother's weight of about 10 lb. It reaches full-size by the following February, and attains sexual maturity three months or so later.

The brush-tail, with its fox-like head, hangs in a bat-like position by its prehensile tail.

△ *Mother and baby coppery brush-tail opossum.*
▽ *The brush-tail has invaded suburbs of towns.*

▽ *The vegetarian diet consists mainly of buds, and it has a taste for Australian mistletoe.*

As with other baby marsupials, the newly-born brush-tail makes its way unaided from the birth-canal to the pouch through the mother's fur. It progresses by an over-arm action of the front legs, which at birth are longer than the hind-legs, the paws being armed with strong claws. The paws can be flexed to grasp the mother's fur. It takes the newly-born brush-tail about 7 minutes to travel the 2½ in. to the pouch. Once inside the pouch the baby seizes one of the two teats in its mouth and hangs on in this way for some weeks, by which time it is beginning to look more like its parents.

Should the baby fail to reach the pouch, or be lost through some other cause, a second will be born fairly soon.

Secrets of survival

The brush-tail population in New Zealand has reached the astonishing number of 25 million. Its only predator, however, is man. About 1 million a year are trapped for their fur and 1 million destroyed under a bounty system. There is an annual increase, despite these losses, of some 25% and this is despite the low birth-rate.

In Australia there are natural predators, but there is the same story of a successful survival despite persecution. The brush-tail is regarded as a pest in orchards and on farms. Moreover, there is continual attrition because of its valuable fur, which has been marketed and exported, and sold under such names as beaver, skunk and Adelaide chinchilla.

During the depression of 1931, unemployed men were encouraged to hunt the brush-tail for its fur. Between June and July of that year over 800 000 skins were marketed from crown lands alone. Previously, in 1906, over 4 million skins were sold in London and New York. The annual toll is not always of that order, and in most years is considerably less.

A more sympathetic attitude is growing, based partly upon the increasing realization of the need for conserving Australia's unique fauna. More particularly, also, is the realization that the brush-tail by feeding on the mistletoe, and thus checking its spread, is not only benefiting the indigenous trees but indirectly helping the honey industry. The economic value of this marsupial in reducing the prejudicial effect of the mistletoe on the flowering gum-trees, which produce the nectar for honey, has been proven by direct experiment.

There is almost certainly, as is so often the case, a need for local control. Given that, there is little to fear, for the brush-tail's ability to survive even under persecution is due, as in the common rat, to its adaptability. It can live in any kind of habitat and, being primarily a vegetarian, it is unlikely to suffer a shortage of food.

class	**Mammalia**
order	**Marsupialia**
family	**Phalangeridae**
genus & species	***Trichosurus vulpecula***

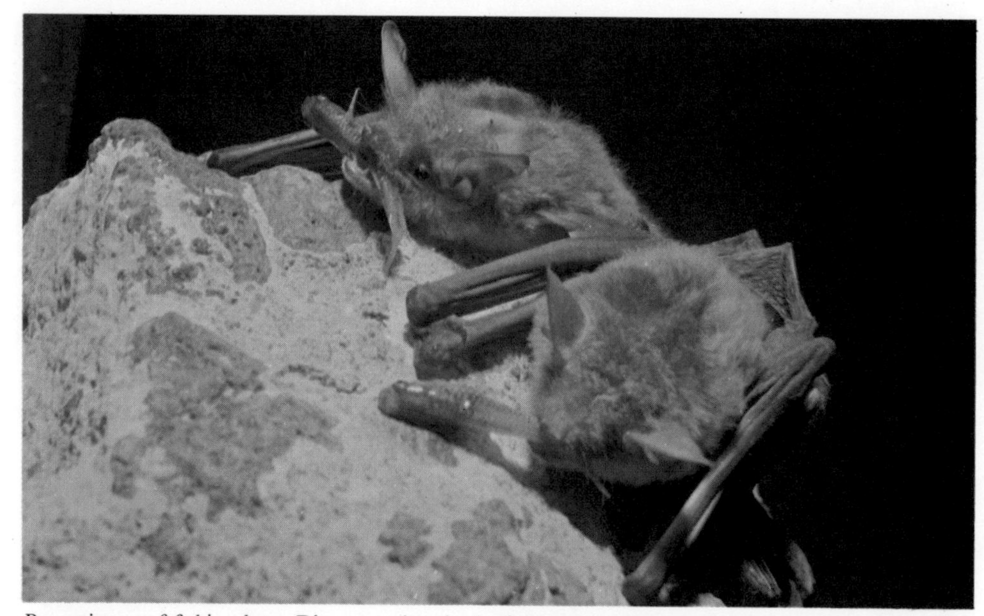

Rare picture of fishing bats ***Pizonyx vivesi*** *feeding.*

The range of ***Noctilio leporinus.***

Bulldog bat

Bats are mammals and the only ones in this class to fly and not just glide. They form the second largest order of mammals —only the rodents have more species. Despite its name, the bulldog bat, also known as the fish-eating bat, is not particularly ugly compared with other bats. Its upper lip is divided into a harelip and there are folds of skin under the lower lip so there is a superficial resemblance to the muzzle of a bulldog. The nostrils project a little beyond the lips giving the bat a rather quaint appearance, but it lacks the complex folds of skin above the nose, the so-called nose-leaves, of more repellent bats.

Bulldog bats have long, narrow wings with a 20 in. span, stretched—as is usual in bats—between the greatly elongated 3rd, 4th and 5th fingers of the hand and running to the ankles. Between the hind legs is the interfemoral or tail membrane. The tail runs down the centre of the membrane. In the bulldog bat it only reaches halfway. The skin is naked except for short reddish-brown fur around the head and shoulders and down the middle of the back.

The bulldog or fish-eating bat is one of three species of bats, belonging to separate families, that prey habitually on fish. The other two are the fishing bat ***Pizonyx vivesi*** *of Baja, California and Sonora, Mexico, and the false vampire bat* ***Megaderma cor*** *of India and southeast Asia which also feeds on insects, birds, frogs and other bats. The bulldog or fish-eating bat lives in America from northwest Mexico southwards to northern Argentina and on the Antilles and Trinidad.*

Impaling fish with their claws

During the day bulldog bats roost in clefts in rocks or in hollow trees. These roosts can easily be found by their powerful and unpleasant smell, which can be detected from 100 yd away. At dusk, or sometimes during the day, the bats come out to feed on fish which they catch from both fresh and sea water.

The feeding habits of these bats have posed a series of problems. First came the question of what they fed on, and this was quite easily settled. A zoologist on an expedition to the Caribbean, organised by the US Fish Commission in 1883, saw some bulldog bats in broad daylight, flying low over the waves in the company of some pelicans that were fishing. The bats also appeared to be fishing as they occasionally dipped down to touch the water. Later some bats were shot as they flew out of a cleft in the cliff face, and their stomachs were found to contain nothing but fish. This is not their exclusive diet, however. They also catch aquatic crustaceans as well as crickets, flying ants and beetles.

The next problem was how the fish were caught. At one time it was claimed that the tail membrane was used as a fishing net, but high-speed photography showed that the bats were using their long, sharp claws, like those of the Mexican fishing bat, as a gaff. They dip their feet in the water trailing them for anything up to 3 ft, and impaling fish about 1–3 in. long and sometimes up to 4 in. They then lift the fish quickly to their mouths and either eat them in flight or store them in cheek pouches until they return to their roosts. Captive bats caught 30 to 40 fish in one night from an artificial pool, but they would presumably catch considerably less than this in the wild.

The only observations on breeding are of female bulldog bats carrying single babies from January to April.

Ultrasonic pulses

Until Donald Griffin, the distinguished American zoologist, showed that bats used echo-location to navigate in the dark, sending out ultrasonic pulses and listening for the returning echoes, it was thought that fish-eating bats merely trailed their claws in the water at random, on the off-chance of catching a fish. Even if they struck a shoal, it is difficult to believe that this method would be very successful. The demonstration that bats used echo-location to detect their prey seemed to show how fish-eating bats could locate fish and spear them accurately. There was one great drawback. The ultrasonic squeaks would be almost entirely reflected back off the surface of the water. Only 0·1% of the sound energy would penetrate the water. Similarly, any sound that might reach the fish and be bounced back would itself suffer a 99·9% loss as it went back into the air. Furthermore, the sound waves would only be reflected by a fish if it had an air-filled swimbladder because flesh offers about the same resistance to sound as does water, so there would be no noticeable echo from it.

Despite these problems it was still argued that a bat might be able to detect fish underwater by flying slowly and very low over the water, as fish-eating bats do, and directing their echo-locating pulses vertically downwards by means of their protruding nostrils.

The problem seems to have been solved by some experiments carried out with tame bats that learnt to catch fish from shallow tanks. They were unable to detect fish or balloons, representing swimbladders, that were just underneath the surface, but they would dip down at any ripples or upwellings. More careful tests showed that they could detect a wire of 0·2 mm diameter sticking $\frac{1}{4}$ in. out of the water from a distance of about 2 ft and a $\frac{1}{2}$ in. cube of fish flesh from 5 ft. So it seems that the bats can only catch fish that bob up to break surface either to catch an insect or for some other purpose. This would explain why the bats are seen fishing with pelicans. The birds are making the fish flee in panic, and break surface in doing so. They have also been seen hawking over shoals that are being forced to the surface by predatory fish underneath.

class	**Mammalia**
order	**Chiroptera**
family	**Noctilionidae**
genus & species	***Noctilio leporinus***

Cacomistle

A relative of the raccoon (page 191), the cacomistle is slender and sleek, up to $2\frac{1}{2}$ ft long of which 17 in. is bushy tail, ringed black and white, and it weighs only about $2\frac{1}{2}$ lb. The coat is greyish-buff, darker along the back and white on the underparts. The face is fox-like, the eyes are large and ringed with white, contrasting with black patches, and the ears are large and pricked. The feet are well furred, and the claws can be partly withdrawn. It ranges the southwestern US and nearby Mexico.

The original Mexican name is cacomixtle, meaning rush-cat, and an alternative Mexican name is tepemixtle or bush-cat. In the United States, it has been given a variety of names: ringtail, coon cat, raccoon fox, band-tailed cat, cat-squirrel, mountain cat, ringtailed cat. Its fur is valuable and is marketed as 'civet cat' and 'California mink'.

Related to the cacomistle is the central American cacomistle, or guayonoche, ***Bassariscus sumichrasti****, about which very little is known. It lives in the forests of southern Mexico and central America.*

The cacomistle has the large eyes and large ears typical of a night hunter.

Elusive ringtail

Nocturnal and secretive, the cacomistle usually manages to keep out of sight, so it is little known, even to the local residents. It sleeps by day in a den among rocks, in holes in trees or at the base of trees between buttress roots. Agile in moving among trees, it uses its tail as a balancer and also curls it over its back like a squirrel. It does this especially when alarmed, at the same time giving a squirrel-like scolding or barking.

Expert mouser

The cacomistle's food is small rodents and birds, lizards, insects and fruit. It is well known around the fruit farms of California for its habit of eating the fallen fruit. At times a cacomistle will come to live in a house, cottage or cabin and will keep the place free of mice and rats. Prospectors gave it the name of 'miner's cat'. However, the cacomistle is a menace to poultry farmers.

Lightweight babies

The gestation period is not known. Births take place in May or June, in a moss-lined nest. The 3 or 4 babies in a litter weigh 1 oz each. They are born blind, with ears closed and the body covered with a downy fur. Some solid food may be taken at 3 weeks. The eyes open at 4 weeks. They are taken hunting at 2 months and are weaned at 4 months. They can live 14 years in captivity.

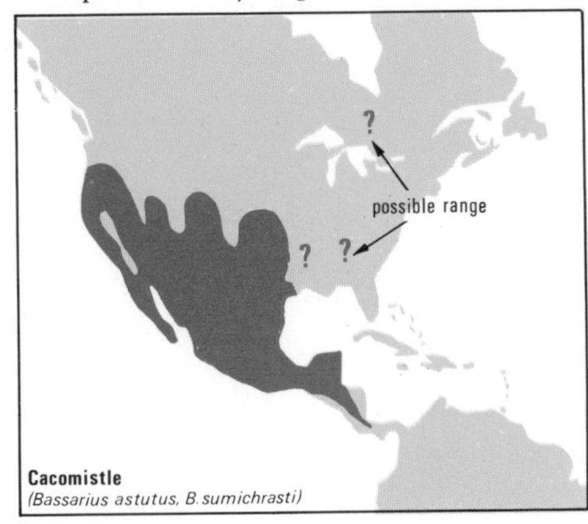

Sunbathers

Animals active by day are termed diurnal, those that come out at night, nocturnal. But even nocturnal animals will sometimes come out during the day, especially to sunbathe. The European badger, one of the most rigidly nocturnal of animals, will do this at times, and so will the red fox. The cacomistle also sunbathes but it is seldom seen doing so, because it does its basking in the tops of trees, crouched along a branch with its ringed tail dangling over the side, the rest of its body harmonizing with the bark.

class	**Mammalia**
order	**Carnivora**
family	**Procyonidae**
genus & species	***Bassariscus astutus***

Camel

There are two species of camel: the Arabian or one-humped and the Bactrian or two-humped. The first is not known as a wild animal, though the second survives in the wild in the Gobi desert. A dromedary is a special breed of the one-humped camel, used for riding, although the name is commonly but wrongly used to denote the Arabian camel as a whole.

Camels have long legs and a long neck, coarse hair and tufted tails. Their feet have two toes united by a tough web, with nails and tough padded soles. The length of head, neck and body is up to 10 ft, the tail is 1½ ft long, height at the shoulder is up to 6 ft, and the weight is up to 1100 lb.

Habits

The wild camels of the Gobi desert are active by day, associating in groups of half-a-dozen, made up of one male and the rest females. They are extremely shy and make off at first sight of an intruder, moving with a characteristic swaying stride, due to the fore and hind legs on each side moving together. Their shyness may be partly due to persecution in former times.

It is often said that a camel cannot swim. Reports suggest they do not readily take to water, but they have been seen swimming.

One-humped camels drinking at a water-hole in the desert. Having taken their fill of water, camels can survive for several days in the desert without drinking, or for several weeks if they have access to succulent desert plants. Water is drawn from the body tissues to maintain the fluid in the blood.

Adaptations to desert life

Everything about a camel, both its external features and its physiology, show it to be adapted to life in deserts. Its eyes have long lashes which protect them from wind-blown sand. The nostrils are muscular so they can be readily closed, or partly closed to keep out sand. The form of the body, with the long neck and long legs, provides a large surface area relative to the volume of the body, which allows for easy loss of heat.

The camel's physiology shows other adaptations which provide protection from overheating, and help it to withstand dessication and to indulge in physical exertion with a minimum of feeding and drinking. These characteristics are often seen in stories of journeys made across waterless deserts. Many of these are exaggerated, but even those that are true are remarkable enough. There is one instance of a march through Somalia of 8 days without water and in Northern Australia a journey of 537 miles was made, using camels which were without a drink for 34 days. Most of the camels in this second journey died, but a few that were able to graze dew-wetted vegetation survived.

Most desert journeys are made in winter, however, and during that season even a man can go without drinking if he feeds largely on juicy fruits and vegetables. Knut Schmidt-Nielsen tested camels in the desert winter and found that even on a completely dry diet, camels could go several weeks without drinking, although they lost water steadily through their skin and their breath as well as in the urine and faeces. Normally, however, a camel feeds on desert plants with a high water content.

Do camels store water?

There are many stories of travellers in the desert killing a camel and drinking the water contained in its stomach. From these arose the myth, which has not yet been completely killed, that a camel stores water in its stomach. Pliny (AD 23-79), the Roman naturalist, first set it on record. Buffon (1707-1788) and Cuvier (1769-1832), celebrated French scientists, accepted it. Owen (1804-1892) and Lyddeker (1849-1915), British anatomists and zoologists, supported it.

In 1801 George Shaw, British zoologist, wrote of a camel having four stomachs with a fifth bag which serves as a reservoir for water. Everard Home, the Scottish surgeon, dissected a camel and in 1806 published his celebrated drawing of alleged water pockets in the first two compartments of the stomach, a drawing which has many times been reproduced in books, and which has served to bolster the story. It was not until the researches of Schmidt-Nielsen and his team, working in the Sahara in 1953-4, that the full story emerged. In the living camel, these pockets are filled with an evil-smelling soup,

the liquefied masticated food, which might be drunk, so saving his life, by a man crazy for water—but not otherwise.

Another of the camel's achievements which served to support the story is its ability to drink 27 gallons of water, or more, in 10 minutes. It will do so only to replenish the body supply after intense dessication. In those 10 minutes a camel will pass from an emaciated animal, showing its ribs, to a normal condition. This is something few other animals can do. But the water does not stay in the stomach; it passes into the tissues, and a camel after a long drink looks swollen.

A camel can lose water equal to 25% of its body weight and show no signs of distress. A man losing 12% of his body water is in dire distress because this water is drawn from his tissues and his blood. The blood becomes thick and sticky, so that the heart has greater difficulty in pumping. A camel loses water from its tissues but not from the blood, so there is no strain on the heart, and an emaciated camel is capable of the same physical exertion as normal. The mechanism for this is not known. The only obvious difference between the blood of a camel and any other mammal is that its red corpuscles are oval instead of being discoid.

The camel's hump

The hump contains a store of fat and it has often been argued that this can be converted to water, and therefore the hump is a water reserve. The hump of the Arabian camel may contain as much as 100 lb of fat, each pound of which can yield 1·1 lb of water, or over 13 gallons for a 100 lb hump. To convert this, however, extra oxygen is needed, and it has been calculated that the breathing needed to get this extra oxygen would itself lead to the loss of more than 13 gallons of water as vapour in the breath. The fat stored in the hump is broken down to supply energy, releasing water which is lost. The hump is thus really a reserve of energy.

Other physiological advantages possessed by a camel are that in summer it excretes less urine and, more important, it sweats little. The highest daytime temperature is 40°C/105°F but during the night it drops to 34°C/93°F. A man's temperature remains constant at just under 39°C/100°F and as soon as the day starts to warm up he begins to feel the heat. A camel starts with a temperature of 34°C/93°F at dawn and does not heat up to 40°C/105°F until nearly midday. A camel's coat provides insulation against the heat of the day and it keeps the animal warm during the cold desert nights.

With all these advantages, camels should be even-tempered, but everyone agrees that they are bad-tempered to a degree. One writer has described them as stupid, unwilling, recalcitrant, obnoxious, untrustworthy and openly vicious, with an ability to bite destructively. There is a traditional joke that there are no wild camels, nor any tame ones.

The power of the bite is linked with the camel's unusual dentition. At birth it has six incisors in both upper and lower jaws, a canine on each side, then a premolar followed by a gap before the cheek teeth are reached. As the young camel grows, it quickly loses all but the outside incisors of the six in the upper jaw and these take on a similar shape to the canines. So in making a slashing bite a camel has, in effect, double the fang capacity of a dog.

Breeding

A baby camel is a miniature of its parents, apart from its incisors, its soft fleece, lack of knee pads and hump. There is a single calf, exceptionally two, born 370–440 days after conception. Its only call is a soft *baa*. It can walk freely at the end of the first day but is not fully independent until 4 years old, and becomes sexually mature at 5 years. Maximum recorded life is 50 years.

Origins of the camel

Camels originated in North America, where many fossils have been found of camels, small and large, with short necks or long, as in the giraffe-like camels. The smallest was the size of a hare, the largest stood 15 ft at the shoulder. As the species multiplied there was one migration southwards into South America and another northwestwards, and then across the land-bridge where the Bering Straits now are, into Asia. As the numerous species died out, over the last 45 million years, the survivors remained as the S. American llamas and Asiatic camels.

A few species reached eastern Europe and died out. None reached Africa. Until 6 000 years or more ago there was only the one species in Asia, the two-humped Bactrian camel. The date is impossible to fix with

In a sandstorm the long lashes protect the camel's eyes, and its nostrils can readily be closed to keep the sand out.

Camels have been used as pack animals since early times, and can carry a load of about 400 lb for long distances.

certainty, as is the date when the one-humped camel came into existence, but the evidence suggests that it is a domesticated form derived from the Bactrian camel. Both readily interbreed, and the offspring usually have two humps, the hind hump smaller than that in front.

Surprisingly, the first record of a one-humped camel is on pottery from the sixth dynasty of Ancient Egypt (about 3500 BC) for the camel was not known in the Nile Valley until 3 000 years later. Its representation on the pottery may have been inspired by a wandering camel train from Asia Minor. Meanwhile, on Assyrian monuments dated 1115–1102 BC, and from then onwards, the camel appears quite often, and when the Queen of Sheba visited King Solomon in Jerusalem, in 955 BC, she brought with her draught camels. The name seems to be from the Semitic *gamal* or *hamal,* meaning 'carrying a burden'.

The one-humped camel was presumably selectively bred from domesticated two-humped camels, in Central Asia, by peoples who left no records. It is also suggested that the nickname 'ship of the desert' is derived from 'animal brought in a ship from the desert' by mis-translation—a reference to the Assyrian habit of naming an animal according to the place from which it came. Presumably this would mean camels were brought by ship across the Persian Gulf.

Feral camels

Today the Bactrian camel is confined to Asia but most of the 3 million Arabian camels are on African soil. Some have, however, been introduced into countries far from Africa or Asia. In 1622 some were taken to Tuscany where a herd still lives on the sandy plains near Pisa. On the plains of the Guadalquivir are feral camels taken to Spain by the Moors earlier still. Camels were taken to South America in the 16th century

◁ *Camel herd of an Arabian caravan. A miniature from a manuscript of about the 12th century illustrates the work of the poet Harari. (Cairo)*

▷ *Camels belonging to the Bedouin tribesmen drinking at a water trough in Saudi Arabia.*

by the Spanish conquistadors but these have died out. Others were taken to Virginia in 1701, and there was a second importation into the United States in 1856. The survivors from these were still running wild in the deserts of Arizona and Nevada in 1915. Camels were taken to Northern Australia, and there also they have reverted to the wild.

For a long time text books reiterated that no camels are now known in the wild state, although they had been mentioned in Chinese literature since the 5th century, and Marco Polo wrote about them. Then, in 1879, Nikolai Przewalski reported wild two-humped camels still living around Lake Lob, southeast of the Gobi desert. The local people told him they had been numerous a few decades prior to his visit but that they hunted them for their hides and flesh. There were reports also of camels in the Gobi, but nobody was prepared to say whether these were truly wild or merely feral camels. In 1945, the Soviet zoologist AG Bannikov rediscovered them and in 1955 a Mongolian film unit secured several shots of them.

These Gobi camels are two-humped but the humps are small. They are swift, with long slender legs, small feet and no knee pads. Their coat is short, the ears smaller than in the domesticated camels, and the coat is a brownish-red.

The Mongolian film shows the Gobi camel to be different from the typical Bactrian and Arabian camels, and it would be not unreasonable to conclude that it represents the ancestral stock from which the other two were domesticated.

class	**Mammalia**
order	**Artiodactyla**
suborder	**Tylopoda**
family	**Camelidae**
genus & species	***Camelus dromedarius*** *1-humped camel* ***Camelus bactrianus*** *2-humped camel*

△ *The camel draws water from the well while his keeper sleeps peacefully.*

▽ *Dromedaries are the riding strain of the one-humped camel, and can travel 100 miles in a day*

Cat

A typical short-haired domestic cat is about 2½ ft long, including a 9 in. tapering tail which, unlike that of the wild cat, is held horizontally when walking. The weight varies considerably, up to 21 lb being recorded. The claws are retractile and are kept in condition on a scratching post. They are used in climbing, and cats readily climb shrubs and trees to escape persecution (from dogs particularly), to rob birds' nests, or to lie along branches to bask in the sun. The muzzle is well-whiskered and the whiskers are used to feel the way in the dark.

The coat colour varies. The most common, the tabby, is of two kinds. One has narrow vertical stripes on the body, similar to those of the bush cat and European wild cat. The other is nearly the same colour, but consists of broad, mainly longitudinal dark lines and blotches on a light ground. In extreme cases, the dark markings are relatively few, strongly drawn, and stand out conspicuously against the lighter background. Such cats are recurrent mutants that parallel the king cheetah.

The names given to cats indicate some of the colour varieties: ginger, marmalade, tortoiseshell, blue, silver and the black cat, traditionally linked with witches, and, paradoxically, good luck. The pure white cat is either a dominant white or, if it has pink eyes, a total albino. The dominant white albino usually has blue eyes, and the popular view is that such cats are deaf.

△ In Great Britain today it is estimated that about 8½ million domesticated cats are household pets.
▽ A cat will readily climb trees to escape persecution, to rob birds' nests or just bask in the sun.

Loss of hearing

Darwin was one of the first scientists to note that white cats with blue eyes are usually deaf, and in 1965 Dr SK Bosher and Dr CS Hallpike, of the Medical Research Council in London, studied the development of the ear in kittens from deaf white parents. They found that for a few days after birth the kittens' ears were normal, then those of 75% of the kittens began to degenerate, and deafness followed. A few of the remainder retained hearing in both ears, and some were deaf in one ear only.

The hearing of other cats extends beyond the range of the human ear into the higher frequencies. This is why a cat responds more readily to a woman's voice. It probably means, also, that a cat waiting beside a mousehole is hearing the rodents' voices when these are inaudible to our ears.

Kittens born blind and deaf

The voice of a cat is the familiar mewing, with the purr being used to express contentment. When hunting, cats are silent, and their loudest vocalizations, usually referred to as 'yowling', must be too well known to need further description. These excessively discordant sounds indicate that the male, or tom, is seeking to impress the female, usually called a she-cat when of the common or garden breed, and a queen when she is a breeding female of pedigree stock.

△ *The distinctive Siamese, the sacred, royal or temple cat of Thailand, is a favoured breed today.*
▽ *Tabby cat asleep in the typical relaxed position shown by most cats – wild and domesticated.*

Domestic cats are sexually mature at 10 months or less, the earliest record being 3½ months. The height of the breeding season is from late December to March but the female may come on heat at intervals of 3–9 days from December to August. She is at her best for breeding purposes from 2–8 years. Males are at their highest potential from 3–8 years. Oestrus (heat) may last up to 21 days and is preceded by 2 or 3 days of excessive playfulness. Gestation is usually 65 days but may vary from 56–68 days. There may be up to 8 kittens in a litter; but 13 are known. Very young mothers have only 1 or 2, and the number drops again as the females approach 8 years.

Kittens are born blind, deaf and only lightly furred. The eyes open between 4–10 days. Milk teeth may appear at 4 days but it may be 5 weeks before all have erupted. Permanent teeth are cut between 4–7 months. Weaning begins at 2 months.

Night hunters

Domestic cats have departed less from their wild ancestors than dogs. Although both are intimate household pets, cats have retained a greater independence and more readily go wild, or feral. Dogs follow their prey mainly by scent and run it down. But cats hunt by sound and sight, stalking with infinite patience and stealth or lying in wait.

Cats are night hunters. Their eyes are protected by an iris diaphragm, used to exclude the bright rays of full daylight, giving a vertical slit pupil. At night, or in full shade by day, as in a room, the diaphragm opens fully, giving a rounded pupil, to take advantage of all possible light. As with most nocturnal animals, the eye has a tapetum, a layer of cells behind the retina which reflects light back across the retina to make the fullest use of it. It is the rays from artificial light at night, reflected back from the tapetum, that cause cats' eyes to glow, or shine in the dark, and to lesser extent this can happen on a starlit night with no extra illumination.

Mixed fortunes of cats

The household or domestic cat has been given the scientific name of *Felis catus*. Although various small cats have been tamed since prehistoric times, the present-day domestic cat seems to have been derived from the cafer cat or bush cat of Africa *F. lybica* perhaps with admixture from the European wild cat *F. silvestris*. It has had a chequered career: venerated and mummified by the Ancient Egyptians, worshipped by the Norsemen, given legal protection when first introduced into Europe, and persecuted with revolting cruelty in the Middle Ages because of its supposed link with witches.

Twenty years ago the dried body of a cat with two rats, one held in its mouth, was discovered in the cavity of a wall in a 17th- or 18th-century house in Southwark, London. Evidently, this was associated with a superstition, but precisely what is unknown. It may have been a talisman to keep out rats, but an octogenarian joiner, interviewed in 1960, declared it was the custom when he was a boy always to put the body of a cat and a rat in the wall of a new house 'to keep out the devil'.

Manx cats and other breeds

There are more than 30 breeds of domestic cat, some long-haired, like the Persian and Angora, but most are short-haired. The majority have long tails, but cats with short tails do occur, especially in southeast Asia. Some are tailless, the Manx cats, and are popularly supposed to be a breed peculiar to the Isle of Man, but probably this breed was originally developed in Japan.

The Manx condition has given rise to strange stories. Occasionally one or more kittens in a litter will be tailless. The story then gets around that the mother had mated with a rabbit or a hare. There have been instances when this story has reached the press, and for a short while accounts of hybrids between a cat and a rabbit create something of a sensation.

In one instance, at least, a litter of four, containing two Manx kittens, belonged to a female domestic cat that had gone wild. The nest was in low undergrowth and all around were rabbits, giving an air of truth to the story, which quickly gained currency locally, that the litter had been sired by a rabbit. Even a fully domesticated pregnant cat tends to seek solitude as the moment of birth approaches. She may even take to woods, where this is possible. In addition, many cats become permanently feral. They are little in evidence as a rule because of the cat's natural wariness, its nocturnal habits and climbing abilities. Also, a cat being a familiar animal, one does not normally stop to enquire whether the cat that disappears into the undergrowth at one's approach is a feral cat, a stray, or a household cat out hunting. Cats have been known to visit woods over a mile away fairly regularly and return to their houses after each hunting expedition.

Feral cats

The ease with which domestic cats become feral was underlined by the story of the Surrey puma. Between August 1964 and November 1967 over 300 people claimed to have seen a puma at large in southern England, the most persistent reports being from the western half of the county of Surrey. Investigation suggested that it was a large feral ginger cat that started the idea. Although feral dogs added to the story, as they did in other southern counties, feral cats were the real instigators and in more than one place the cat was killed and could be measured.

The number of feral cats in Britain, especially in woods on high ground, must be very high. It is usually the larger cats which take themselves off into the wild, but many unwanted cats are taken out into the countryside and abandoned. Such cats seem to grow larger than usual, possibly the result of a more athletic life as well as the abundance of natural food. Mayne Reid, writing in 1889, tells of a domestic cat that went wild and in 4 years doubled in size. The maximum size of which we have records is 42 in. overall, and eye-witness reports suggest that feral cats might attain the maximum dimensions for the wild cat of 45 in. and 30 lb weight.

Their numbers can only be surmised, but the game records for the Penrhyn Estates

△ The cat's eyes at night. In dim light the muscles of the iris relax to open the pupils wide to admit as much light as possible.

△ The cat's eyes during the day. In daylight certain muscles contract to make the pupils slit-like so not too much light enters.

△ Cat's typical hostile reaction to a dog.
▽ A frightened kitten ready to take flight or fight. The hairs are erected, and the claws are out.

▽ Tommy, a grey-brown tabby, the pet of Mervin Bedell, Long Island, USA, was taught to swim when he was about 4 months old.

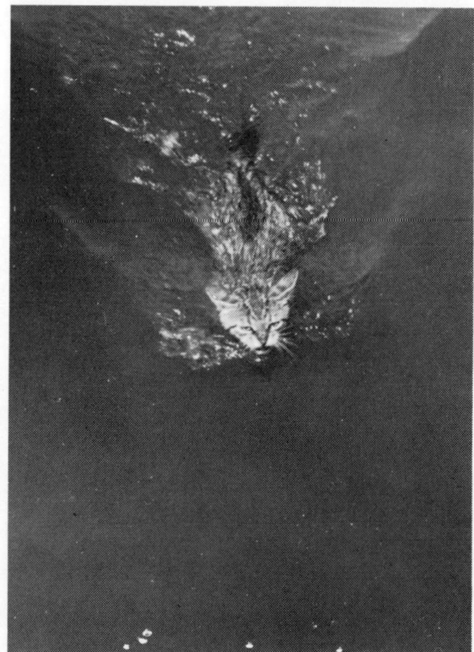

for 1874 to 1902 show that 98 polecats were shot as well as 13 pine martens and 2 310 cats. Even discounting household cats out hunting being shot by keepers, this is an extraordinary figure—one cat per week killed on average, in an area not densely populated by people. While some of these may have been truly wild cats it is fairly certain that many were feral cats and it suggests large-scale breeding in the wild. This is supported by conversations with present-day gamekeepers.

Ferocious cats

In *The Field* for 1871 there is an account showing how readily the tame cat can revert to the wild. It tells of a man who kept a number of tame birds and a cat. For a long time all lived amicably together. Then the cat began to kill the birds, and one day it leapt at its owner, landing on his chest, knocking him over backwards and scratching and biting him as he lay on the ground. His thick overcoat took the brunt of the attack, giving him time to beat off the cat, which ran off and was not seen again.

In breeding it is found that tabby is dominant over black and short hair is dominant over long hair. In practice it is found that feral cats not only achieve a size and ferocity uncommon in tame cats but that all revert to the tabby, whatever the colours of their forbears. All of the well known colour variation in domestic cats has arisen by mutation after they were domesticated.

Outstanding breeds

The long-haired Persian was formerly much sought after, but one of the most favoured breeds today is the Siamese, the sacred, royal or temple cat of Thailand. It was first introduced into England in 1884, and into the United States 10 years later. Related to it is the Burmese, which is probably the older breed, seal brown, with yellow or golden eyes, and without the dark points of the Siamese. Originally reputed to be owned by Burmese aristocracy and priests, each cat had its own servant and could not be bought.

The Abyssinian is probably the nearest to the cats of Ancient Egypt. Grey-brown to reddish in coat, with large, yellow, hazel or green eyes, lithe, with long slender legs and a weak voice, the Abyssinian has come into western Europe in relatively recent times. Its ancestors may well have been those associated with the cat-headed goddess of Ancient Egypt, whose chief seat of worship was the town of Bubastis. Some people claim that 'Puss' is a corruption of 'Bast', but the more likely explanation is that the word is onomatopoeic, in imitation of the cat's hiss. The word came into use first in 1605, when the study of Egyptology was not even in its infancy.

Mummified cats

It is generally accepted that the cat was first domesticated in Ancient Egypt or, if not, that the domestication reached a high peak there. The evidence for this rests on pictures of cats in Egyptian antiquities, and mummified cats deposited in tombs, especially those around Bubastis, during the 2 000 years before the birth of Christ Many statuettes, figurines and amulets of cats have also been collected from the same sources. That some of these, at least, represent fully domesticated cats can be seen in the tomb paintings of Thebes (1250 BC). In two of these, cats are depicted sitting under a chair, and in one of these the cat is wearing a collar and gnawing a bone.

The mummified cats in particular ought to enable us to decide their wild ancestors, yet this was for a long time in doubt—and for a quite unexpected reason. During the 19th century and the early years of the 20th, mummified cats were excavated in very large numbers and spread over the land as manure. They were also shipped abroad to be converted into fertilizers. One consignment alone, which reached England, contained 19 tons of mummified cats. All that was salvaged from this for study purposes was one skull sent to the British Museum.

In 1907, Professor WM Flinders Petrie. the distinguished Egyptologist, presented a collection of skulls of mummified animals to the British Museum. It consisted of 192 cat, 7 mongoose, 1 fox and 3 dog skulls. They were examined by Mr Oldfield Thomas who came to no firm conclusion about the cats.

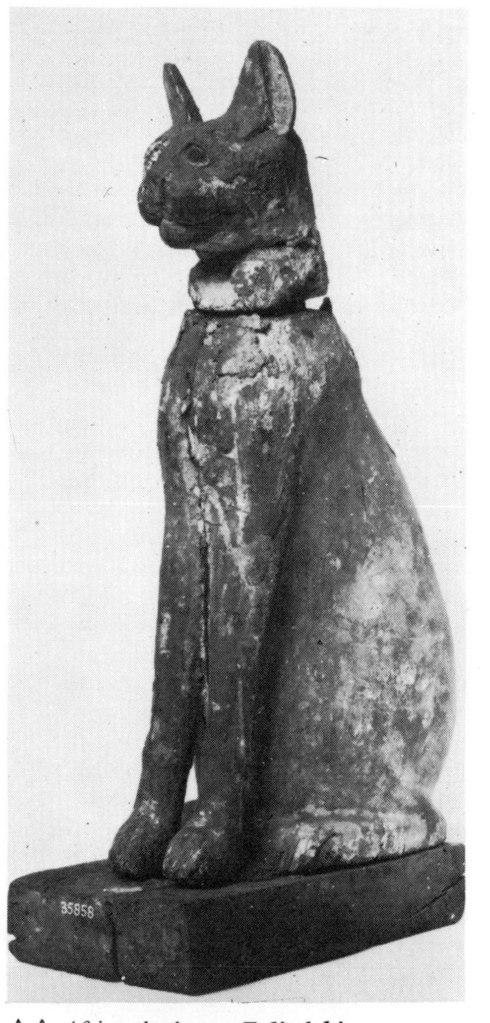

△△ *African bush cat,* ***Felis lybica*** *hunting. The present day domestic cat seems to have been bred from this species with admixture from the European wild cat,* ***F. silvestris.***
△ *Case for holding the mummy of a cat, which was regarded as sacred in Ancient Egypt.*
◁ *The black cat has been linked with evil and witches for centuries, but some say it is lucky.*

Then they were put in store and forgotten. In 1952 they were studied by Dr (now Sir) Terence Morrison-Scott. He concluded, after comparing his results with those of several authors who had examined collections of mummified cats during the 19th and early 20th centuries, that two species were represented. One was the jungle cat *Felis chaus,* the other a form of the African bush cat *Felis lybica.*

From other evidence, it seems that the jungle cat, which is the larger of the two, was not domesticated but foraged around the human settlements. It was only occasionally mummified. The most frequently mummified was the bush cat, but unlike its wild ancestors, judging from the pictures and statuettes, was ginger-coloured, with rather long ears and legs and with a long ringed tail.

class	**Mammalia**
order	**Carnivora**
family	**Felidae**
genus & species	***Felis catus***

Cattle

Two species of large hollow-horned ruminants, belonging to the family Bovidae, are called cattle. The name, from the Latin for 'head' indicates property, and meant originally much the same as 'capital' and 'chattel', words having the same origin. Although we sometimes speak of wild cattle this is a contradiction. Strictly speaking, cattle must be domesticated, and the two species are the western cattle (dealt with here) and the zebu of southern Asia and Africa.

Western cattle were derived from the aurochs, ***Bos primigenius****, which once lived in the forests of Europe and Asia and became extinct in 1627, the last survivor dying in a Polish park. The bull stood 6½ ft at the shoulder, with long curving horns. Its coat was black with a white stripe down the middle of the back, with white curly hair between the horns and white or greyish muzzle. The cows were smaller, brownish-red with some black or fawn patches. The calves were red.*

The colour, size and shape of horns varied from one part of the range to another, and this is reflected in the types of cattle seen over Europe today.

The date and place of the first domestication are unknown. By 2500 BC *several distinct domesticated breeds were already in existence. For example, Egyptian drawings of that date show a hornless breed, as well as long-horned and short-horned cattle.*

Invaders bring their own cattle

While prehistoric man was still a hunter, he killed the aurochs for its meat and hide. When it was domesticated he could get milk from it as well, and as man became a farmer, he could use it for pulling the plough. British cattle are generally thought to be of mixed ancestry, arising from different strains coming in from continental Europe. In settlements of Bronze Age and Iron Age people have been found remains of long-faced cattle, smaller than aurochs and with shorter horns. This particular species is now called Celtic shorthorn. Later, the Romans brought in their own cattle, larger than the Celtic, with white coats, some with long horns and, probably some hornless. These, it is thought, were the ancestors of the English white park cattle, which include herds such as the Chillingham and the Chartley.

The Anglo-Saxons brought red cattle, larger than the Celtic, and these became established more especially in southern and eastern England, and there was a marked division for a long time. In the 17th century for example, there were mainly all-black cattle in the north of England, the Midlands and in Wales, reflected today in the Welsh Black, Aberdeen Angus, and others. The Danes brought dun-coloured, hornless cattle, the Normans nothing new, but in the 16th to 18th centuries Dutch cattle were imported. These had broken colours, black and white and red and white.

Resurrecting the aurochs

The various breeds of cattle in Britain today reflect these admixtures, in their colours as in their names. Other breeds on the continent of Europe also tell of the course of events there. Some of this is contained in the account of experiments carried out in Germany aimed at 'resurrecting' the aurochs by breeding back.

In 1921 it occurred to Lutz and Heinz Heck that it might be possible to cross selected modern breeds of cattle to reconstruct their wild ancestor. Heinz Heck, director of the Munich Zoo, crossed Hungarian and Podolian (SW Russia) steppe cattle, Scottish Highland, Friesian and Corsican cattle and several grey and brown Alpine breeds. After some years he obtained a bull and a cow both of which had the characteristics of the aurochs, and these two bred true to type.

Lutz Heck, at the Zoological Gardens in Berlin, used Spanish fighting cattle, Corsican and Camargue cattle and English park cattle. These seem to be closer to the aurochs in type and he obtained quicker results. His stock in Berlin was lost during the Second World War although some of his specimens, sent to other zoos, survived.

The reconstituted aurochs not only had the physical characters of the wild ancestor but also its agility and wildness. It is of interest to recall in this connection that bones from Neolithic settlements, through the Bronze and Iron Ages, show that domesticated cattle decreased in size until the Iron Age, and then increased again until modern times. The original diminution in size may have been due to deliberate selection of less agile and less wild individuals, for in domestication, docility is preferable to aggressiveness and a marked agility makes for difficulty in control. Once these two ends were achieved later selection could be profitably directed towards increase in size.

Cattle ask to be domesticated

Two factors which made domestication of cattle relatively easy are their social nature and ability to do well on a wide variety of foods. Cattle naturally live in groups, or herds, which means they readily form social bonds with others of their kind. They also live readily with individuals of other species, which makes it easy for them to live in association with man.

The aurochs was widely distributed over Europe and Asia, which indicates a species ready to adapt to a wide range of climate and to accept a diversity of food plants. Both these were important in the early years of domestication, and also in the spread of cattle by humans to all parts of the world.

The whole process of domestication is so sophisticated that one is constantly surprised that primitive peoples should have accomplished it. Their knowledge of natural history may have been rule-of-thumb, but it was none the less profound. For example, primitive peoples today dealing with semi-wild cows quickly learn that the sight of a calf, or of a boy dressed in a calf skin, will cause the let-down of milk in an otherwise refractory cow. Moreover, in the absence of these the let-down can be stimulated through inflating the cow's vagina by blowing air into it. This gives the animal a stimulus similar to that which it has when a calf is born.

The gestation period of cattle varies with the breed and may be as little as 270 days or as much as 439 days.

The use to which cattle are put has varied

▷ *Highland cattle keeping cool.*

Aurochs, the wild ox of Europe and Asia, died out in 1627. It stood 6½ ft at the shoulder and the many modern breeds are descended from it.

Reconstituted aurochs, one of a herd produced in the Munich zoo by several years of interbreeding among modern varieties of cattle.

at different times and still varies in different countries.

Farmers usually divide cattle breeds into three types: beef, dual-purpose and dairy. In Britain they may be grouped respectively: beef–Shorthorn, Devon, Sussex, Hereford, Aberdeen Angus, Galloway, Belted Galloway, Highland; dual-purpose–Ayrshire, Friesian, Kerry, Jersey, Guernsey; dairy–Shorthorn, Lincoln Red Shorthorn, South Devon, Red Poll, Welsh Black, Dexter.

Beef cattle are primarily intended for meat production and their milk yield is of secondary importance. A good beef animal is deep, thickly and evenly fleshed, and blocky in appearance. In some parts of the world—for example, the La Plata river basin of South America and the western ranges of the United States—beef animals predominate. This is because they are well adapted to sparse grazing and can be farmed in arid areas. They may also be kept to convert surplus corn to protein.

A good dairy cow has a more angular shape, lighter in the fore-quarters and widening out backwards to give a wedge shape. Backbone and hips are less well fleshed but the chest is wide and deep, and the udders are well developed. The average milk yield of a British Friesian cow is 900 gallons in one lactation; many cows give over 20 000 lb of milk a year, and some give over 30 000 lb.

In western societies, milk, meat and leather, in that order, are the desired products. In some less sophisticated societies today none of these is important. In one place the dung may be valuable as fuel, or as a building material. In another place cattle may be kept for their meat and for sacrificial purposes. The Masai of East Africa drink the blood of their cattle, periodically bleeding them for this purpose. Leather may once have ranked first, for covering shields or for body armour. The bulls have been used for sport since the Minoan age in Ancient Crete.

It seems highly likely that in the early days of domestication and, in some countries, persisting even to this day, the primary use of cattle was as draught animals, for pulling carts and ploughs. In parts of Africa they are an index of wealth, a means of purchasing wives—cattle or chattels in the original sense of the word.

class	**Mammalia**
order	**Artiodactyla**
family	**Bovidae**
genus & species	***Bos taurus*** *western cattle*

Swiss cow in a mountain meadow, complete with cowbell (used in herds the world over).

Cheetahs qualify as 'big cats' by the way they crouch. They have blunt claws which can only be partly retracted.

Cheetah making a kill. Having caught its prey it despatches it by biting the neck and severing the jugular vein.

Cheetah

The cheetah is one of the 'big cats', distinguishable from other cats by the way it crouches. The big cats hold their feet out in front of them when crouched, like the lions in Trafalgar Square, London, while the small cats tuck their feet in, like a domestic cat. Big cats are more noisy than small cats, some of them being noted for their roaring. The cheetah utters a cry that is described as a 'barking howl'.

The legs of a cheetah are disproportionately long when compared with other cats, and the head is small. The length of head and body is just over 4 ft and that of the tail 2 ft. Cheetahs stand just over 3 ft high and weigh over 100 lb. Altogether, the physique suggests a lithe and speedy creature, and the cheetah is, indeed, the chief claimant to the title of the world's fastest land animal.

The ground colour of the coat is tawny to light grey with white underparts. Most of the body is covered with closely-spaced black spots, merging into black rings on the tail. On each side of the face there is a black stripe running from eye to mouth. In 1927 a second species of cheetah was described from Rhodesia. It was called the king cheetah and had black stripes replacing some of the spots, longitudinal on the back and tail and diagonal on the flanks. The species is based on only a few specimens and may be only a local aberration rather than a true species, or even a recurrent mutant.

Unlike other cats, the cheetah has claws that are blunt and can be only partly retracted.

Diminished range

The range of the cheetah extends from India westwards across to Morocco and Rio de Oro and southwards through the African continent to South Africa. It prefers open country. In places, for instance some parts of East Africa, it is still relatively abundant, but elsewhere it is very rare or extinct.

In India it is very unlikely that any survive. EP Gee, the authority on Indian animals, considers, in his book *Wild Life of India,* that they have been extinct for some years. Apart from unconfirmed reports, the last record seems to be of three males being shot in the same place at night using artificial light. This is a sad state of affairs when one considers that the 16th-century Emperor Akbar kept 1 000 cheetahs in captivity for hunting. Now, scarcity of native cheetahs has forced the Indian potentates to import hunting animals from Africa. The staggering decline of the Indian cheetah was not just due to the trade in coats, rugs and trophies; the spread of agriculture robbed it of its habitat, and its staple food, the blackbuck and axis deer, have been wiped out in many places.

Cheetahs are solitary but sometimes hunt in pairs. Occasionally, small bands of up to a dozen may be seen. These are composed of males and females or males alone. Bands made up of females only have not been seen.

Like other cats, cheetahs have regular scratching posts. This habit is used by trappers who set nooses at the post, with a reasonable certainty of success if the scratch marks are fresh.

Cheetahs can make good, but expensive pets, being affectionate and playful. Many stories in popular magazines attest to their good qualities. However, ownership should be restricted to people experienced in handling animals. An animal weighing a hundred pounds or more, with non-retractile claws, can inflict injury without meaning to do so, and there is always the fear that a cheetah may revert to its normal hunting habits of chasing animals that are moving away from it.

Sprinting hunters

The main prey are small-hoofed animals, chiefly axis deer and blackbuck in India, and Thomson's and Grant's gazelles in Africa. They will also attack large animals such as wildebeeste and zebra, and small animals such as hares probably form a substantial part of the diet. Ostriches and game birds, such as bustards and guinea fowl, are also eaten.

Unlike other cats, who tend to lie in wait for their prey and pounce with a single leap or a short rush when they are close enough, cheetahs will stalk their prey and then race after them, for some distance. In a short sprint they can easily overtake their prey, but if the latter gets a good start, the cheetah will drop out of the chase, exhausted from its burst of violent energy. Having caught up with its prey, the cheetah is said to knock it over and quickly despatch it by biting the neck to sever the jugular vein.

It seems that cheetahs attack only frightened animals. One story tells of four cheetahs trying to frighten a young warthog. The warthog refused to flee and the frustrated cheetahs left it, to kill an impala. If this idea is true, it is good evidence that pet cheetahs should be treated with caution. It would not do to run away from one.

Another story, recounted by a woman who lived in East Africa for some years, illustrates the cheetah's hunting habits. She came across a cheetah sitting within a few hundred yards of some Thomson's gazelles. Having made up its mind to catch one, the cheetah trotted towards them, upwind and making no effort at concealment. The gazelles immediately became alarmed but instead of fleeing they jinked about. Only when the cheetah suddenly bounded towards them did the gazelles break and flee. By this time it was too late, and the cheetah caught its intended victim. The strange part about this story is the careless way in which the cheetah approached its quarry. It certainly suggests that the spotted coat is not employed for camouflage, and it seems strange that the gazelles, which are built for swift running, did not flee immediately.

Captive breeding success

There are few records of breeding in the wild, which appears to take place all the year round. The litter ranges from 2–5 cubs. Observations over a period of years in East Africa suggest that there is a 50% mortality in the first year of life.

It is amazing that the cheetah was the last of the large cats to be bred successfully in captivity. Even today it has been bred in zoos only in a few instances. The first record of successful breeding in captivity was at Krefeld, Germany, in April 1960.

Dr Luciano Spinelli in his private zoo in Rome succeeded in breeding his cheetah 'Beauty'. She gave birth to two litters: one in January, 1966 and one in December, 1966. These two litters were the first recorded cases of cheetahs born in captivity, being successfully reared by their mother. The gestation period was 13 weeks, the cubs weighing $8\frac{1}{2}-9\frac{1}{2}$ oz at birth.

Since then several zoos have succeeded in breeding cheetahs in captivity. These include Milwaukee, Toledo and San Diego in the US; Whipsnade in England; and Montpellier in France.

How fast is the cheetah?

The cheetah is traditionally the fastest animal on land, but it is very difficult to find good evidence as to how fast it can run. One difficulty is to know over what distance should the speed be timed. A cheetah is fast over only very short distances. After a few hundred yards it gives up, so although a cheetah will outclass a human athlete in the 200-metres sprint, it is unlikely to complete the course in the 1 500-metres event.

One record that has been widely accepted is for a cheetah running at 71 mph, over 700 yards while timed with a stop-watch. It had also been claimed that it can accelerate to 45 mph in 2 seconds. Later writers have put the maximum speed at 60 mph, but doubt is often cast on this by people who have chased cheetahs with cars. They have suggested speeds of 30–50 mph only, but chasing an animal over, presumably, rough ground while watching a speedometer is not the best way to record speed. At the same time one wonders whether the cheetah is really making an effort to get away, under these circumstances.

The only satisfactory method of settling the dispute about the cheetah's capabilities is to time it over a properly measured course. This has already been tried. In the 1920's cheetah versus greyhound races were staged at Harringay, in London. The results were disappointing, the spectators were more interested in straightforward betting and the cheetah clocked up only 45 mph. But for those who would like to think cheetahs can really run faster, there is the consoling thought that this cheetah was the equivalent of the man who prefers to sit by the fireside on Saturday afternoon. Having failed to catch the hare in the first dash, the cheetah sat down to watch the fun.

class	**Mammalia**
order	**Carnivora**
family	**Felidae**
genus & species	***Acinonyx jubatus***

Cheetah slaking its thirst. This and the other picture on this page clearly show the animal's disproportionately long legs; with about the same body length as a leopard, it stands a foot higher.

Cheetah at rest, but still watchful, on the East African plains.

Chimpanzee

One of the great apes and the nearest in intelligence to man, the chimpanzee is one of the most studied and popular of animals. Scientists have examined its mental capacities and sent it into space in anticipation of man. To the general public, the chimpanzee is the familiar clown of circus acts and tea parties at the zoo. Yet despite all our knowledge of the chimpanzee's capabilities in the laboratory, it is only recently that its habits in the wild have been studied, and these are proving to be more remarkable than its antics in captivity.

Chimpanzees need little description. Being apes and not monkeys, they have no tail. Their arms are longer than their legs and they normally run on all fours, but they can walk upright, with toes turned outwards. When erect they stand 3–5 ft high. The hair is long and coarse, black except for a white patch near the rump. The face, ears, hands and feet are naked and, except for the black face, flesh coloured.

Forest families

The single species of chimpanzee lives in the tropical rain forests of Africa, roughly from the Niger basin to Angola. They are at home in the trees, making nests of branches and vines each night to sleep in, but they often come down to the ground to search for food. Whereas their normal gait on the ground is all fours, they will run on three legs, leaving one free to hold food, or on their hind legs, in an amusing waddling gait, when carrying an armful of food.

Chimpanzees live in small parties, occasionally numbering up to 40, but the bonds between the members of a party are weak. There is no fixed social structure like that found in baboon troops. A chimpanzee party is constantly varying in size as members leave to wander off in the forests by themselves or return from such a wandering. The only constant unit of social life is a mother with her young. She may have two or three of different ages with her at any time because they stay with her for several years. The usual size of a group is from 3–6, but numbers increase as chimpanzees gather at a source of plentiful and tasty food, or if a female comes on heat when the males will gather round her for several days.

Within a party, the males are arranged in a social order, the inferior ones respecting the superior ones. Dominance is related to age; a chimpanzee gradually rises in social position from the time he is physically mature and leaves the protection of his mother. The status of a male seems to be partly determined by noisy displays, charging about waving branches or rocks or drumming the feet on the plank-like buttresses of the forest trees. This behaviour is sometimes sparked off by frustration brought on by seeing more dominant males enjoying food without sharing it. Yet the chimpanzees recognize the right of ownership sufficiently to prevent a dominant male from wresting food from one of his inferiors.

The gamut of emotions

△ *'I'm content'—relaxed and normal.*

△ *'Hello'—the greeting pout.*

△ *'I'm happy' smile, showing bottom teeth only.*
▽ *Tantrum, showing top and bottom teeth.*

First aid and affection

When chimpanzees meet after having been apart they greet each other in a very human way, by touching each other or even clasping hands and kissing. The arrival of a dominant male is the signal for the rest to hurry over and pay their respects to him. The members of a party also spend a considerable amount of time grooming each other, and themselves. Mothers carefully go through the fur of their babies for any foreign particles, spending more and more time on the task as the babies grow older. Dirt, burrs, dried skin and ticks are plucked off and splinters may be removed by pinching them out with forefingers or lips. Such mutual help may lead to further first aid. A captive female was once seen to approach her male companion, whimpering. She sat down while the other chimpanzee sat opposite her and, holding her head steady with one hand, pulled the lower lid of one eye down with the other. After a short inspection he removed a speck of grit from her eye with his finger, to her evident relief.

Hunting for meat

About 7 hours a day may be spent feeding either up trees or on the ground. The chimpanzees investigate any source likely to produce food. Crevices in logs are searched for insects and nests are robbed of eggs and chicks, but their usual food consists of fruits, leaves and roots. Ripening fruit crops, of bananas, pawpaws or wild figs, are a special attraction to them and they are sometimes a nuisance when they attack plantations. A big male chimpanzee can eat over 50 bananas at one sitting.

Until recently it was thought that the only flesh eaten by chimpanzees was that of insects and occasionally birds and small rodents. They have now been found to hunt larger animals, some individuals apparently being particularly fond of meat. Young bushbucks and bushpigs have been seen caught by chimpanzees, as well as colobus monkeys and young baboons. Jane Goodall, the British naturalist who spent several years in Africa studying chimpanzees in the wild, has given a very graphic description of a chimpanzee catching a young baboon and killing it by holding its back legs and smashing its head against the ground.

Obedient children

Chimpanzees are promiscuous. When a female comes on heat the males gather round her, bounding and leaping through the branches. All of them mate with her, no matter what their social standing. She remains on heat for several days, then the males lose interest.

A single baby—twins are rare—is born after about 230 days. If it is the female's first baby, she does not at first seem to know what to do with it, but by a combination of instinct, knowledge gained from having seen other babies, and learning, she soon starts to care for it. For 2 years the baby will be completely dependent on her. At first she carries it to her breast, but as it grows larger it rides pick-a-back.

The standard of baby care shown by

Top: *Young chimps, Primrose (left) and Peter.*
Right: *Chimpanzees are the best tool-users apart from man, using natural objects to gather food, crack nuts, and drive off enemies.*

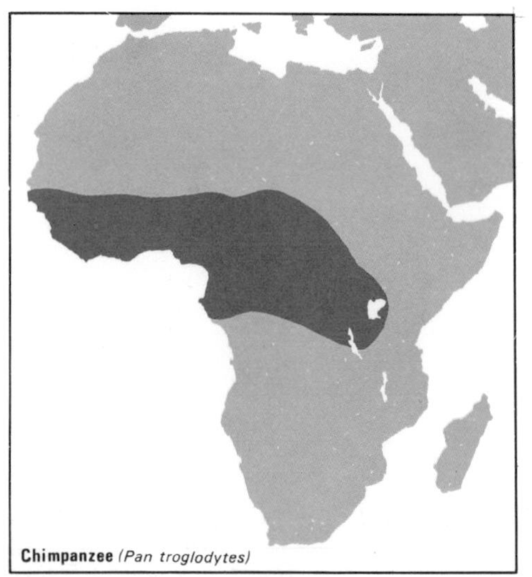

female chimpanzees varies considerably. Some are ideal mothers, caring for their babies zealously and caressing and kissing them. Others are over-attentive, and the babies are 'spoilt'; and yet others neglect their children. The standard of care and education, however, is on the whole exemplary. The babies are not usually bullied or spoiled, yet they obey the parents' orders instantly. When they leave their mother's back they have considerable freedom, and can climb over dominant males without fear.

The babies are carried for varying periods. Sometimes they are still riding on their mothers when 4 years old. By this time the mother will have another baby and the elder one has to fend more for itself, but chimpanzees have been seen hand feeding young that are 6 or 7 years old.

Tools for chimpanzees

Man is sometimes called the toolmaker to distinguish him from other animals. It is difficult to decide when our ancestors became human-like rather than ape-like,

and toolmaking is one factor used as a line of contrast. Upright gait and speech are others, but it is difficult to make rigid pronouncements about features that must have evolved gradually.

Tools can be regarded as extensions of the body used to help with certain tasks. Few animals are known to use tools, but the real difference that separates man and the rest of the animal world is that he not only uses a variety of tools, he makes them, fashioning natural objects to suit his purpose. In this way, opening a nut with a stone is tool using, but shaping the stone into an axe is toolmaking.

Chimpanzees are the best tool-users apart from man. In captivity, they have been seen to throw stones and brandish clubs when put in a cage near a leopard and they are mentally well equipped to work out how to use tools, which are used by some other animals more or less instinctively. Chimpanzees have solved such problems as fitting two sticks together or balancing boxes on top of each other to get at otherwise inaccessible bananas.

The observations by Jane Goodall and others on wild chimpanzees have shown that they also use a variety of tools. The most common use is to extract honey, ants or termites from nests. Sticks 2–3 ft long are picked off the ground or broken from branches and pushed into nests, then withdrawn, and the honey or insects licked off. Stones are used to crack nuts, or as missiles to drive humans or baboons away from the chimpanzees' food. The stones, which sometimes weigh several pounds, are thrown, overarm, not very accurately but definitely aimed. Another material used for tools is leaves. Chimpanzees have been seen plucking leaves, chewing them up, and using the resultant mass as a sponge. Water, in a natural bowl in a tree, was soaked up into the sponge and squeezed out into the chimpanzee's mouth. Whole leaves have also been used for wiping sticky lips and hands after eating bananas.

The variety of tools used by the chimpanzees is made more interesting because they actually make some of their implements. To make a suitable rod to extract insects, the chimpanzees will strip the leaves off a twig or tear shreds off a grass stem to make it narrower. These are clear signs of modifying natural material for a specific use, as is the chewing of leaves to make a sponge. So man is not the only toolmaker, merely better at it than his relatives.

A final point arises from these observations on wild chimpanzees. Babies were seen to play with tools discarded by their elders after having watched them being used. At first their efforts at imitation were clumsy but by 3 years of age they were using them competently. Here is the beginning of a culture in which individuals learn skills passed on from generation to generation.

class	**Mammalia**
order	**Primates**
family	**Pongidae**
genus & species	***Pan troglodytes***

△ △ *Chimpanzee first-aid: incredible instance of male removing grit from his mate's eye.*
△ *A mother suckles her baby. Chimpanzees show great affection to their young, and discipline is good without the need for bullying.*

Chipmunk

There are 2 genera of chipmunks among many kinds of ground squirrels. The eastern chipmunk is the larger of the two, with head and body measuring 5–7 in. and the tail 3–9 in. The fur is reddish-brown with dark stripes on the back, alternating with two lighter stripes. The tail is not as bushy as that of tree squirrels. The western or Siberian chipmunk has a smaller body than the eastern chipmunk, being about 4 in. long, but the tail is as long. Its fur is lighter and there are 5 lightish stripes between the dark ones.

The two genera are quite easily distinguished by the rufous rump of the eastern chipmunk and by the teeth. The eastern chipmunk has one upper premolar (grinding tooth) on each side of the jaw, and the Siberian chipmunk – a north Asiatic ground squirrel and a proficient climber – has two.

The names eastern and western chipmunks refer to their distribution in North America. The single species of eastern chipmunk lives in most of the eastern United States and south eastern Canada, where it thrives in regions of deciduous forest and shrub, being found around fallen logs, rocks or outbuildings. The western chipmunks, of which there are about 17 species, live in North America from the Yukon to Sonora in Mexico, and are spread across Asia from northern Russia through Siberia to northern China and northern Japan. There are 16 American species and one, the Siberian chipmunk, extends across Asia. They are widely distributed within their ranges, except in dense forests. On the whole, the western chipmunks prefer more open country, and are abundant in pasture land and on rocky cliffs.

▽ *Chipmunks feed mainly on berries, fruit, nuts and seeds – but they never store any fruit or flesh likely to go bad. Food that is not to be eaten immediately is carried in the cheek pouches and hidden until winter.*

Permanent residents

Chipmunks are active by day and are good climbers, but prefer to stay on the ground, although the eastern chipmunks sometimes rear their young in trees. They make a complicated system of burrows underground, often running under logs and stones, or delving several feet under the turf. Each burrow is owned by one chipmunk who continues digging throughout its life, so that the burrows may reach lengths of 30 ft or more, and have more than one entrance and perhaps several side chambers, one of which probably contains a nest of leaves and grass.

Although common animals, chipmunks rarely become pests of agriculture or forestry, and when they do they can be controlled easily. Apart from this damage, they are popular animals and readily become tame, visiting campsites to steal food or to accept it from the hand.

Chipmunks do not hibernate in the strict sense, but during bad weather they go into a state of torpor, awakening every now and then to feed from their caches of supplies.

Wide variety of food

The main food is berries, fruits, nuts, and small seeds, which are collected after they have fallen to the ground, or harvested by climbing trees and shrubs to pick them. Fungi, grass and leaves are also eaten. The chipmunks are also carnivorous, taking slugs, snails, aphids and other insects. Small birds, eggs, mice and small snakes are also taken, and in some places the eastern chipmunk is considered one of the chief enemies of the rosy finch.

Food that is not immediately needed is carried in the cheek pouches and cached for use in the winter. The cheek pouches are loose folds of skin, naked inside but not moist, that open into the side of the mouth. To fill its pouches, a chipmunk holds a nut in its paws, neatly bites off the sharp point on each end and slips it into one pouch. The next nut is placed on the other side and the pouches are filled alternately so that the chipmunk's face, although looking extremely bizarre, is at least balanced. It can take up to four nuts in each pouch and another between the teeth.

△ *One for the road; a chipmunk can hold about four nuts in each cheek pouch with another in its teeth. A chipmunk's winter store is made up of many caches; there are reports of hoards of up to 32 quarts of nuts.*

Breeding in burrows

Mating takes place from February to April, the males seeking out the females in their burrows. The young are born 31 days later. The babies, 2–8 in number, spend a month in the nest then begin to accompany their mother on foraging trips above ground, venturing further each time. They stay together for 6 weeks then go off on their own. There may be two litters in one year.

Chipmunks' enemies

Coyotes, bobcats, foxes, hawks and owls prey on chipmunks. Weasels and snakes are their worst enemies as they can follow them into their burrows, where they are safe from other predators. When danger threatens, the chipmunks alert each other with an alarm call, and dash to cover. The alarm call is a guttural scolding or a whistle.

Misers' hoards

To compare someone with a squirrel is to condemn him as an inveterate hoarder who fills his house with all manner of objects, not so much with any end in mind, but because something might come in useful sometime, and it would be a pity to get rid of it. Chipmunks are probably the most expert hoarders of all the squirrel family but, unlike their human namesakes, they are selective about the things they hoard. When a chipmunk is collecting its winter store, it selects only nuts and cones, never any fruit or flesh that would go bad. The chipmunk takes first prize, however, for the sheer bulk of its stores. Reports have been made of caches containing '8 quarts of acorns' or 'a bushel (32 quarts) of nuts', and one cache does not form a chipmunk's complete winter store. More than one cache may be made in the burrow, and small caches are made all over the chipmunk's home range. In this way it combines the behaviour of the chickaree that makes one or two large stores, and the grey squirrel that makes many small ones. Like the grey squirrel, the chipmunk forgets the position of its small, scattered stores which consist of just a mouthful of nuts buried under leaves or turf. During the winter it may find some of them by smell, otherwise they remain hidden until they germinate and contribute to the growth of the woodland. So the chipmunk does not have to venture into bad weather to feed, and to save further trouble, the store is usually placed in the legendary hideaway of human misers—under the bed!

class	**Mammalia**
order	**Rodentia**
family	**Sciuridae**
genera & species	***Tamias striatus*** *eastern chipmunk* ***Eutamias sibiricus*** *Siberian chipmunk* *others*

△ *A troop of coatis. Outside the breeding season the males are solitary, but the females and young live in bands of up to 20, sharing nests to sleep and taking turns to groom each other.*

▷ *An inquisitive coati standing on its hind legs to get a better view.*

Coati

The coati is a small carnivorous mammal related to the raccoon (page 191), and the cacomistle (page 39). The head and body measure $1\frac{1}{2}$–2 ft long. The ears are small and the forehead flat, running down to a long, mobile snout that extends beyond the jaw. The black nose at the tip is moist which helps give the coati an excellent sense of smell. The general colour is reddish-brown to black, with yellowish-brown underparts and black and grey markings on the face. The tail is up to $2\frac{1}{2}$ ft long and is banded. It is generally held vertically with the tip curled over.

The three species of coati are found in the forests of South and Central America. They are also found in the southwestern parts of the United States, in Arizona, New Mexico and Texas and here they are extending their range northwards.

Social forest dweller

Young coatis and the females live in bands of up to 20. The males lead a solitary life outside the breeding season. Each band has a home range, the borders of which overlap with those of other bands, but they rarely meet because they tend to stay near the centres of the ranges. When they do meet, there is usually some threatening and arguing but rarely any fighting. The band forages together, retiring just after sunset to a favoured tree where the coatis remain until sunrise. They sleep curled up with their tails over their faces, in nests of twigs and creepers placed in a fork or on a mat of branches. Several coatis may share a nest. There is no fixed hierarchy, or 'peck order', and there is very little aggression between members of a band. Neither is there any system of sentinels to keep guard while the band is feeding.

Although very active throughout the day, the coatis being very inquisitive, there are periods of rest, when the members of a band will groom themselves. This is an all engrossing habit and one was seen to scratch so hard with her forepaws that she rolled off a log and lay on the ground still scratching. Coatis also groom each other. Sitting head-to-tail they work over each other's bodies, gently nibbling with their teeth. Not only will this keep them clear of parasites, it probably strengthens the bonds between the individuals in the group. While they are out foraging, the band may get split up, but before long the parties will search for each other, with chittering calls, the laggards running to catch up and the others doubling back to find them.

When travelling through the forests coatis walk or gallop with fore- and hindfeet working together, presenting an amusing sight as the band bounds over the ground with tails held high. They also adopt two gaits in climbing, either ascending hand-over-hand or galloping up wide trunks with forefeet and hindfeet clutching the bark together. To descend, they come down head-first with their hindfeet held backwards, like a squirrel, but they are not so completely at home in the trees as a squirrel or pine marten. They will jump from branch

to branch, but not over any great distance, and they seem to fall quite often, although they land without hurting themselves.

Foraging in the forest

Coatis forage both in the leaf litter on the forest floor and up the trees. On the ground they run about, sniffing to locate any small animals hidden underground or in rotten logs. Here many kinds of invertebrate can be snapped up, including millipedes, earthworms, termites, snails and tarantulas. Lizards and mice are chased when flushed and dug out of their holes, as are land crabs and caecilians. The coatis are very persistent once they have discovered some small animal's lair, digging down after it like a dog. They have been seen to spend half an hour digging out lizards, burrowing so far down that their bodies have disappeared from sight. Land crabs are dealt with by deftly flicking them into the open and ripping off their claws. If their prey escapes and bolts, the whole band will set off in pursuit of it.

Once the prey is caught it is killed either by being bitten in the neck in the case of lizards and rodents, or by being rolled under the front paws until well mangled. This may be to crush the shell or wipe off hairs, stings and so on.

Coatis eat fruits such as wild bananas, figs, mangoes and papayas, either eating the fallen fruit or climbing the trees to pluck it. Sometimes a band splits up, one part waiting under the trees to pick up fruit dropped by the others.

Young reared in nests

Mating takes place during the dry season. During this time the bands of females and juveniles are joined by the males, who are aggressive towards each other, each male jealously guarding his own band, although additional males are also discouraged by the females. The male sleeps and grooms with the band. Gestation lasts 10 or 11 weeks and just before the litter is due, the female leaves the band and makes a nest in a tree. The 2–6 young are born in the nest and remain there for 5 weeks, by which time they can run and climb well enough to keep up with their mother when she rejoins the band.

The mother sits to suckle her babies

The babies are suckled by their mother who is in a sitting position, with her infants lying on their sides, holding on with their forepaws. Sometimes the young coatis become too adventurous and have to be rescued from a precarious position among the branches by their mother, who carries them in her mouth.

After the female rejoins the band the young stay with her, only gradually becoming more independent. The young males leave the band when they become sexually mature at 2 years.

The longevity of the coati in captivity is around 10 years. A red coati lived for 14 years, 9 months, 1 day in the Philadelphia Zoological Gardens.

Segregation of the sexes

'Coati' is derived from South American Indian words 'cua' (belt) and 'tim' (nose), but why they should be called 'beltnoses' is obscure. The Latin name *Nasua* or 'nosy one' is more appropriate, when one considers the way they thrust their long snouts through leaf litter and down holes in search of food. Another English vernacular name is coatimundi, but this should refer specifically to the lone males, for the inhabitants of Central and South America thought that there were two kinds of animal—the coati living in bands, and the coatimundi, leading a solitary life. It also has native names of 'pisote', or 'pisote solo' in Spanish American, the latter referring to the solitary animal. Before anyone had studied the animal this distinction was accepted by zoologists and two species were named: *Nasua sociabilis* and *Nasua solitaria*.

Coatis certainly have an unusual social life with the segregation of the sexes. This is more usual in the hoofed animals, but in the bighorns, for instance, the males flock together outside the breeding season, whereas the male coatis are strictly solitary. Each male has its own range and is aggressive towards any other male it should meet. The bands are also aggressive to the males, driving them away by threats or actual attacks. One adult female can send a male packing and sometimes the whole band will descend on him and drag him down. Even the immature males will face up unafraid to the old males because they have the support of the rest of the band.

class	**Mammalia**
order	**Carnivora**
family	**Procyonidae**
genus & species	***Nasua nasua*** ***N. nelsoni*** ***N. narica***

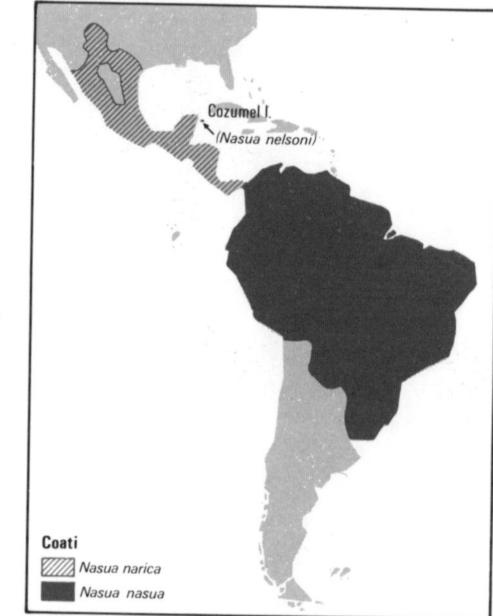

On the track of a meal. Coatis are inveterate diggers—especially when hunting lizards—and will often dig themselves underground.

In the Amazon River jungle. As well as foraging in the leaf litter on the jungle floor, coatis readily climb for their food.

Caught by the camera: a cottontail surprised while resting prepares to make itself scarce. These rabbits fall prey to every flesh-eater in their region, and often need all their flashing speed to escape.

Cottontail

The cottontail is a small rabbit varying from dark grey to reddish-brown. The upper parts are brown and sides grey, with a rufous nape and legs. The ears are short and rounded. The tail is brown above and white below and gives rise to the name cottontail from its resemblance to a cotton boll. There are about 13 species whose head and body length varies from 10–18 in. and weight from 14 oz–5 lb.

Abundant everywhere

Cottontail rabbits of various kinds range from southern Canada to South America, as far as Argentina and Paraguay. Over this range they live in a wide variety of habitats. Most species prefer open woodland and brush or clearings in forests. Consequently some species flourished in early colonial times when European settlers were first opening up the forests. Later, complete destruction of forests acted against these species. The New England cottontail of the Appalachians preferred open woodland with undergrowth, but with the spread of agriculture this species has been replaced by the Eastern cottontail that inhabits open country such as pastureland.

Other species of cottontail live in more extreme habitats. The smallest species, the Idaho cottontail, lives in deserts, and the largest, the swamp or marsh rabbit, sometimes called the 'canecutter', lives in swampy areas. It has large feet with splayed, slightly furred toes, and swims well. When alarmed it is said to make for water where it hides with only nose and ears showing.

Cottontails are timid animals, ready to bolt for cover at the slightest hint of danger, finding a suitable place where they crouch motionless, their neutral colour blending in so well with the vegetation that they are very difficult to find. In this they are aided by their nocturnal habits. Most species are active when the light is dim, but during the summer months, when nights are short, the cottontails are more active and are more likely to be seen in broad daylight.

Each cottontail has a range of several acres that is crossed by regular runs through the ground vegetation. In these runs the animal knows its way so well that it can hurtle at a full speed of 20–25 mph when frightened. Cottontails usually use the burrows of other animals, such as woodchucks. Only the Idaho cottontail makes its own burrow.

Unusual digestion of food

Cottontails eat grass and broad-leaved annuals and can severely damage crops and gardens. The damage is made the more severe because more plants are injured than eaten. Some are trampled and others only nibbled, sufficient to spoil them as a food crop without satisfying the cottontail's hunger. If this is done to young plants they grow stunted and deformed. In winter the cottontails feed on buds and soft twigs, and young saplings can be found neatly cut off at the level of the snowline.

Digestion of coarse herbage and twigs is difficult, as they contain large quantities of indigestible cellulose, which has to be assimilated. Other grazing animals, such as cows and sheep, chew the cud, a process in which food first goes to the rumen, the first compartment of the multiple stomach, where the cellulose is broken down, then brought up to the mouth and back to the true stomach. In rabbits another process is used. Food is passed through the digestive system twice, to ensure complete digestion, for they eat the faeces containing food undigested after the first time through.

Breeding like other rabbits

The breeding season is long, from February to September in temperate parts of North America, but in snowy areas such as the Sierra Nevada it starts earlier, for in summer the vegetation will have shrivelled and there will be little food for the young. In some parts where conditions are always favourable, breeding may take place all the year round.

Fighting sometimes occurs between males, with fur flying, and courtship dances may take place in which one rabbit leaps in the air while the other runs under it. After mating the male is driven away and the female rears the litter alone. The young cottontails, called fawns, are born a month after mating, naked and blind. There may be up to seven of them, weighing less than 1 oz each. They are placed in a shallow nest that the mother has scraped in the ground, perhaps by enlarging the hoofprint of a horse or cow, and lined with grass and fur plucked from her breast. When she leaves the nest she carefully covers it with grass. The nest is so small that the mother cannot lie in it. Instead she crouches over it and the fawns have to climb up to suckle.

When a fortnight old the fawns leave the nest and feed nearby during the day. They finally disperse when 3 weeks old. Their mother will have mated again within a few hours of their birth and the next litter will soon be due. She may have up to 5 litters in a year.

Every man's meat

Cottontails fall prey to every flesh-eating animal. Skunks, foxes and crows search out and kill the young in the nests and owls, hawks, snakes and chickarees take the young when they leave their nests. Other predators such as bobcats and eagles take the adults. Man also hunts and traps cottontails and in the eastern United States they are the chief game animal, sometimes several million cottontails being killed annually in one State. Their fur is of little use except in the manufacture of felt, but they are eaten, although care is needed in handling and cooking them as a rabbit disease called turalemia is infectious to man. Proper cooking is sufficient to kill all germs.

Staple food for many

The cottontail, whose only defence is not to be seen as it crouches motionless, half hidden by greenery, forms the staple food of many American predators, much in the same way as the European rabbit once was the mainstay of many carnivores and birds of prey on the other side of the Atlantic. The importance of the cottontail in the natural economy has been shown in studies of the feeding habits of the predators.

One method of finding out an animal's food preferences is to examine the undigested remains passed out in the excrement. Bones of small mammals and birds are often easily identifiable, and while fish-bones are usually digested, their earbones, or otoliths, may pass through unchanged. The shells of snails and other molluscs may also be present, along with feathers and fur. This method has to be used carefully because it can give misleading results. For instance, fish remains are rarely found around herons' nests because the bones are so easily digested.

In a study made in the Sierra Nevada, cottontails were found to be the staple diet of coyotes, grey foxes, bobcats, horned owls and gopher snakes. The horned owl's diet consisted of 61% cottontails. Rattlesnakes and red-tailed hawks also took large numbers of cottontails, but preferred ground squirrels. These predators are only the main natural enemies of cottontails in this region, and, as well as other animals which take cottontails only occasionally, we must add man, floods and disease as agents that significantly reduce their numbers.

The point of cataloguing the sources of dangers to cottontails is to demonstrate the vast numbers of cottontails that must exist. In the study of the predators' feeding preferences it was calculated that they captured over 7 lb of cottontails each year on every acre of ground. So more than the actual cottontail population is being eaten by only some of their enemies, but the apparent contradiction is solved by the extremely rapid rate of breeding. Each female cottontail is producing litters, averaging four young, 3–5 times a year. If those struck down by a predator only weigh 1 lb, there is still a good surplus left to continue the species.

▷ *Getaway: surprised from the rear, a cottontail erupts from a hollow in the snow.*

▷△ *Hotly pursued by a labouring beagle, this cottontail has a good chance of escape.*

▷▽ *After clearing the snow off its front doorstep, a cottontail surveys the scene.*

class	**Mammalia**
order	**Lagomorpha**
family	**Leporidae**
genus	***Sylvilagus***

Coyote

The coyote belongs with domestic dogs and wolves in the genus **Canis.** *The name comes from the Mexican* **coyotl** *a [illegible] can be pronounced with or without the 'e' sounded. A coyote weighs 20–50 lb and measures about 4 ft from nose to tail-tip. The fur is tawny and the tail, bushy with a black tip, droops low behind the hind legs, instead of being carried horizontally as wolves do. Another difference from wolves is that coyotes are smaller, and they hunt smaller game than wolves.*

Coyotes used to live on plains and in woods of the western part of North America, being known as the brush wolf in forested regions and the prairie wolf in open lands. Within the last century their range has increased despite persecution and they are now found from north Alaska to Costa Rica. They have also spread eastwards to the Atlantic seaboard. At the turn of the century they had reached Michigan; they were seen in New York State in 1925 and in Massachusetts in 1957. The northward spread carried them to the southern shores of Hudson's Bay by 1961.

The prairie wolf's spread

In the face of man's persecution most carnivorous animals have been retreating. Their habitat has been destroyed and they are hunted mercilessly as vermin or as valuable fur bearers. The coyote is extending its range. There is no market for its fur but the coyote has long been shot on sight, or trapped and poisoned, because it has been regarded as an enemy of livestock and a competitor against man for game. Many thousands have been killed – 125 000 a year according to one estimate – yet the coyotes flourish. Their powers of survival seem to lie in their proverbial wariness and their adaptability. They are difficult to trap unless a ruse is employed. One coyote avoided every trap set for it until the trapper buried an alarm clock near a trap and the coyote, overcome by curiosity, walked right into it.

The spread into the northeast United States is probably linked with widespread tree-felling and the decline and extinction of the timber wolf. This left a gap in the wildlife of the area which the adaptable coyote was able to fill. Even urban development has not deterred coyotes. They have moved into suburbs where, like the red fox in Britain, they can supplement their diet with gleanings from dustbins and other sources. There is a story of a Californian who wondered why his dog was not gaining weight despite being very well fed. He later discovered that a pair of coyotes were stealing its food. One lured the dog away while the other bolted the contents of the feeding bowl.

The coyotes' diet

Coyotes are persecuted because of their reputation as killers of livestock and deer. While sheep, goats and deer are occasionally killed, the reputation has probably been encouraged by the coyote's carrion-eating habits. A half-eaten carcase of a cow or sheep with coyote tracks around it leads to the assumption that coyotes killed it, and revenge is exacted without thought that the animal might have died for some other reason, such as thirst.

As proof that coyotes are not major threats to livestock, several thousand dead coyotes have been examined. Their stomachs contained mainly jack rabbits and cottontails, together with mice, voles and other small rodents. Poultry and livestock made up about $\frac{1}{8}$ of the sample. It is probable that, as with other animals with a varied diet, coyotes will eat whatever is most available. If rabbits are abundant, then poultry runs are left alone, but if a square meal in the form of a weak calf is found, then it is not overlooked. Many other items are eaten; insects, birds, trout and crayfish have been found in coyote stomachs. Beavers, domestic cats, skunks and even grey foxes have been known to be attacked and eaten. Sometimes coyotes eat large amounts of vegetable matter, including prickly pears, grass and nuts.

Coyotes hunt singly or in pairs, running down their prey with speeds of over 40 mph. Sometimes they chase deer in relays, one coyote taking over the pursuit as another becomes tired. Another habit is to sham dead, waiting for inquisitive, carrion-eating birds such as crows to land and examine the 'corpse', when it leaps up and grabs them.

Coyotes are model parents

Breeding begins when coyotes are a year old and they pair for life. They mate during January to March and the pups are born 63 days later. The den is usually made in a burrow abandoned by a woodchuck, skunk or fox, which is enlarged to form a tunnel up to 30 ft long and 1–2 ft in diameter, ending in a nesting chamber which is kept scrupulously clean. Nests are sometimes made on the surface—for instance, in marshlands where tunnels would be flooded.

Up to 19 pups may be born in a litter, the average being around 10. They are born with their eyes shut and stay underground for over a month. The father stays with the family, bringing food first for the mother, then for the pups; this is regurgitated to them as a partly-digested mess. Later the family go out on communal hunting trips and the pups learn to hunt for themselves. Although hunters themselves, coyotes are still not immune to attack by larger predators, and coyotes are known to have been killed by wolves, golden eagles and pumas.

Distinctive call

Scientifically the coyote is *Canis latrans*, barking dog, so-called because apart from the domestic dog it is the only member of the dog family that habitually barks. Foxes, wolves and jackals only bark at specific times. The call of the coyote has become part of the background to the Wild West, necessary to produce 'atmosphere' for any night scene in a Western film. Coyotes can be heard all the year round, usually at dawn and dusk. In the evening coyotes sing in chorus. One starts with a series of short barks, gradually increasing in volume until they merge into a long yell. Other coyotes join in and the chorus continues for a minute or two. After a pause, the chorus starts again.

Two or three coyotes may meet each night to sing and the eerie effect of the songs of several such groups ringing over the countryside on a still moonlight night is very impressive. Not surprisingly, there are numerous legends connected with the coyotes' song. The faraway sound is supposed to be made by the coyote barking into a badger hole to produce a hollow echo, while the quavering howl is said to be produced by the coyote making its chest vibrate by bouncing on rigid legs.

Many Indians claimed to understand the coyotes' language. The Comanches had their equivalent to Kipling's Mowgli, who was brought up by coyotes and later taught his tribe to understand them.

class	**Mammalia**
order	**Carnivora**
family	**Canidae**
genus & species	***Canis latrans***

Winter feast: coyotes clean an elk carcase while ravens wait for their turn.

△ *Coyote pups. Up to 19 may be born in one litter, but the average is 10. After a month underground, the pups begin to sally out on communal hunting trips until they learn to fend for themselves.*

▽ *Suspicion: a swimming coyote approaches the bank. Hunters themselves, coyotes are known to have been attacked by larger predators.*

▷ *A coyote heads in to intercept a deer. When it comes to a straight chase, coyotes often find themselves outpaced by deer, and have been known to wear down their quarry by chasing in relays until it tires.*

▽▷ *A young coyote emerges from its lair. Coyotes live in scrupulously clean nesting chambers at the end of tunnels which can be up to 30 ft long.*

▽ *A coyote in its classic pose as the spirit of the North American prairie—but within the last century they have spread across North America from Alaska to Costa Rica.*

Australian spotted cuscus. Not a monkey but a pouched mammal related to the kangaroo. In both, the second and third toes of the hindfoot are bound together by skin and the claws are used for combing the fur.

Cuscus

The cuscus is an unusual phalanger that is often mistaken for a monkey. It has a rounded head, small ears almost buried in the fur, protruding yellow-rimmed eyes, short muzzle and yellow nose. About the size of a large domestic cat, $3\frac{1}{2}$ ft long including a long prehensile tail, it makes good use of its voice. It has been suggested that its name refers to its scolding or 'cussing', but in fact it was originally **couscous,** *the French rendering of the aboriginal New Guinea name. There are about 7 species living in the forests of Queensland, Australia, and the jungles of New Guinea.*

Although monkey-like, especially in the face, a cuscus shows its relationship to kangaroos not only in the female having a pouch but also in the hindfoot. This is used for grasping branches and is more like a human hand, the long first toe being thumb-like and particularly strong. But the second and third toes are bound together by skin (that is, syndactylous) and their claws are used, as in kangaroos, for combing the fur.

Bouts of cussing

Cuscuses live in trees and are adept at clinging to branches with the hind feet and the prehensile tail, which can be wrapped round a branch for extra support when climbing. At rest, the tail is tightly coiled, like a watch spring.

When quarrelling among themselves or challenging another animal coming near it, the seemingly gentle and inoffensive cuscus will snarl and bark, in a guttural voice, in bursts of scolding. As it begins to bark it raises a forepaw in a menacing gesture, and if further provoked, strikes out with its front feet, perhaps even biting savagely, and all the time 'cussing'.

They live in thick cover, well hidden among foliage, keeping still by day. At night they move about among the trees feeding on leaves and insects, sometimes eating eggs and small birds. When feeding, a cuscus grips a branch with its back feet and tail, which leaves the front feet free for holding food. In captivity they will eat chopped fruit and vegetables.

Tightest sitter

The best-known species are the spotted cuscus, found in Queensland, and a second species of spotted cuscus in New Guinea in which the males and females are so unlike they were first thought to belong to different species. The female's fur is grey, brown or

Only rarely is a cuscus seen during the day, as it keeps still among thick foliage. At night it moves about branches feeding on leaves, insects and sometimes eggs and small birds. When feeding it grips the branch with its back feet and tail so the front feet are free for holding food.

fawn, a little lighter on the sides of the head. There is also a patch on the rump which is lighter still, and this is found in all cuscuses of whatever species and of both sexes. As the cuscus gets older this patch gets lighter because it is where the fur is worn down through the animal sitting so tight and so continuously through the day. There can be few animals to rival a cuscus for sitting tight.

By contrast with the female, the male spotted cuscus in New Guinea is a creamy white with many ½in. black spots uniformly spaced all over when young. As it grows older the black spots run into each other to form irregular grey blotches. The hind limbs also change to a uniform grey.

Little more is known about the life history or growth except that breeding must be fairly continuous. It is rare for a female to be found without at least one baby in the pouch.

There is little information on enemies either. One cuscus in captivity panicked at the approach of a python but this means little. Captive animals have often been seen to panic at the sight of strange objects which they could not possibly meet in their natural habitat.

Repulsive odour

Whether it is a defence or not, a cuscus has a strong and repulsive odour, and it is this very often that betrays its presence to the hunter. David Fleay, the distinguished Australian naturalist, has described how he returned from New Guinea with several cuscuses. The odour they gave out clung to every part of the ship the animals had occupied for a long time after they had been disembarked. Fleay himself handled them no more than he was compelled to and then very gingerly. Yet he tells how people edged away from him in cinemas even after his suits had been twice cleaned.

In spite of the odour, Aborigines relish cuscus flesh, and they have a singularly callous way of ensuring a continuous supply of it fresh. When one of their hunters finds a cuscus he breaks its hindlegs so it can feed but cannot move far—a repulsive substitute for the kitchen 'fridge'.

class	**Mammalia**
order	**Marsupialia**
family	**Phalangeridae**
genus & species	***Phalanger nudicaudatus*** *Australian spotted cuscus* ***P. maculatus*** *New Guinea spotted cuscus* *others*

Meal among the toadstools: a deer mouse strikes a Disneylike pose while taking a snack.

Deer mouse

Deer mice are American, very similar to the European long-tailed fieldmouse, both in appearance and in habits, but the two belong to different families. There are about 20 species, varying in colour from sandy or grey to dark brown. Some are almost white and others nearly black, but in general those living in woods are darkish, and those living in open or arid country are pale. The underparts and feet are white, hence the alternative name of white-footed mouse. A deer mouse measures 5–15 in. from nose to tip of tail, the tail varying from 1½–8 in. in different species.

They are found over most of North America, from Alaska and Labrador southwards, and one species extends into South America, reaching the extreme north of Colombia. They inhabit many kinds of country from swamps and forests to arid, almost desert, country, but each species usually has only a limited habitat and consequently is found only in a relatively small part of the total deer mouse range.

Overlapping territories

Deer mice are nocturnal, coming out during the day only if they are very hungry, or if there is a cover of snow that allows them to forage under its shelter. During the evening they can be heard trilling or buzzing, a noise quite unlike the squeaks of other mice, and in some parts of the United States this has led to their being called vesper mice. They also drum with the front feet when excited.

Each deer mouse has a home range which it covers regularly in search of food. The extent of the range varies considerably and depends on the amount of food available. In the grasslands of south Michigan the average size of the ranges of male deer mice is $\frac{3}{5}$ of an acre, while those of the females are slightly smaller. The home range of a mammal is not strictly comparable with the territory of a bird. Only a few birds keep a territory all through the year, but more important, a mammal does not defend its range so vigorously. The borders of neighbouring ranges overlap, sometimes considerably, and the ranges of two females may be almost identical, but it is only the inner parts of the territory around the nest that will be defended vigorously.

Within its range, a deer mouse may have several refuges in abandoned burrows or birds' nests, under logs or in crevices. Sometimes a deer mouse will come indoors and make its nest in an attic or storage room. Each nest is used for a short time, being abandoned when it becomes soiled, for deer mice limit hygiene to cleaning their fur.

Burrows with a bolt hole

The nest is an untidy mass of grass and leaves, lined with moss, fine grass or feathers. Sometimes the deer mice make their own burrows. The Oldfield mouse, a species of deer mouse living in Alabama and Florida, makes a burrow leading down to a nest which is 1 ft underground. Then from the other side another burrow leads up again but stops just short of the surface. This presumably serves as a bolt hole in case a snake or other narrow-bodied enemy finds its way in. A traditional way of catching these mice is to push a pliable switch or wand down the hole, twiddle it about until it can be pushed up the escape burrow, and catch the mouse as it breaks out.

Are they a pest?

Seeds and berries are the main food of deer mice but they also eat many insects such as beetles, moths and grasshoppers, which are chased and bitten or beaten to death. Insect larvae, snails and slugs are devoured, and deer mice also eat carrion such as dead birds and mammals, and they will gnaw cast antlers.

Deer mice are something of a problem in plantations or on farms, where they eat seeds of new-sown crops, which they smell out and dig up. But even when abundant, they are not as much of a pest as meadow voles and other small rodents. To even the score, deer mice are helpful because they eat chafer grubs that damage the roots of young trees.

Hanging on to mother

In spring the males search for mates, perhaps finding females whose ranges overlap theirs. At first their advances are repulsed but the males eventually move into the females' nests, staying there for a few days only but sometimes, it is thought, forming permanent pairs.

The female gives birth to a litter of 1–9 young after 3 or 4 weeks. At birth the young mice are blind, deaf, and apart from their whiskers, naked. They hang firmly to their mother's teats and she can walk around with them trailing behind. If the nest is disturbed she will drag them in this manner to a new site. Any baby that does fall off is picked up and carried in its mother's mouth.

Litters of deer mice can be found from spring to autumn but more are born in spring and autumn than during the summer, and if the winter is mild, breeding will continue through it. Females begin breeding at 7 weeks, only a few weeks after leaving their mothers, and have up to 4 litters a year.

Many nocturnal enemies

Most deer mice live less than two years, and many never reach maturity but provide food for the many predators that hunt at night. Foxes, weasels, coyotes, bobcats, owls and snakes, all feed on deer mice, and even shrews will occasionally eat them.

Racial 'segregation'

Although similar, the 20 different deer mice can easily be told apart by the specialist in classification, and, one must presume, by the mice themselves. Otherwise they would mix and interbreed and their differences would disappear, especially when different kinds live in the same habitat. Experiments by an American scientist using a Rocky Mountain deer mouse and a Florida deer mouse, which are closely related, showed how the deer mice are segregated.

Special cages were made, each with two side compartments. In preparation for the experiment a Rocky Mountain mouse was put in one compartment and a Florida mouse in the other, and this was repeated for all the cages. After these had remained long enough to impart their smell to the compartments they were taken out. Now, into each cage were put either a Rocky Mountain mouse or a Florida mouse, and these naturally made full use of the available space, including wandering into each compartment. By timing the period each mouse spent in each of the two side compartments of its cage the scientist found that in all cases the test mouse was very obviously drawn to that compartment which carried the smell of its own species. This almost certainly is how mice of the same species, even when sharing a habitat with another species, would be drawn together to breed, for it was noticed that males reacted particularly strongly to the smell of females of their own species on heat. So although it may sometimes appear that there are mixed populations of deer mice, the different species are really living separately.

There was, however, one difference between the Rocky Mountain mice and the Florida mice: the latter were much more likely to spend time in compartments smelling of Rocky Mountain mice. The reason for this seems to be that in Florida there is only one species of deer mouse, and discrimination is no longer necessary; but in the west there are many species and if their strains are to be kept pure, they must be able to distinguish between their own fellows and those of closely related species.

class	**Mammalia**
order	**Rodentia**
family	**Cricetidae**
genus & species	***Peromyscus maniculatus*** *others*

Deer mouse cleaning and drying its fur—its only hygienic habit.

Dingo

The dingo is the wild dog of Australia, about the size of a collie, standing about 20 in. at the shoulder. The ears are erect and pointed and the tail bushy, often with a white tip. It is popularly supposed that dingoes always have the yellow-brown coat immortalised in the Yellow-dog Dingo of Kipling's ***Just So Stories****, but they vary from light red to brown and some may be brown with black streaks. Albinos are known and in southeastern Australia there is a whitish breed. The darker dingoes are often assumed to be the offspring of matings with domestic alsatians, but this is not necessarily so, although dingo × dog crosses are quite common.*

▽ *A dingo family, which can run to 8 pups. Dingoes hunt in family parties, running down their prey in a long, tenacious chase and worrying it until it tires.*

Dingo bitch with wobbly pup.

Adults often lead packs to kill scores of sheep and cattle.

The Aborigines' companion

The animals of Australia are unique because the continent was cut off from the main land mass of Europe, Asia and Africa before the placental, or true, mammals arose, and the marsupial, pouched mammals were able to survive in large numbers. When the Europeans arrived in the 18th century the only true mammals apart from man were bats and rats, which are thought to have floated over from southeast Asia on driftwood, and the dingo. The most likely explanation of the dingo's presence is that it was brought over in one of the Aborigine invasions from Asia as a domestic dog, and later went wild. The New Guinea singing dog is thought to have had the same origin. Remains of dingoes dating back 6 000 years have been found, and present-day Aborigines use them for hunting. They are captured when young, and, if necessary, suckled by the Aborigine women.

Support for this idea lies in the similarity between the dingo and the Asian pariah dogs. They may have both descended from the dhole. Neither the dhole, the dingo nor the singing dog can bark. Instead they howl or whine, the howl of the singing dog being a remarkable yodel that gives the dog its name.

For thousands of years dingoes have flourished in Australia. Their only competitors were the Tasmanian devil, thylacine and, perhaps, the tiger cat. It is probably the competition with the dingo that caused the Tasmanian devil and the thylacine to become rare and finally extinct on the mainland. Their final stronghold, if they still survive, is Tasmania, which the dingo never reached.

Dingoes live alone or in small family parties. They are found all over Australia and appear to have regular migrations along definite tracks. There is evidence that many dingoes breed in inland parts of Australia and migrate to the coastal strip in winter.

Chasing their prey

Like all dogs, dingoes chase their prey, wearing them out in a long chase, for they are not fast runners. Large animals such as kangaroos, sheep or cattle are chased until the dingoes can catch them, or if there is a number of sheep or cattle they are harassed and chivvied about until the weaker ones drop back. As the prey fails in strength, the dingoes worry it, slashing at its head and legs but keeping clear of the hooves, until it collapses.

Sometimes the dingoes meet their match, as when a kangaroo turns at bay. Leaning back on its tail it can deliver kicks powerful enough to rip open the dingoes' bellies.

Breeding

The male dingo marks its territory with urine, like a domestic dog. It mates in winter and the pups are born in spring, some 9 weeks later. The litter consists of up to 8 pups, which are sheltered in a den where they are suckled for 2 months. After that they stay with their parents for at least a year, and hunt as a pack.

Eagle fodder

Apart from man, and prey brought to bay, dingoes are in danger from crocodiles and snakes in the tropical parts of Australia, and from wedge-tailed eagles. These are the world's largest eagles, and two working together have been seen to kill an adult dingo, but this is exceptional; it is young dingoes and the old or infirm that usually fall prey to them.

Well-earned bounties

Since Europeans began farming in Australia, the dingo has been an ever present problem. Thousands of sheep and cattle are slain each year, a family of dingoes sometimes killing a score or more in one night, apparently out of sheer blood lust, but more likely through the inexperience and excitement of the young dingoes. Consequently, firm steps are taken against them. Thousands of square miles of sheep country have been fenced, at a cost of millions of pounds.

As a result of the bounty thousands of dingoes have been shot or poisoned, but despite a prediction 80 years ago that they would soon become extinct, they are still common, even in fairly well-populated areas. But a greater problem than the ordinary dingo is the rogue, perhaps one that is wounded and so unable to hunt wild prey, or a dingo-collie cross, that makes a speciality of killing sheep. Bounties of over $A200 may be offered for the scalp of such a rogue, and they may be well earned. Ellis Troughton, the distinguished Australian zoologist, tells of a wily rogue that evaded all the efforts of a dogger, as dingo hunters are called, to trap or poison it. Eventually the dogger noticed that the dingo followed regular tracks so he spread a trail of poisoned golden syrup. The dingo, finding his paws becoming sticky, licked them, and succumbed to man's greater cunning.

class	**Mammalia**
order	**Carnivora**
family	**Canidae**
genus & species	**Canis familiaris dingo** *dingo*

Dog

In both legend and fact, the dog was probably the first animal to be fully domesticated. Even as far back as 8 000 BC there seem to have been at least two breeds of dog. The ancestral wild animal is still in doubt and may never be known for certain. Orthodox opinion has long held that the domestic dog was derived from the northern wolf, perhaps with a later admixture of jackal. We now have more extensive information available from dog skeletons unearthed in excavations of prehistoric sites. In addition there have been modern studies of the behaviour of members of the dog family. Together they suggest, as some scientists have suspected for the last 20 years, that the domestic dog's ancestor may have been a wild dog, of a species distinct from wolf or jackal, that has long been extinct.

An infinite variety

No domestic animal shows so much variation in size, colour, coat and behaviour as the dog. The smallest is the chihuahua (pronounced chi-wa-wa), a breed developed from the Mexican hairless dog: 6–9 in. high, weighing up to 6 lb–the smallest weighed $1\frac{3}{4}$ lb. The largest are the St Bernard and the mastiff, 28 in. high, 200 lb weight and 30 in. high, 165 lb weight respectively. In between we have such diverse animals as the great dane and the Scotch terrier, the greyhound and the bulldog, the German shepherd dog, also known as the Alsatian, and the dachshund, to mention only a very few.

When is a dog a wolf?

It seems that in any species of wild dog colour and size tend to vary widely, and the more a population is spread over a continent the more it breaks up into local races. The typical North American wolf, for example, is yellowish or brownish-grey brindled with black. In the Arctic tundra, however, wolves are white, on the plains farther south they are usually grey, and in Florida they are often black. The wolves of Europe and Asia are no less variable, in colour as well as size, being smaller in more southern regions.

The remains of wolf skeletons from prehistoric times, when wolves ranged from the British Isles to Japan, show endless differences in size. The earliest remains of domestic dogs are also variable in size, although all are intermediate between the northern wolf and the jackal. In addition, all wild dogs and domestic dogs differ so little in the shapes of their bones that it is often a matter of individual opinion whether the bones found on a particular prehistoric site are those of a wolf that had wandered in to scavenge and had come to an untimely end, or of a tamed dog kept as a pet that had died a natural death.

From the many races of wolf, the Indian wolf has been selected by some students as the probable forerunner of the domestic dog. Halfway between the larger, more typical northern wolf and the southern jackal for size, it is said to bark occasionally but not to howl, and it has many of the

△ *Team mates fighting. The husky on its back, although apparently snarling viciously, lies in a submissive posture to its dominant team mate.*

▽ *Harlequin great danes. Of the mastiff family, these dogs were originally used to trace and fight wild boars.*

qualities of both the pariah dog and the dingo, both of which are usually regarded as primitive domestic dogs.

All this is so much guesswork, however, for under domestication many things can happen. For example, juvenile characters tend to be retained in the adult. This is a form of neotony and results in several marked changes. The size of the animal, for example, tends to be reduced. The leg length tends to change, and the muzzle becomes either shorter or, sometimes longer. These changes have happened with sheep, cattle and pigs as well as dogs, even in the first generation born in captivity.

The oldest breeds

The puzzle is increased by the fact that even 10 000 years ago, according to the finds made by archaeologists and others, there were domesticated dogs in Asia, Egypt and Europe, possibly also in America, although there is less evidence there. In each of these there was already more than one breed.

In early Europe, by at least the Bronze Age, four types can be recognised, a wolf-like dog similar to the Eskimo dog of today, a sheepdog, various breeds of hounds, and a small house dog from which terriers, Pomeranians and others were later derived. In Ancient Egypt mummified dogs as well as paintings and figurines leave no doubt that there were various types of hounds, mastiffs, sheepdogs and Pomeranians, and more especially greyhounds. Between these, as always, were the mongrels. On Assyrian sculptures are depicted large dogs recalling the mastiff or the bloodhound.

The sociable dog

In addition to its intelligence and stamina, the character that made the dog so useful to man was its sociable nature. It is often said that the domestic dog is a pack animal, and that this alone points to a wolf ancestry. In fact, modern researches on the North American wolf tend to discredit this. Stories of wolves hunting in packs are now believed to be incorrect. The truth seems to be that wolves normally associate in groups no larger than family parties. Only occasionally, especially when food is scarce, do several family parties come together for a temporary period, the families still retaining their identity within the group.

Domestic dogs are promiscuous breeders, the bitch being in season usually once a year, for 1–3 weeks, beginning at 7–10 months of age. After that the details are much the same as for wolves (see page 237), though they do not differ much from those of other species of wild dog, so far as these are known. The gestation period is 53–72 days. A litter may be anything up to a dozen or more, the largest recorded being 22. Puppies are born blind and deaf. The eyes open at 9 days, hearing begins at 10–12 days. Weaning is at 4–8 weeks.

Feral domestic dogs, where present in numbers, as for example in Australia, will form packs, apparently similar to those of the Cape hunting dog.

Whatever the ancestry of the domestic dog may be – and domestication may have occurred independently in several parts of the world – it acts as a sociable animal. More especially does it attach itself to one person,

△ *Hunters with hounds, c. 1680.*

△ *St Bernard – little truth in rescue tales.*

△ *German shepherd dog or Alsatian in police work is trained to guard, track and detain.*

▽ *Golden retriever guide-dog is intensively trained to take her owner around all obstacles.*

▷▷ *Dutch Krothas griffen retrieves a duck.*
▷ *Pointer—a clean-cut gun dog known the world over, used to find concealed game.*
▽ *Bloodhound—tracker and police dog but known to Britain since the Norman invasion.*
▽▽ *An Italian breed of bull mastiff.*
▽▷ *Shih-tzu—quite a popular ornamental dog.*

and will then show the highest degree of co-operation. Conversely the domestic dog readily suffers loneliness. This too suggests that its ancestor was not solitary by nature.

One of the habits of wild dogs of all kinds is that of marking a territory with urine. A dog will pay periodic visits to these sign-posts, re-marking them, having used its nose to determine what other dogs may have visited it. This marking of trails and territories is a nuisance under civilised conditions, but is probably inseparable from the ingrained sense of territory which gives a dog its most valuable traits: as guardian and watch-dog.

Types of breed

The classification of breeds of dogs tends to vary according to the author. Dr Johannes Caius (John Keys) physician to Tudor monarchs during the second half of the 16th century, divided all breeds into three classes: high-bred, country and mongrel. Linnaeus, in the 18th century, listed 35 breeds. Today we can reckon at least a hundred distinct breeds, the number rising considerably if all named varieties are included. They can be grouped as: sporting dogs, hounds, working dogs, terriers, toy dogs and non-sporting.

Sporting dogs include pointers, retrievers, spaniels and setters. Hounds are of two kinds; those like the bloodhound, foxhound and beagle that use their noses, and those like the greyhound, Afghan, borzoi, whippet and dachshund that use sight. Working dogs include sledge dogs, guard dogs, sheep dogs and others such as St Bernard and Newfoundland. Terriers need little explanation and toy dogs, all of small size, include the long-haired Yorkshire terrier and the Mexican hairless, the Pomeranians, Pekingese, pug, chihuahua and toy poodle. Non-sporting dogs constitute those difficult to classify otherwise, like the bulldog, Dalmatian, chow-chow and poodle, although originally the poodle was a sporting dog. Clipping its coat was not for decoration but in order to allow the dog to move more freely in water, when it was used as a water spaniel.

This classification will not satisfy everybody, nor is it the best that might be given. Its virtue is that it indicates broadly the uses to which the various dogs are put. Moreover, these uses vary from place to place as well as in time. The dachshund (German: badger hound) was formerly used on the Continent for sport. It is now more a toy dog. Welsh corgis, cattle dogs to the Celts, today enjoy favour with the British Royal family. The Dalmatian, originally a sporting dog, later a guard dog on coaches, then a running dog with horsedrawn carriages, is more commonly kept today as a house pet.

Help for the blind

One of the latest uses to which dogs have been put is as guide-dogs to the blind. This

was started in Germany at the end of the First World War and soon became world-wide. The first breed used was the German shepherd dog–only bitches are used to cut out the risk of fighting–and although other breeds–boxers, Labradors, Border collies–have been tried, the Labrador is now the most favoured.

One breed that has acted as a guide, but in another way, is the St Bernard. A long-standing legend has, however, masked its services. The legend is that this dog was specially bred at the Swiss monastery during the lifetime of the saint, in the 11th century. The dog is supposed to rescue travellers lost in deep snowdrifts in winter. There is no truth in the first part of this story and little in the second. In fact, this breed of dog was unknown in the monastery before the second half of the 18th century and was then used as a guide to the monks travelling up and down the pass. Where a heavy dog could tread in deep snow with safety the route was then known to be safe for a man.

This does not mean that dogs could not show the compassion needed to carry out the work popularly ascribed to the St Bernard. There are many stories of the loyalty of dogs, of their faithfulness unto death. Even wolves will look after their own kind and there are stories, seemingly well-authenticated, of wolves supplying food to an injured or senile member of their species.

Dogs of war

Another breed which has played a variety of parts during a long history is the mastiff. A mastiff-type, used in hunting and as a guard dog, figures on Assyrian sculptures. The mastiff was also used in the arenas of Ancient Rome, but Caesar was impressed by the size of the mastiffs used by the Ancient Britons; and Britain's 'clever hunting dogs' were famed even before the Roman conquest. Mastiffs are the one breed used in war, other than for rescue work, as sentinels or for light transport work. The Romans used them for attacking the ranks of their enemies, and in the Middle Ages they were used against mounted knights. The mastiff, itself wearing a small coat of armour, was equipped with a cauldron of flaming sulphur and resin fastened on its back. From this a vicious iron spike projected forward over the dog's head. Added to these were the dog's great weight and fearsome jaws. It was a formidable combination to let loose among heavily encumbered cavalry.

class	**Mammalia**
order	**Carnivora**
family	**Canidae**
genus & species	***Canis familiaris***

Donkey

The donkey is a domesticated ass. It deserves separate treatment because of its long and chequered history, and because of the marked, if subtle, changes in appearance and temperament it has undergone.

Wild asses are confined to northern Africa and southern Asia. They are a dying race, and as their numbers dwindled, so the numbers of donkeys increased. Donkeys were called asses until the end of the 18th century, by which time they had been taken over a large part of the world. They were important beasts of burden, were used for riding, also for ploughing, turning wheels and any form of menial task. To a large extent the donkey has been the helpmate of the poor man throughout the ages.

Origin of the donkey

The Persian wild ass was domesticated by the Sumerians, about 3 000 BC, but horses soon displaced it. A race of African wild ass was domesticated at about the same time, possibly earlier. It may have been drawn from a race in northeastern Africa which has long been extinct. It seems certain that the donkeys so widely used in Asia were not domesticated forms of the Asiatic wild ass but were imported, possibly from Egypt. Moreover, it is very likely that donkeys used in Europe were imported from Asia, not direct from Egypt. When this took place is uncertain. In fact, almost everything connected with the ancestry of the donkey is guesswork, except that its original home was Africa. So far as the European donkey is concerned, Homer, the Greek poet, who lived nearly 3 000 years ago, made no mention of it, although he wrote of a mule from Asia Minor. The Romans, however, knew of the donkey. There is a Saxon word for ass, so we can suppose the donkey was taken across Europe during the Roman occupation of Britain and was established in Britain before the Norman Conquest.

Breeds of donkey

Breeds vary in colour from nearly white to nearly black. They are of different sizes, markings and performance. Darwin reported four distinct breeds in Syria alone. There was a lightly built graceful breed, with easy gait, used by ladies of high rank. The so-called Arab breed, well fed and carefully groomed, was used for the saddle. A stouter, more heavily built donkey was used for ploughing, usually harnessed with a camel, as well as for general purposes. The fourth and largest was the Damascus breed, usually white, with a long body and particularly long ears. Donkeys of similar breed to this, but greyish-white, are found all over the Middle East, and a similar breed was favoured for centuries in Baghdad. The main feature of these types was their endurance. They could trot or canter for hours without sign of fatigue, and uphill or over broken ground their performance was better than that of a horse.

The Mahratta breed, of Pakistan, western India and Ceylon, is the smallest, 7½ hands high, and can carry loads out of proportion to its size. The largest is the Poitou breed of France, up to 16 hands, grey to black, more like a carthorse for size, with a large ungainly head, stout limbs, broad hoofs, large ears, a long heavy coat and a hanging instead of the usual erect mane. It is kept almost entirely for breeding mules.

The vitality of donkeys is reflected in their long life. Wild asses kept in zoos have commonly lived to 20–24 years. The record for a mule is 37 years. For the donkey there are many records for above 20 years, some for 37 or more, and a white donkey is said to have lived 50 years. There are claims for 50 years or more, even to as much as 80. Whatever the truth, the mere frequency of these reports suggests that donkeys are traditionally held to be long-lived.

Healthgiving donkey milk

Apart from the uses made of donkeys for riding and as beasts of a variety of burdens, their flesh has sometimes been eaten. There is yet another service, recalled by Biblical references to a man's possessions including so many she-asses. The milk of a she-ass is said to be highly nutritious because it contains more sugars and less cheesy matter than cows' milk and is reputed to be particularly good for tuberculosis. Indeed, it was more valued in the past as medicine than as food, which may be why Cleopatra bathed in asses' milk. A 16th century antidote to poisoning was to drink the broth from donkeys' bones bruised and steeped in water. Doubtless this would produce the necessary vomiting. It was also held to be an antidote to scorpion poison. There was even the belief that anyone stung by a scorpion had only to look into a donkey's ear to be cured—and we say a donkey is stupid!

Other curative properties were ascribed

The donkey, a domesticated ass—seen here in Greece—is still used as a beast of burden in many countries. Throughout the ages the donkey has been the helpmate of the poor.

to donkeys. Hair cut from the black cross on its shoulders prevented fits and convulsions if put in a bag and hung around a child's neck. In parts of England, notably Dorset, it was the custom to put a child astride a donkey to prevent whooping cough. If the child already had it, the cure was to pass the patient three times under the belly and three times over the back of a donkey. The skin of a donkey hung over a small boy would prevent him from being frightened.

Ass, not donkey

The word donkey appears only very late in history; it was a nickname given to the animal in England in the 18th century. Nobody knows how the name arose. Some say it was derived from *dun*, the colour of the animal's coat, with the addition of the word *kin* meaning small. But the colour of the ass in Britain is usually grey. Another suggestion is that it had something to do with the name Duncan which comes from a Gaelic word meaning a brown warrior. Other nicknames for a donkey are a dicky, a neddy and a cuddy; and, here again, nobody knows how or why these came about. Finally there is the word 'moke', a slang name which seems to have been mentioned first by Thackeray. Although its origin, too, is obscure, it is tempting to recall that the crest of the Dymoke family, of Lincolnshire, dating from 1377, is a pair of ass's (later called donkey's) ears. It is customary to use a pun wherever possible in heraldry, and Dymoke could be loosely translated as two mokes, and represented by a pair of ass's ears. If this be so, the name moke, for a donkey, must have come before the late 14th century.

Whether called ass, donkey or moke, the animal is usually linked in our minds with stupidity. Many stories could be told to counter this. One of the briefest and best is told by Canon Tristam, the celebrated divine and traveller, who wrote: 'One of our donkeys which had been severely beaten for misconduct by a member of our party, never forgot the circumstances, but, while ready to sniff and caress any of the others, would stand demurely whenever his old enemy was near, as if unconscious of his presence, until he was within reach of his heels, when a sharp sudden kick, with a look of more than ordinary asinine stolidity, was the certain result.'

class	**Mammalia**
order	**Perissodactyla**
family	**Equidae**
genus & species	***Equus asinus***

△▷ *Mother and foal. The milk of a she-ass is said to be highly nutritious because it contains more sugars and less cheesy matter than cows' milk and it is said to be extremely good for tuberculosis. In the past, it was more valued as medicine than as food.*
▷ *Relief is just a nibble away.*

△ *Family group – the youngsters stay with the adults until their teens.*

◁ *Feeding time. At birth the baby is 3 ft high and weighs some 200 lb. It uses its mouth when suckling from the mother's nipples, situated between the cow's forelegs.*

▷ *Largest land animals alive today – African elephants feeding and drinking on the river bank. Their vegetarian diet includes grass, foliage and branches of trees and fruit. The mobile trunk is used to gather and carry food to the mouth.*

Elephant

The elephant is the largest living land animal and there are two species, the African and the Indian. During fairly recent geological times elephants of many species making up six families ranged over the world except for Australia and Antarctica. The African elephant, the larger of the two surviving species, is up to $11\frac{1}{2}$ ft high and weighs up to 6 tons.

Elephants have a massive body, large head, short neck and stout pillar-like legs. The feet are short and broad with an elastic pad on the sole and hoof-like nails, five on each foot except for the hind foot of the African elephant, which has three. The bones of the skeleton are large, and instead of marrow cavities they are filled with spongy bone. The outstanding feature of elephants is that the snout is remarkably long, forming a flexible trunk with the nostrils at the tip. The trunk is used for carrying food and water to the mouth, for spraying water over the body in bathing or spraying dust in dust-bathing, and for lifting objects, as well as being used for smelling. The single incisor teeth on either side of the upper jaw are elongated and form tusks.

The main differences between the two living species are the larger ears and tusks of the African, its sloping forehead and hollow back, and two 'lips' at the end of the trunk compared with one lip in the Indian elephant.

The African elephant is found in most parts of Africa south of the Sahara, in savannah, bush, forest, river valley or semi-desert. It lives in herds of bulls and cows, each herd being led by an elderly cow, while the older bulls live solitary and join the herd only to mate. The Indian elephant is also native to Ceylon, Burma, Thailand, Malaya and Sumatra, living in dense forests. More correctly it should be called Asiatic, not Indian, but the use of 'Indian elephant' is now too deeply rooted for change. The social structure of its herds is much the same as in the African species.

Keeping its skin in condition

Elephants are sometimes grouped with rhinoceroses and hippopotamuses under the loose heading of pachyderms (thick-skins). In all the skin is thick and only sparsely haired, and all need to keep the skin in condition by wallowing. An elephant will bathe in water, almost completely submerging itself and will also spray water over itself with its trunk. It indulges in dust baths, too, and if water is scarce it will wallow in mud. The African elephant at least is adept at finding water in times of drought, boring holes in the ground using one of its tusks as a large awl. The requirements of the two species differ because the Indian elephant keeps mainly to dense shade. This also influences other aspects of their behaviour. The African elephant, for example, must seek what shade it can from the midday sun and cool its body by waving its large ears. The enormous surface these present allows for loss of body heat, which is helped by waving the ears back and forth. The Indian elephant, with much smaller ears, keeps itself to dense shade.

Asleep on their feet

A vexed question of long standing is how elephants sleep. Both species can sleep standing, or lying on one side. To lie down an elephant uses similar movements to a horse, but it does what no horse will do: it will sometimes use a pillow of vegetation pulled together on which to rest its head. When standing asleep an elephant breathes at the normal rate. Lying down it breathes at half this rate. When 17 elephants were kept under observation it was found they usually slept for 5 hours each night, in two equal periods. Of this 20 minutes were slept standing, the rest lying down.

Dangers of over-population

The diet is entirely vegetarian and includes grass, foliage and branches of trees and fruit. The trunk is used to gather these and convey them to the mouth. African elephants, living where bushes and trees are scattered, will use the forehead to push over small trees to get at the top foliage. When an area becomes over-populated the loss of trees can be serious. In national parks in Africa the populations of elephants, under protection, tend to increase so much that their ranks have to be thinned out by selective shooting, usually spoken of as culling, to prevent destruction of the habitat. Otherwise all the elephants in the area would be in danger from starvation.

Under free conditions elephant herds trek from one area to another, often seasonally in search of particular fruits. Long distances may then be covered, and this relieves the strain on the vegetation, which can regenerate in their absence.

The molars of elephants have broad crushing surfaces for chewing fibrous vegetation. The wear on them is considerable. Every elephant in its lifetime, assuming it dies of old age (70 years in the Indian, 50 years in the African elephant) has 7 teeth in each half of both upper and lower jaws, exclusive of the tusks. The first are 4 milk teeth which are soon shed. After that a succession of 6 teeth moves down each half of both jaws on a conveyor-belt principle. The first is in use alone but as its surface is getting worn down the next tooth behind it is moving forward, to push out the worn stump and take its place. When the last teeth have come forward and been worn down the elephant must die from starvation, if nothing else.

Purring from the stomach

For a long time big-game hunters and naturalists were perplexed by one feature of elephant behaviour: their tummy-rumblings. Nobody was surprised that these abdominal noises should be so loud and persistent, in view of the enormous quantities of food the huge pachyderms must eat. What puzzled people was that the elephants could apparently control the noises, stopping suddenly when someone approached. Within the last few years it has been discovered that these noises have nothing to do with digestion. When elephants are out of sight of each other they keep up this sort of purring. When danger approaches one of them, it becomes silent. The sudden silence alerts the rest of the herd, which also grows silent. Only when danger has passed is the purring resumed, by which the elephants tell each other that all is well.

Trumpet Voluntary

Apart from these sounds elephants will 'trumpet'. The sound is as startling and as loud, if less pure in tone, as that from the brass wind instrument. In paintings of elephants made in the Middle Ages, or even later, the trunk was always given a trumpet-shaped end, the artists being influenced by travellers' stories of the elephants' trumpeting.

Elephant 'midwives'

Mating is preceded by affectionate play, especially with the bull and the cow entwining trunks or caressing each other's head or shoulders with the trunk. The gestation period is 515–760 days, mostly about 22 months. The single baby—twins are rare—is about 3 ft high and weighs about 200 lb. On several occasions hunters or naturalists have seen a cow elephant retire into a thicket accompanied by another cow. Some time later the two come out again accompanied by a baby. Nobody knows whether the second cow acts as midwife or merely

◁ *Woe betide those who ignore this warning notice.*

▷ *Dressed up for a reception at Bahawalpur, West Pakistan, an Indian elephant looks very decorative. It is distinguished from its larger African relative by its smaller ears, arched back, domed forehead and smoother trunk which has only one 'finger' or lobe at its end compared with the African's two (below). In general, the Indian elephant appears to be an animal of jungle or bush country, although it is found in grassland areas.*

▽ *Enjoying a dustbath, an elephant uses its hose-like trunk to snort dirt over its body.*

stands guard while the calf is being born. The baby is able to walk soon after birth and can keep up with the herd in two days.

Hefty train-stoppers

Such large and powerful animals have few enemies. In India a tiger may kill a baby and in Africa the large predators, such as the lion, may do the same. The power of an elephant in defence can be gauged by the several stories told of a bull elephant meeting a train on a railway and charging the engine head on. In all reports it is stated how the engine driver drew the train to a halt and the elephant charged the engine repeatedly, doing itself great injury yet persisting in the attack. Another feature of elephant defence is the close co-operation between members of a herd. Hunters have reported seeing a shot elephant being helped away by two others ranged either side of it, keeping it upright on its feet. On one occasion the herd combined to drag the carcase of one of their fellows throughout the night, in an abortive attempt at rescue. In 1951, in the Johannesburg *Star,* Major JF Cumming was reported as having seen some elephants dig a grave to bury a dead comrade!

Do they fear mice?

In contrast with the elephant's comparative freedom from large enemies is the long-standing belief that elephants are afraid of mice. Lupton, in his *A Thousand Notable Things,* published in 1595, wrote: 'Elephants of all other beasts do chiefly hate the mouse.' The idea still persists, helped no doubt by such stories as that of the elephant in a zoo found dead from a haemorrhage and with a mouse jammed in its trunk.

In 1938 Francis G Benedict and Robert C Lee, American zoologists, tested zoo elephants with rats and mice in their hay, and by putting rats and mice in the elephants' house. The pachyderms showed no concern even when the rodents ran over their feet or climbed on their trunks. White mice were also put in the elephants' enclosure, again without result. There was, however, one moment when a rat ran over a piece of paper lying on the ground. The unfamiliar noise of rustling paper set the nearest elephant trumpeting and before long all the others were joining in the chorus.

class	**Mammalia**
order	**Proboscidea**
family	**Elephantidae**
genera & species	***Elephas indicus*** *Indian elephant* ***Loxodonta africana*** *African elephant*

△ *Bulls contest for the cow who appears to be rather disinterested in the combat.*

▽ *African elephants were thought to be untameable but the Belgians succeeded at the turn of the century by training immature ones for work using kindness and patience rather than brutality.*

Elephants playing in Lake Edward. The thick sparsely-haired skin is kept in condition by wallowing. When bathing an elephant will almost completely submerge itself and also spray water over itself with its trunk.

Elephant seal

Ponderous mountains of flesh, male elephant seals are the largest of all seals and may reach a length of 22 ft and weigh up to 8 000 lb. The females are smaller, growing up to 12 ft and weighing 2 000 lb. They were given the name because of their great bulk and the large drooping nose of the adult males.

There are two species, both very alike in size and appearance. The northern elephant seal lives on islands off the coast of California and Mexico. The southern elephant seal is found around the Southern Ocean. It breeds on the mainland of Argentina and on the many islands scattered around the cooler waters of the southern hemisphere, including the Falklands, South Georgia, South Orkneys, South Shetlands, Kerguelen, Heard and Macquarie. Outside the breeding season, the southern elephant seal wanders considerable distances, reaching the coasts of Antarctica, South Africa, Australia and New Zealand. Breeding has occasionally taken place on the coast of South Africa and Tasmania.

The southern elephant seal is the more numerous of the two. The population is estimated to be 600 000, of which half live around South Georgia.

Bellowing beachmasters—the biggest bulls, perhaps 12 or more years old, threaten each other at the start of a battle to win control of a section of the beach and custody of some of the females. The nose of the bull becomes enlarged during the breeding season, being inflated by muscular contraction and blood pressure to form a big cushion. At the height of the battle the bulls have reared up to face each other, with their loud roars echoing around the beaches.

Moulting en masse

In the latter half of the summer the elephant seals come ashore in their hundreds to moult. They congregate on low-lying ground just behind the beach where they remain throughout the period of moult. Hundreds of elephant seals gather in 'wallows' where the vegetation has been killed and the ground hollowed out to form muddy pools. The seals lie in these pools, sleeping most of the time and rarely, if ever, going into the sea. As a result they soon become unpleasant to both eyes and nose, as they lie in the increasingly foul wallows, sometimes several layers deep. As they can hold their breath for a long time they even go to sleep submerged in mud and under their fellows' bodies. Occasionally, however, one may be trapped under too great a crush and drowned.

During this time the seals are very placid. They sleep very soundly and it is possible to sit down on an elephant seal without awakening it. Even if awake they will take little notice of man; one is conscious of hundreds of pairs of big, limpid eyes following every movement, but the seals make no move unless provoked. Then they will rear up and roar, and flop back to sleep if the source of annoyance does not move. It takes gross provocation to make an elephant seal retreat down the beach to the sea. At first it slowly moves backwards like a caterpillar in reverse, then turns round and races seawards, the foreflippers held out from the body to act as crutches levering the heavy body over the sand.

The actual process of moulting is startling. Patches of hair come away along with the underlying skin, together with a liberal covering of noisome mud, so moulting elephant seals present a repulsive sight. It seems likely that the peeling skin irritates the elephant seals and that the mud helps to soothe this. When lying on a beach elephant seals will scoop damp shingle or sand over their backs with their foreflippers, probably for the same reason.

Outside the breeding season, elephant seals spend most of their time feeding at sea. The southern seals may wander some distance from their home waters. Seals marked with metal tags at South Georgia have been found at the South Orkneys and Argentina. The longest recorded journey was by a one-year-old that reached South Africa, 3 000 miles away.

Huge beast, small teeth

Elephant seals' teeth are surprisingly small. Although the canine teeth of a fully-grown male are 6 in. long no more than $1\frac{1}{2}$ in. stick through the gums and its cheek teeth are very much smaller. They feed on fish and squid, which are soft-bodied and easily chewed. The northern elephant seal is known to feed on sharks, skates and ratfish. As the latter lives more than 300 ft down, elephant seals must feed at a considerable depth. While ashore to moult or breed elephant seals often eat sand and stones, probably to ease hunger pangs.

Guarding the harem

In the spring the mature bulls come ashore to take up territories on the breeding beaches. Only the biggest bulls—perhaps 12 or more years old, called beachmasters—are able to win control of a section of beach; the others have to hang around the shores, or wait farther inland. During this time there is a good deal of rivalry. The bulls challenge each other with loud roars that echo around the beaches and can be heard for several miles. The nose of the bull becomes enlarged during the breeding season, being inflated by muscular contraction and blood pressure to form a big cushion. The northern elephant seal has the larger nose, with the tip hanging over the mouth like an elephant's trunk. It snorts down its nose and the sound waves are directed into the mouth which acts as a resonating chamber. The southern elephant seal's nose is not so well developed and its roar is generated in the mouth with the nose acting as a resonator.

Fighting also breaks out. Two bulls face each other and rear up so that over half their vast bulk may be held off the ground. Then they throw themselves against each other, chest to chest, their bodies quivering at the impact like giant jellies. At the same time horrible-looking injuries are inflicted. The nose may be badly torn and chunks of blubber gouged out, but the

Harem beauties on Annenkor Island, South Georgia. Since the closure of the whaling factories in 1965 the shooting of seals has stopped.

power of healing seems to be very strong and the wounds heal up, leaving large scars.

Quite soon after the bulls have come ashore, the cows arrive and lie on the beach ignoring the bulls. After about a week, each female gives birth from last year's mating to a single, black woolly-coated pup, about 4 ft long and weighing between 80 and 100 lb. The pup grows very rapidly, doubling its weight in 11 days and quadrupling it in a month. At this age the pup is left to fend for itself. At first it joins the other pups on one side of the territory where they moult their black coats, exposing shiny grey adult coats. They then go down to the shore where they feed on crustaceans they find among the rocks and finally swim out to sea.

During the pupping period the beachmasters will have been defending their territories and the harem of cows within the boundaries. After the cows have abandoned their pups they become receptive to the advances of their beachmaster, who mates with all his harem. The young bulls are kept away from the harem and can mate only with the young, pupless cows that are waiting about offshore. After mating, the harem breaks up, the seals go to sea to feed before coming back to the shore to moult.

A hazardous youth

The first hazard in an elephant seal's life is the danger of the beachmaster unwittingly squashing it. If the harem area is covered in deep snow the young seal also runs the risk of becoming trapped in a pit of its own making. If it lies in one place it gradually melts the snow under its body until it is lying in a pit too deep to climb out of and it eventually starves to death. Even adults sometimes die in the same way.

The young seals may fall prey to leopard seals or killer whales, but it is doubtful whether a healthy adult would be attacked by any but a very hungry killer whale. Adult elephant seals have been found with wounds apparently caused by killer whales and northern elephant seals have been seen with scars from wounds caused by sharks.

Carnage from the hunters

The pelt of an elephant seal is of no commercial value but under it lies 1–6 in. of blubber which practically led to their extinction. The northern elephant seal was almost annihilated by 1890 but with strict protection its numbers have risen to 15 000. The southern species was similarly persecuted. Shortly after Captain Cook discovered South Georgia in 1775, sealers made their way there first to slaughter fur seals for their pelts, then the elephant seals for their blubber, the blubber from each seal providing about 80 gallons of oil.

By the turn of this century both seals were very nearly extinct in the Southern Ocean. Then the British Government, that claimed South Georgia and many other sub-Antarctic and Antarctic islands, placed them under protection. The elephant seals soon began to recover and sealing was allowed under licence.

Sealing was usually carried out by the whalers who ran the whaling factories on South Georgia. Using old whalecatcher ships they would set off in parties around the island and land on the beaches where the seals were moulting. The seals were driven down to the water's edge by beaters where they were shot. The skin and underlying blubber were cut off in one sheet and towed out to the waiting ship.

To prevent the seals from being almost wiped out again, government inspectors supervised the sealing. In the early 1950's the coastline of South Georgia was divided into four sections and each year one section was left untouched and there were also two reserves with complete protection. In the other sections only a limited number of seals could be shot, usually 6 000 in all. Furthermore, only bulls more than 10 ft long were killed. As elephant seals have a 'harem system' of breeding so only a few bulls take part in mating, large numbers of males can be shot without affecting the birthrate.

Under this system the elephant seals continued to flourish. Since 1965 the whaling factories have been closed down so the elephant seals are left in peace.

class	**Mammalia**
order	**Pinnipedia**
family	**Phocidae**
genus & species	***Mirounga angustirostris*** *northern elephant seal* ***M. leonina*** *southern elephant seal*

Ermine

The ermine, also known as the stoat, a relative of the larger polecat, is up to 17 in. or more in length including 4½ in. of long-haired tail. The males are larger than the females, weighing from 7 oz to 1 lb, while the females weigh only 5–10 oz. The fur is reddish-brown with the white underparts and throat tinged with yellow. The tail is the same colour as the back except for the tuft of long black hairs on the tip. Like the polecat, the ermine can secrete an objectionable odour from its scent glands, but this is not quite so offensive as in the larger animal.

In winter in the northern parts of its range the ermine fur becomes white all over, with the exception of the tip of the tail which always remains black. It is then very valuable to the fur-trade. In Great Britain the traditional ceremonial robes of royalty and nobility have always been made from ermine.

The ermine is widespread in Europe from the Alps and Pyrenees to the Adriatic shores and east into Asia. It is also found in North America where sometimes known as the short-tailed weasel. The ermine is found throughout the British Isles and a smaller local race, varying somewhat in colour, is found in Ireland where it is known as the weasel. Another local race, called the Islay stoat, is found on the islands of Islay and Jura on the west coast of Scotland.

Snake-like hunter

The ermine is found in most types of country, hunting along hedgerows, across fields, by rivers and brooks or wherever there is a chance of food. It moves characteristically in a succession of low bounds, its long, lithe body assuming an almost snake-like appearance. It can swim and climb well. Its senses of smell and hearing are acute but its sight is poor. Whether hunting or not, the ermine is alert, agile and energetic, with a natural ability to take advantage of cover. A common trick is to use the runs of moles or rats, either to escape enemies or to hunt prey. There are a number of accounts of ermines playing together, twisting and turning like snakes, zigzagging over the ground, rolling over each other, somersaulting on the ground or in mid-air, leaping anything up to four feet into the air and, finally, sitting up on their hind-legs and boxing furiously with their fore-paws.

Although largely nocturnal in its habits, there is a good chance of seeing an ermine hunting in broad daylight.

Truly carnivorous

An ermine hunts largely by scent, picking up the trail of its prey and following this relentlessly. Truly carnivorous, it rejects little that is flesh. Until the disease myxomatosis reduced the numbers of rabbits in Europe, they formed the ermine's main food. A rabbit will cry out in terror, apparently paralysed with fear, even while the ermine is some way off. Similarly, a hare, which can outwit a fox or a pack of trained hounds, becomes so terrorised that it hardly tries to escape.

Given the opportunity an ermine can be destructive to game and poultry, which has led to its persecution by gamekeepers and poultry farmers from early times. The fact that it also destroys vermin is not so commonly stressed. Moles, rats, mice and voles are killed by a bite at the back of the neck. It also takes fish, small birds, eggs and reptiles. It will sometimes employ tactics known as 'charming' (see red fox p. 193).

Family parties on the hunt

Fertile matings take place in March and again in June and July and because the males are partially sexually active until October, infertile mating may take place after July. After fertilisation in the spring and summer, implantation is delayed until the following spring, after which there is a gestation period of 20–28 days. The nursery is made in a hole in a bank or the hollow of a decayed tree and in April or May 4 or 5 young are born, occasionally 6–9. The female alone tends the young, which she will defend fiercely against all dangers. She has only one litter a year. The babies are covered with fine white hair at birth, the black tip appears on the tail at 20 days and the eyes open at 27 days. Weaning is at 5 weeks of age.

The young remain with their mother after weaning and hunt with their parents in a family party. Two or more family parties may join up, like some of the larger carnivores, to form the well-known packs of ermine that are reputed to attack dogs and even men. When, through an increase in their numbers, the food supply of a district is largely reduced, the ermines sometimes migrate in large numbers. There are reports of several scores of them moving across country in a column, but these stories are viewed with caution by many zoologists. Apart from man the ermine has few enemies, but young ermines and a few adults are taken by owls and hawks.

Change in colour

It was believed for a long time that the change in the colour of the coat in autumn was caused not by the loss of hairs but by the loss of pigment in the hairs. This is now known to be incorrect. An ermine moults twice a year in spring and autumn and, as has already been said, its coat turns white in northern latitudes. The change from brown to white is very rapid because the

Except for the black tip on its tail, which always remains black, this ermine has moulted to its winter coat, having discarded its brown summer coat in a matter of days.

◁ *A first look at life: two young ermines gaze inquisitively at the world about them.*

△ *An ermine on the look-out, its body sprung for action. It is a remarkably fast and agile animal. An expert hunter, it will climb trees and pursue its prey into small holes and burrows.*

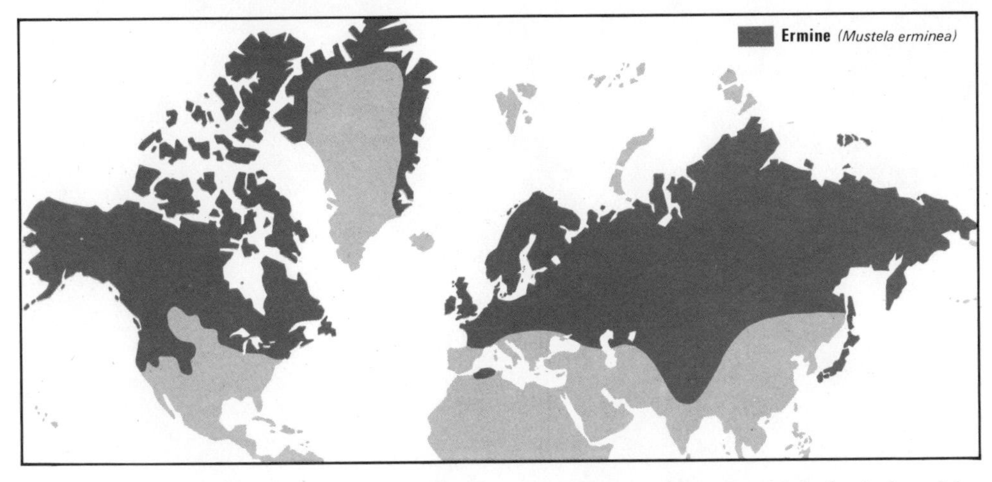

new white coat grows underneath the old one. The old coat may be shed in as little as three days in very cold conditions.

The accepted idea is that the summer russet coat is protective in that it harmonises generally with the colour of the leaf litter over which the ermine is moving, and that on snow-covered ground, as in Alpine districts, the change to a white fur enables the ermine to steal up on its prey unseen. It is not easy, however, to accept this when we recall the behaviour of rabbits and hares in the presence of an ermine. Their terror must almost certainly be induced by the sight of the animal. Even if it were induced by smell, it would still suggest that any coat colour is not of primary value as camouflage. More likely, a white coat cuts down the loss of body heat, as in the plumage of the ptarmigan.

Although the matter has not been fully investigated, experiments have shown that the change in colour of the ermine's coat seems to be dependent partly upon temperature and partly upon day length. Since both these factors are variable many permutations can result. In North America, for instance, all ermines turn white in the north, and in the south all remain russet, but there is a zone between where white, particoloured or russet ermines are found in winter. Again, in the south of England, an occasional ermine will turn white in the autumn, whereas others may be partly white and partly coloured. Both these forms may occur in a mild winter when there is little or no snow. One explanation for this is that temperature may also have a delayed action, so that an ermine experiencing lowered temperatures in one autumn may turn white in the following autumn even if temperatures are high. But the fact that white or partly white ermines in southern England tend to be localised suggests that the change is genetic.

class	**Mammalia**
order	**Carnivora**
family	**Mustelidae**
genus & species	***Mustela erminea*** *ermine, stoat or short-tailed weasel* ***M. e. hibernica*** *Irish weasel* ***M. e. ricinae*** *Islay stoat*

Flying phalanger

Sometimes called flying squirrels in Australia, because they look outwardly alike. Flying phalangers are, however, not even distantly related to squirrels but are true marsupials like kangaroos and opossums.

There are five species distributed over the eastern half of Australia, one species overlapping in the south into Tasmania, and one species of sugar glider in New Guinea. They belong to three types: the pigmy or feathertail glider, the 3 species of sugar glider and the greater glider. The first is mouse-sized, just over 6 in. long, of which a half is tail, olive-brown above, white below. The sugar gliders are nearly 16 in. long, of which one half is bushy tail, with a fine silky fur, grey to brown with a dark line along the back and lighter underparts. The greater glider is over 3 ft long, of which the tail is over 1½ ft, grey to dark-brown with yellowish underparts. All live in trees and take gliding flights from tree to tree. The gliding membrane of the sugar glider is narrow, fringed with long hairs and stretches from the fore to the hind limbs. In the others this 'parachute' is a broader web of furred skin stretching from the 5th toe on the forefoot to the ankle of the hindleg. In the sugar glider the tail is feathered—that is, fringed either side with long hairs; in the remainder it is bushy.

△ *Takeoff; a well-judged launch into space . . .*
▽ *. . . and landing, gripping with sharp claws.*

▷ *A sugar glider takes a snack. These pretty marsupials eat almost anything they can get.*

Hidden in the tree tops

The feathertail flying phalanger, or glider, has the large eyes and ears typical of a nocturnal animal. It is seldom seen except when a tree is felled or a domestic cat brings one home. During the day it rests curled up in holes in trees, up to 50 ft from the ground, lying hidden in a nest of shredded eucalyptus bark. At night it takes gliding flights from tree to tree. It is said to be common wherever there are eucalyptus trees and especially those with a white smooth bark. In the trees it runs quickly over bark using its claws, and it can run over smooth leaves with the pads on the tips of the toes giving a sure grip.

The habits of sugar gliders and greater gliders are similar. When in the tops of the trees there is little to indicate they are moving about except a faint scratching on bark or the rustle of leaves. Their glides also are sudden and swift, usually seen only by accident. The gliding feats are most spectacular in the greater glider; one is recorded as having covered 590 yd in 6 successive glides, an average of nearly 100 yd between each pair of trees. During a glide the phalangers lose height, and having landed on the next tree they run rapidly up the trunk for the next takeoff. Sometimes one will land on the ground, over which it runs awkwardly.

All have a sweet tooth

The various flying phalangers differ in one respect: the teeth. The pigmy gliders have insectivorous teeth, recalling those of shrews. They eat insects and especially plant lice, such as aphides and scale insects, that give out honeydew. The sugar glider also eats insects, and small birds as well, but its food is mainly flowers, fruit, buds, nectar and sap. The greater glider feeds only on leaves and flowers, mainly those of gum trees. Both these have the kind of teeth associated with a vegetarian diet.

Useful mobbing

Being marsupials, the females carry their young in a pouch, and when the babies are large enough to leave it they ride on the mother's back; at least this is true for the smaller species. This is known from only chance observation, as when a flying phalanger out in broad daylight was mobbed by a crowd of birds, including Australian magpies. One of these swooped and drove the phalanger hard against a tree. It hit its head on a branch and fell to the ground, where the baby fell from its back. Otherwise little is known of the breeding habits of these shy creatures. The pigmy glider has 2–4 young at a birth. The sugar glider has 1–3, usually 2, babies after a gestation of

3 weeks, the young becoming independent at 4 months. The greater glider has one young in July–August, which leaves the pouch at 4 months, but remains with the parents until fully grown.

Powerful owl enemy

The greatest hazard to flying phalangers lies in the steady felling of eucalyptus or gum trees. A phalanger occasionally falls victim to the introduced red fox when it lands on the ground. Otherwise the main enemies are owls, especially the one known as the powerful owl.

Bundles under the tail

Several marsupials use their tails for carrying nesting materials, and so does that other primitive mammal, the platypus. This is the more remarkable since the tail of a platypus is not long and slender but broad and flat—less suited, one would have thought, to being wrapped around a bundle of leaves. The rat kangaroos of Australia do the same, but their tails are prehensile, anyway. The American opossum brings its tail forward under the body, passes leaves and grass—or similar building materials—backwards under its chest, then with its hind legs arranges these for the tail to grasp. The greater glider has been recorded as carrying a bundle of twigs and leaves for a nest with its prehensile tail, and sugar gliders have been seen to carry out the same operation.

David Fleay, the Australian naturalist, watched a captive sugar glider hang by its hindfeet, bite leaves off eucalyptus boughs and, using the forepaws, transfer them to its tail. When it had a bundle about 6 in. long and 3 in. across, the phalanger ran along to its nesting box holding the burden with its tail wrapped round it.

◁▽ *Sugar glider feeding on foliage. The folds of skin between the limbs give little idea of the massive 'parachute area' so dramatically illustrated below. The neatly-curled tail can be used for carrying nest material.*
▽ *Study in unpowered flight. With only a leap from a treetrunk and the gliding effect of the outstretched skin, a flying phalanger can cover 100 yd a trip, landing with remarkable accuracy some way up the trunk of a selected tree.*
▷ *Baby phalanger, blind, naked, and completely helpless, nestles in its mother's pouch. It will not become completely independent until about 4 months old.*

class	**Mammalia**
order	**Marsupialia**
family	**Phalangeridae**
genera & species	***Acrobates pygmaeus*** *pigmy glider* ***Petaurus australis*** *sugar glider* ***Schoinobates volans*** *greater glider*

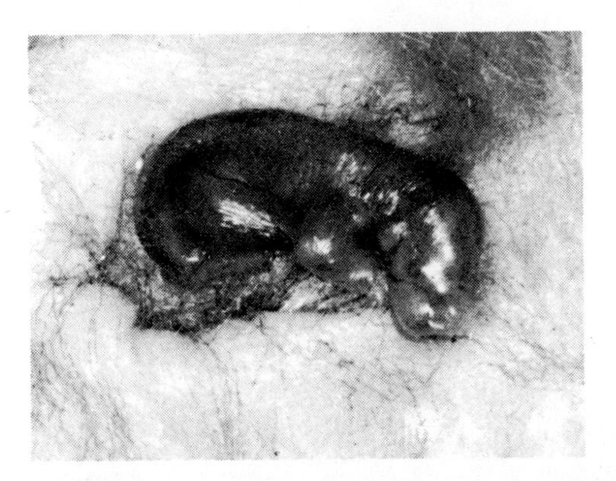

Giant Panda

This black and white bear-like carnivore has leapt from obscurity to worldwide fame in less than a century. Also called the panda and, by the Chinese, **beishung**, *the white bear, it was first made known to the western world in 1869, by the French missionary, Pere David.*

The giant panda is stockily built, with a 6 ft long body and a mere stump of a tail and weighs 300 lb. Its thick, dense fur is white except for the black legs and ears, black round the eyes and on the shoulders. There are 5 clawed toes on each foot and each forefoot has a small pad which acts as a thumb for grasping. The cheek teeth are broad and the skull is deep with prominent ridges for the attachment of strong muscles needed in chewing fibrous shoots. It lives in the cold damp bamboo forests on the hillsides of eastern Tibet and Szechwan in southwest China.

Habits unknown . . .

Giant pandas are solitary animals except in the breeding season. They live mainly on the ground but will climb trees when pursued by dogs. They are active all the year. Little more is known of the habits in the wild of this secretive animal which lives in inaccessible country. When live giant pandas were first taken to zoos it was thought they lived solely on bamboo shoots. Later it was learned that during the 10—12 hours a day they spend feeding they eat other plants, such as grasses, gentians, irises and crocuses, and also some animal food. This last includes small rodents, small birds and fishes flipped out of water with their paws.

Breeding unknown . . .

Little is known about the giant panda's breeding habits in spite of attempts to induce a mating between An-an, the male giant panda belonging to the Moscow zoo, and Chi-chi, the female in the London zoo. In 1966 Chi-chi was taken to Moscow but no mating took place, and An-an was brought to London in 1968 with no more success. It is believed that giant pandas mate in spring, and that probably one or two cubs are born in the following January, each cub weighing 3 lb at birth. Several cubs have been born in Chinese zoos. On September 9, 1963, a male cub Ming-ming was born to Li-li and Pi-pi in Peking zoo, and a female cub, Ling-ling, was born on September 4, 1964, to the same parents. A third cub Hua-hua, a male, was born to Chiao-chiao on October 10, 1965. According to Mare Ribaud, a French photographer writing in *Natural History,* April 1966, Ming-ming and Ling-ling were produced by artificial insemination. Presumably the same is true for Hua-hua.

Bad treatment

In 1869 Père Armand David of the Lazarist Missionary Society, and an experienced naturalist, came upon the skin of an animal

◁ *Chinese mother love. Although breeding has not been achieved in the western world, Chinese zoos have bred pandas.*

in a Chinese farmhouse in Szechwan which he did not recognise. He sent it to Paris and later sent more skins. Not until 1937, however, was the first live giant panda seen outside China. Theodore and Kermit Roosevelt had shot one in the 1920's and in 1936 two other Americans, Ruth and William Harkness, with the animal collector Tangier Smith, captured several. They quarrelled, presumably over the spoils, and all the giant pandas died except one, which Ruth Harkness delivered to the Chicago zoo where it was named Su-lin. Another, given the name Mei-mei, reached the same zoo in 1938. In December the same year a young female, Ming, aged 7 months and two young males, Tang and Sung, reached the London zoo. The two males died before the female reached maturity, and she died in December 1944. In May 1946, the government of the Szechwan Province presented a male, Lien-ho, to the London zoo and he lived until 1950. By 1967 there were a score of giant pandas in various zoos, 16 or more in Chinese zoos, An-an in Moscow and Chi-chi in London.

Although the species is now protected it was formerly hunted by the local Chinese, and the history of western animal collectors does nothing to offset this. The story of Chi-chi gives point to this. In 1957, Heini Demmer, then living in Nairobi, was commissioned by an American zoo to negotiate the exchange of a collection of East African animals for a giant panda. He reached Peking zoo with his cargo, was given the choice of one of three giant pandas, chose Chi-chi, the youngest, and took charge of her on May 5 1958. Chi-chi had been captured by a Chinese team of collectors on July 5 1957, and was reckoned then to be 6 months old. She had been taken to Peking zoo and cared for night and day by a Chinese girl. By the time Demmer had taken charge of Chi-chi the United States had broken off diplomatic relations with the Chinese People's Republic, so she became automatically a banned import.

Bamboo shoots are not the sole food of giant pandas—other plants and some animal food is eaten.

Demmer took her on a tour of European zoos during the summer of 1958, reaching the London zoo on September 26.

After such treatment perhaps it is not surprising she refused to be mated! She died on July 21 1972.

London Zoo now has two 3 year-old pandas, Ching-Ching and Chia-Chia, which were presented to Mr Heath during his visit to Peking. Mr Nixon was similarly honoured on his visit there in 1973. In 1974 Tokyo had two pandas, and Korea and Paris owned one each.

class	**Mammalia**
order	**Carnivora**
family	**Procyonidae**
genus & species	***Ailuropoda melanoleuca***

Giraffe

Tallest animal in the world, the giraffe is remarkable for its long legs and long neck. An old bull may be 18 ft to the top of his head. Females are smaller. The head tapers to mobile hairy lips, the tongue is extensile and the eyes are large. There are 2–5 horns, bony knobs covered with skin, including one pair on the forehead, a boss in front, and, in some races, a small pair farther back. The shoulders are high and the back slopes down to a long tufted tail. The coat is boldly spotted and irregularly blotched chestnut, dark brown or liver-coloured on a pale buff ground, giving the effect of a network of light-coloured lines. A number of species and races have been recognised in the past, differing mainly in details of colour and number of horns, but the current view is that all belong to one species. The number of races recognised, however, varies between 8 and 13 species depending on the authority.

The present-day range of the giraffe is the dry savannah and semi-desert of Africa south of the Sahara although it was formerly more widespread. Its range today is from Sudan and Somalia south to South Africa and westwards to northern Nigeria. In many parts of its former range it has been wiped out for its hide.

A leisurely anarchy

Giraffes live in herds with a fairly casual social structure. It seems that males live in groups in forested zones, the old males often solitary, and the females and young live apart from them in more open country. Males visit these herds mainly for mating.

Giraffes do not move about much, and tend to walk at a leisurely pace unless disturbed. When walking slowly the legs move in much the same way as those of a horse. That is, the right hindleg touches the ground just after the right foreleg leaves it, and a little later the left legs make the same movement. The body is therefore supported on three legs most of the time while walking. As the pace quickens to a gallop the giraffe's leg movements change to the legs on each side moving forward together, the two right hoofs hitting the ground together followed by the two left legs moving together.

The long neck not only allows a giraffe to browse high foliage, the eyes set on top of the high head form a sort of watch-tower to look out for enemies. In addition, the long neck and heavy head assist movement by acting as a counterpoise. When resting crouched, with legs folded under the body the neck may be held erect or, if sleeping, the giraffe lays its neck along its back. To rise, the forelegs are half-unfolded, the neck being swung back to take the weight off the forequarters. Then it is swung forwards to take the weight off the hindlegs, for them to be unfolded. By repeated movements of this kind the animal finally gets to its feet.

◁ *Dappled freaks of the African veld: a group of giraffes rear their extraordinary necks against the skyline of a pale sunset.*

Necking parties

The habit of 'necking' has been something of a puzzle. Two giraffes stand side-by-side and belabour each other with their heads, swinging their long necks slowly and forcibly. Only rarely does any injury result, and the necking seems to be a ritualised fighting, to establish dominance, and confined exclusively, or nearly so, to the male herds.

Not so dumb

One long-standing puzzle concerns the voice. For a long time everyone accepted the idea that giraffes are mute—yet they have an unusually large voice-box. During the last 25 years it has been found that a young giraffe will bleat like the calf of domestic cattle, that the adult female makes a sound like 'wa-ray' and that adult bulls, and sometimes cows, will make a husky grunt or cough. Nevertheless, there are many zoo-keepers who have never heard a giraffe utter a call and there is still the puzzle why there should be such a large voice-box when so little use is made of it. Some zoologists have suggested the giraffe may use ultrasonics.

Controlled blood pressure

In feeding, leaves are grasped with the long tongue and mobile lips. Trees and bushes tend to become hourglass-shaped from giraffes browsing all round at a particular level. Acacia is the main source of food but many others are browsed, giraffes showing definite preferences for some species of trees or bushes over others.

Giraffes drink regularly when water is available but can go long periods without drinking. They straddle the front legs widely to bring the head down to water, or else straddle them slightly and then bend them at the knees. Another long-standing puzzle concerns the blood pressure in the head, some zoologists maintaining a giraffe must lower and raise its head slowly to prevent a rush of blood to the head. In fact, the blood vessels have valves, reservoirs of blood in the head and alternative routes for the blood, and so there is no upset from changes in the level of the head, no matter how quickly the giraffe moves.

Casual mothers

Mating and calving appear to take place all the year, with peak periods which may vary from one region to another. The gestation period is 420–468 days, the single calf being able to walk within an hour of birth, when it is 6 ft to the top of the head and weighs 117 lb. Reports vary about the suckling which is said to continue for 9 months, but in one study the calves were browsing at the age of one week and were not seen suckling after that. The bond between mother and infant is, in any case, a loose one. Giraffe milk has a high fat content and the young grow fast. Captive giraffes often live for over 20 years.

Defensive hoofs

Giraffes have few enemies. A lion may take a young calf or several lions may combine to kill an adult. Even these events are rare because the long legs and heavy hoofs can be used to deadly effect, striking down at an attacker.

Symbol of friendliness

Rock engravings of giraffes have been found over the whole of Africa and some of the most imposing are at Fezzan in the middle of what is now the Sahara desert. The animal must have lingered on in North Africa until 500 B.C. Some of the engravings are life size, or even larger, and many depict the trap used to capture giraffes, while others show typical features of its behaviour, including the necking. The engravings also show ostriches, dibatag, and gerenuk. Giraffes were also figured on the slate palettes, used for grinding malachite and haematite for eye-shadows, in Ancient Egypt, similar to that believed to portray the dibatag. The last giraffe depicted in Egyptian antiquities is on the tomb of Rameses the Great, 1225 BC.

There are references to the animal in Greek and Roman writings and a few pictures survive from the Roman era, but from then until the 7th or 8th century AD the principal records are in Arabic literature. The description given by Zakariya al-Qaswini in his 13th-century *Marvels of Creation* reflects the accepted view, that 'the giraffe is produced by the camel mare, the male hyaena and the wild cow'. The giraffe was taken to India by the Arabs, and from there to China, the first arriving in 1414 in the Imperial Zoological Garden in Peking. To the Chinese it symbolised gentleness and peace and the Arabs adopted this symbolism, so a gift of a giraffe became a sign of peace and friendliness between rulers.

In medieval Europe, and until the end of the 18th century, knowledge of the giraffe was based on descriptions in Greek and Roman writings and on hearsay accounts. It was at best a legendary beast.

class	**Mammalia**
order	**Artiodactyla**
family	**Giraffidae**
genus & species	***Giraffa camelopardalis***

Wiped out for its hide in many parts of its range, the present day distribution of the giraffe is much reduced. A number of races are recognised within the single species.

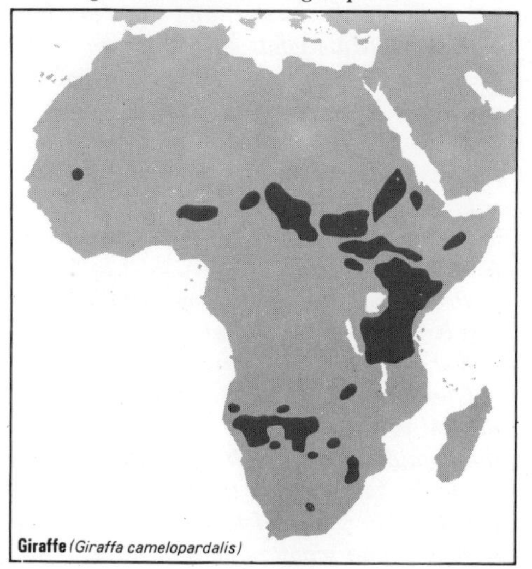

Goats browsing in semi-desert. When presented with more lush conditions they will soon eat and spoil them until the land is like this.

Goat

Scientifically, it is not easy to sort sheep from goats. The distinguishing features are that the horns of sheep grow to the sides of the head, those of goats curve upwards and backwards and are worn by both sexes. Male goats have a strong smell and wear beards. In goats the forehead is convex, not concave as in sheep.

There are five species of wild goat, including the ibexes and the markhor. The wild goat ***Capra hircus*** *from which the domesticated goat was derived, ranges from southeast Europe through Asia Minor to Persia and Pakistan. Domestication can be traced to 6–7 thousand years ago and it may have been earlier.*

Goats are 4½ ft long in body and head, the tail is 6 in., they are 3 ft high at the shoulders and weigh up to 260 lb, the males being larger than the females. The horns of males are sweeping and scimitar shaped, up to 52 in. long, compressed sideways and ornamented along the inner front edge with large knobs. The horns of the females are shorter and more slender. The coat is typically reddish-brown in summer, greyish-brown in winter with black markings on the body and limbs.

Desert-making goats

Goats usually live in rugged, rocky or mountainous country, but sometimes on lowland plains. Where hunted they become extremely wary, and difficult to stalk, as their sure-footed skill as they progress from rock to rock is legendary. They generally move about in herds of 5–20, led by an old female. When living on mountains they may go up almost to the snow-line but in winter migrate down to lower levels, returning in spring to the fresh pastures. Goats do not sleep; they merely have periods of drowsiness.

Goats will eat straw, and have been seen to scratch their backs with straws held in the mouth. Like sheep, goats chew the cud, but whereas sheep take mainly grass, goats browse chiefly on leaves and twigs as well. They will eat desert scrub and climb into trees to browse, and goats have been seen to jump onto the backs of donkeys to reach the lower boughs, and from there move to higher and higher boughs by jumps. They readily take bark, will eat paper and are notorious for eating linen cloth. In this they are helped by protistans living in the gut which pre-digest cellulose. Domestic goats will eat the foliage of yew, which may be fatal to horses and cattle, and suffer only a temporary diarrhoea. Released on oceanic islands, goats have reduced earthly paradises to barren soil with only low vegetation. In the Near and Middle East herds of goats have contributed to the formation of deserts.

Life-cycle

A female goat over two years old is known as a nanny-goat, the male is a buck or billy-goat. Both are relatively recent names, the first having been used since 1788, the second since 1861. Mating is normally in autumn and the kids, born 147–180 days later, are able to run shortly after birth and soon become adept at climbing. There are one, sometimes two at a birth, exceptionally three or four. Sexual maturity is reached in about 12 months, when the male is known as a buckling, the female as a goatling. The life span is up to 18 years.

Indestructible spoilers

Goats have probably always been more useful for their milk than for their hair or flesh. Their flesh is somewhat rank and the hair short, but sometimes used for spinning, especially from longhaired breeds such as the Angora and the Kashmir. In the days of sail, ships took goats on board to provide fresh milk as well as meat. Ships' captains would put goats ashore on oceanic islands for the use of castaways, or to get rid of surplus. The marooned goats multiplied and, as on St Helena and other islands, denuded the flora. In 1773 Captain James Cook put goats ashore in New Zealand. These went wild and multiplied. Later goats were taken there for other purposes, to feed those building roads and railways, for use in miners' camps, and also to prevent introduced bramble, gorse and bracken running amok.

In fact, the goats barked trees, ate shrubs, brought many native plants to the verge of extinction and cleared the ground of mosses that not only held water but protected the topsoil from wind erosion. Their hoofs cut the turf so that it was washed away by rain, so adding to the erosion. The natural home of a goat is the barren hillside and wherever goats go they convert the landscape into their natural habitat.

The speed at which goats multiply is also an embarrassment. In 1698 an English ship put into the harbour of Bonavista. Two Negroes went aboard and offered the captain all the goats he cared to take away. There were only 12 people living on the island and not only were the goats eating everything but they were so tame nobody could go anywhere without a crowd of goats following.

Holinshed, in his *Chronicles of England*, 1577, wrote: 'Goats we have in plenty, and of sundry colours, in the west parts of England; especially in and towards Wales, and among the rocky hills, by whom the owners do reap no small advantage.' What the advantage was he did not say but it is believed that goats were deliberately allowed to go wild in these regions by the sheep farmers. In the Welsh mountains grass grows lush in inaccessible places. Sheep attracted up by the grass cannot get down and have to be retrieved. Wild goats, better climbers than the sheep, climb the high rocks and eat the grass (so removing temptation for the sheep) and have no difficulty in descending.

△ ▷ *The sure-footed agility of young goats.*
▷ *Every kid needs its mother now and then, especially if he is the only one.*

class	**Mammalia**
order	**Artiodactyla**
family	**Bovidae**
genus & species	***Capra hircus***

Gorilla

The gorilla is the largest of the man-like apes. The males average 5 ft 8 in. high and may exceed 6 ft, and the females are about a foot less. An adult male may weigh 400–450 lb but in a zoo he tends to get fat and may weigh 100 lb more. Unlike the chimpanzee whose skin normally turns black only at maturity, the gorilla's skin is jet black from a few days after birth. The hair is grey-black or brown-black in western gorillas, jet black in the eastern race. The adult male in both develops a silvery white back and this makes a strong contrast with the jet black of the eastern gorilla. He also has a large sagittal crest (a bony crest on the top of the skull) to which the jaw muscles packed with connective tissue are attached. This gives a helmet-like effect to the head. The nostrils are broad and the ear is small, in contrast with that of the closely related chimpanzee. The chest is broad and the neck short and muscular. The hands and feet are broad and strong, the great toe being less widely separated from the other toes than in the rest of the apes. A gorilla walks normally on all fours, with knuckles to the ground, in a semi-erect posture because the arms are longer than the legs. Adult gorillas seldom climb trees.

There are three very well-marked races. The western gorilla lives in lowland rain forest, from sea level to about 6 000 ft, in the Congo (Brazzaville), Gabon, Equatorial Guinea, Cameroun, extreme southwest of Central African Republic, and in the extreme southeast of Nigeria. The eastern lowland gorilla is found in a similar habitat in the eastern Congo (Kinshasa), ascending the mountains in the Central African Lakes region to about 8 000 ft. There are no gorillas in the vast lowland forest area between the ranges of these two races, and it is somewhat of a mystery why this should be. Finally, the mountain gorilla is found between 9 000 and 12 000 ft in the Virunga Volcanoes and Mt Kahuzi. All eastern gorillas are blacker than western, with larger jaws and bigger teeth; mountain gorillas are distinguished, in addition, by their comparatively short arms, long silky hair, and strikingly manlike feet.

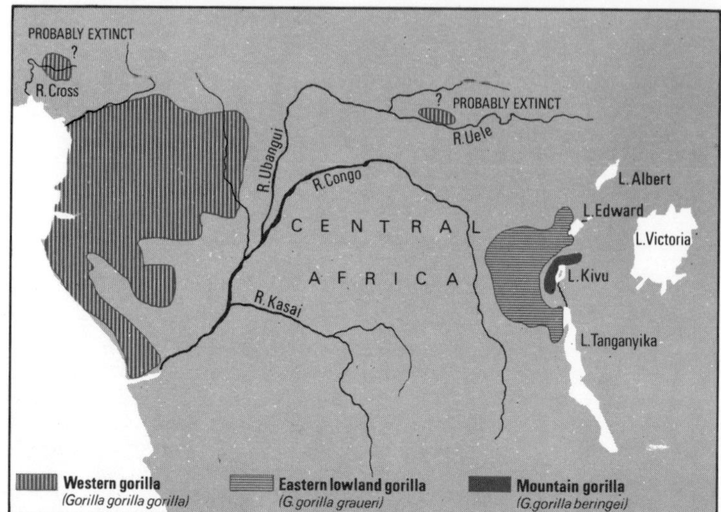

△ A relaxed snack: having torn away the tough outer covering of a plant stem, a young gorilla tucks into the juicy centre.
◁ The three gorilla races come from three distinct areas in Central Africa. Mountain gorillas have shorter arms, longer hair, and almost human feet. Eastern races are darker than western, with big jaws and teeth.
▷ Young gorillas are very like human children in their temperament.

Peaceful co-existence

Gorillas live in troops of a single adult male and several females with their young. Other males wander alone, sometimes travelling along with a troop for a while. There is thus a much tighter social organisation than with chimpanzees, with much smaller troops and only one adult male to each. The home ranges of the troops overlap extensively;

there is no defended territory. A meeting between two troops may result in their mingling temporarily, or the two more or less ignoring one another. No fighting has been recorded, although some gorillas have been seen with wounds, especially bruises and cuts about the eyes, which suggest they do occasionally come to blows. A troop tends not to wander over more than 10–15 sq miles. Their wandering is irregular.

Aversion to water

In the lowlands, gorillas feed on fruit and leaves, and raid banana plantations. They strip the stem from the banana tree and eat the marrow. In the mountains they eat much more tough, fibrous vegetation, such as bark, stems and roots. They get their moisture from their food, rarely drinking in the wild. When they do they soak the fur on the back of the hand and suck the water from it. They are afraid to cross even small streams, which limits their wanderings and makes it possible for them to be kept in by moats in zoos. They have not been known to eat eggs, insects or other animal protein. In lowland forests the density of gorillas is about 1 per sq mile; but in the open *Hagenia* forests at high altitudes in the Virunga Volcanoes, with dense ground cover and abundant fodder, the density of gorilla populations reaches 6 per sq mile.

Precocious babies

There is no special birth season and it is not necessarily the troop male who mates with the females in the troop. Often a wandering male who has joined the troop for a while mates with a female in it. Gestation is about 255 days—a little shorter than in the human. The young weigh about 4–5 lb at birth, which is less than a human baby, but the baby gorillas develop twice as fast: their eyes focus in the first or second week, they crawl in 9 weeks, and can walk a few steps bipedally at 35–40 weeks. The infants play among themselves and with the adult male of the troop, who tolerates a great deal of nonsense from them. As they mature they become less playful. Females stay in the troop but the males leave it before they have developed the silver back of full maturity, to become wanderers. It is not known how a male becomes troop leader. The female is sexually mature at 6 or 7 years, the male a year or two later, but he does not reach full size until he is 12 or 14. Gorillas have been known to live 37 years.

No enemies but man

Leopards may take young gorillas and have even been known to kill adults, but this must be uncommon since gorillas show no particular uneasiness when leopards are around. There seem to be no other enemies except man, who kills them when they raid plantations and some tribes eat gorilla meat.

The gentle giant

The gorilla is not the savage, untameable ogre of popular imagination. He is normally a gentle peaceful creature. When disturbed gorillas of both sexes and all ages after about a year, beat their chests. This is a kind of tension-releaser, but in the male it is part of a full display of roaring, rising on hindlegs, beating the chest, a quick sideways run and, finally, tossing vegetation into the air. This is all very terrifying to a human being, and may precede a bluff charge, but an actual attack is rare. The male may, however, give chase if the intruder runs away, biting and scratching. Among gorilla-hunting tribes it is considered a disgrace to be wounded by a gorilla, for the man must have been running away! Unprovoked attacks are not known to have occurred.

Young gorillas have been kept as pets, but they are difficult to keep as they are subject to the same tantrums and fits of exhilaration as human children, and being much stronger they are more destructive. Moreover, they need constant companionship and affection and if the favourite person is not at hand the young gorilla will sulk, refuse food, become ill, even die. In zoos, keepers go in and play with young gorillas and every effort is made to obtain a companion to avoid the misery of loneliness and to prevent the animal becoming 'humanised'.

Gorillas are at least as intelligent as chimpanzees, less volatile in temperament and more patient and methodical. This stands them in good stead in the performance of intelligence tests, but they tend not to do well in tests involving manipulation. This makes sense, as they do not make use of natural objects as 'tools' in the wild as chimpanzees do.

class	**Mammalia**
order	**Primates**
family	**Pongidae**
genus & species	***Gorilla gorilla gorilla*** *western gorilla* ***G. gorilla graueri*** *eastern lowland gorilla* ***G. gorilla beringei*** *mountain gorilla*

▽ *Freak of nature: a baby albino gorilla, Snowflake, at the zoo in Barcelona. He would have had little chance of survival in the wild, lacking the gorilla's normal protective colouring.*

The bushy-tailed, forest-dwelling grey fox. Shy and nocturnal, it often passes undetected—and one reason is because its voice resembles the familiar call of the coyote. Unlike many other foxes, it is not a fast runner, nor can it cover long distances at full speed.

Grey fox

Sometimes called the tree fox, the short-legged grey fox is noted for its ability to climb and it uses trees much more than other foxes. Up to 27 in. head and body length and 15 lb weight, the grey fox has a bushy tail up to 17 in. long. The general colour of the fur is grey with underparts white but there is a rusty tinge along the sides of the neck, lower flanks and underside of the tail. There is a black line along the middle of the back, continuing along the tail, and black lines on the face. There is a noticeable ridge of stiff hairs along the top of the tail. Its size and colour vary from one region to another. In the north-east of its range its coat is a dark grey; in the southwest it is paler and slightly redder.

The range is from southern Canada through the United States to Mexico, Central America and northern South America. A smaller animal with shorter ears living on certain islands off southern California is regarded as a second species. It scavenges the beaches and makes its den among the cacti.

Climbing fox

Grey foxes live in forests, especially of southern pines, or brush country in the dry areas of the southwestern United States and Mexico. It is difficult to assess numbers because the animal is not only about mainly at night but is also adept at keeping out of sight at all times. It is therefore comparatively seldom seen and even its yapping bark often passes unrecognised, even if heard, partly because it is somewhat like the call of the coyote. During the day it rests in thick vegetation or among rocks, or in a tree hollow. Much of its food is caught on the ground but the fox will not only go up into trees when pursued but will also do so of its own will, especially to find fruits in season. It will run up a leaning trunk or climb a straight trunk gripping it with its forelimbs and pushing upwards with the hindfeet, the long claws on the toes of the hindfeet acting as climbing irons. Once in the tree it may leap from one branch to another. In descending the fox backs down the tree. It is not a fast runner, nor can it run long distances. The difference between the crude climbing of the red fox and the skill of the grey fox can best be illustrated by an accident. A grey fox was found dead in a tree its tail caught by the tip in a forked twig and further held by having passed through a second fork. From the scarring on the bark of the nearby twigs the fox had made desperate efforts to free itself. The important point is that all the twigs around the fox were no more than $\frac{3}{4}$ in. thick and most were nearer $\frac{2}{5}$ in.

Fox and grapes

Its diet is wide and takes in mice, squirrels, small birds and eggs, as well as insects. It also includes more plant food than is usual in the dog family. Grain and fruits, especially wild grapes and wild cherries, form the bulk of the food at certain seasons and in particular areas. With such a wide diet the grey fox readily takes to farmland and can be a nuisance, especially where there is poultry. It is also established in some built-up areas, for instance, the outskirts of New York City. The actual requirements of grey foxes were worked out by Richard F Dyson, Curator of Large Mammals at the Arizona-Sonora Desert Museum at Tucson. Because some of the mammals were overweight and had shaggy coats he tested grey

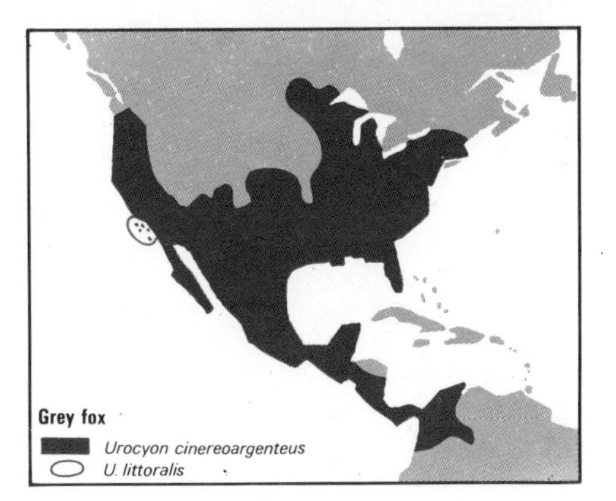

foxes for 6 months and found they kept in excellent health on 3·8% of their own body weight of food (flesh and fruit) per day. Later it was shown that this held good for other carnivores.

United families

The cubs are born in spring after a gestation period of about 2 months, the litter averaging 3 or 4, but it may be from 2 to 7. At birth the cubs are black, blind and helpless, about $3\frac{1}{2}$ oz weight. They are weaned at 6 weeks. The male helps in bringing up the family, the cubs finally leaving the parents at the age of 5 months. Grey foxes have lived up to 12 years in captivity.

The grey fox may be killed by wolf, coyote, bobcat and lynx but today its main enemy is man. Because of its habit of going quickly to ground or up into trees it is not hunted but trapped. In this the trapper takes advantage of the regularity with which a grey fox uses a run through the vegetation and sets his traps accordingly. The pelts make only second-rate furs.

Tree-climbing

But for its habit of tree-climbing the grey fox would hardly be noticed by zoologists. Yet tree-climbing foxes are no novelty, even among those whose coats are red. Many a fox has outwitted the hunt by running up the trunk of a leaning tree and hiding among the foliage. Others have ascended by using low branches but at least one red fox in England denned up in the crotch of a large tree and had her cubs there. The crotch was 15 ft from the ground and the vixen reached it, judging by the scratch marks on the bark, by jumping up from a buttress root and scrabbling the last few feet. This is highly unusual, but ordinary tree-climbing by red foxes seems to be more common than we suppose. One thing they never do is cling by the forelegs, as the grey fox does. That is a cat-like action, but it is probably also a result of the grey fox's short legs. Domestic dogs will sometimes climb trees. Those that do this most successfully are the small breeds with short legs.

class	**Mammalia**
order	**Carnivora**
family	**Canidae**
genus & species	***Urocyon cinereoargenteus*** *grey fox* ***U. littoralis*** *beach fox*

Bundle of innocence: a grey fox cub. At 5 months, however, when it becomes independent, it will be a wily, hard-to-catch, farmland predator.

Grey whale *Eschrichtius glaucus.*

Grey whale

At one time the Californian grey whale lived in the Atlantic Ocean, for its remains have been found in reclaimed land in the Zuider Zee. Now it is confined to the North Pacific where there are populations on both sides of that ocean.

It is a rather unusual whale, having points in common with both the rorqual (family Balaenopteridae) and right whales (family Balaenidae). It is about the size of the right whales, reaching 45 ft long and 20 tons in weight. The flukes of the tail are proportionally longer and more delicate than those of right whales, but more stubby than those of rorquals. The dorsal fin is replaced by 8–10 small humps along the tail just in front of the flukes. On the throat the grey whale has 2–3, rarely 4, grooves extending a short distance as compared with the 40–100 grooves extending to the belly in rorquals and the complete absence of grooves in right whales.

As the name implies, the grey whale is usually dark slate-grey but it may sometimes be blackish. It is lighter on the belly than on the back, as is usual in marine animals. Many grey whales have crescent-shaped marks or patches on the skin, especially on the back. These are caused either by lampreys or by barnacles.

Sluggish swimmers

Grey whales are very slow, usually swimming at 2–3 knots with bursts of 6–7 knots when alarmed, compared with 20 knots of a fin whale. As they also come very close inshore this makes them very vulnerable to hunters. In early spring the grey whales migrate down the west coast of North America. In 1840 there were estimated to be around 25 000 grey whales but soon after this there was very intense hunting all along the coast. By 1875 it was unusual to see more than 50 migrating whales at a time, although they used to be seen by the thousand. While the whales were in the Arctic Ocean they were hunted by Eskimos; in the Bay of Vancouver and around the Queen Charlotte Islands they were attacked by Indians from canoes, and farther south the Yankee whalers chased them in sailing boats.

Dog food or tourist bait?

By 1936 the world population was thought to be as low as 100–200. Then the governments of America, Japan and Russia came to an agreement on the future of the grey whale and declared it a protected species. This protection, together with the animal's fairly high rate of reproduction, has resulted in a gradual build-up of the population and it is now thought that they number between 5 and 10 thousand. But the grey whale's future is still dubious: it has recently been reported that the Mexican government is planning to kill grey whales when they migrate south to Mexican water, the idea being to use the carcases as dog food. This is a short-sighted plan: far more could be made out of tourism, for each year thousands of tourists gather on the west coast to watch the grey whales come down from the Bering Sea to give birth to their calves in the shallow and sheltered coastal waters of California and Mexico.

Straining out their food

Like the rarer blue whale the grey whales collect their food by means of rows of baleen plates in their mouths. Various crustaceans and molluscs floating in the sea are eaten in this manner.

Swimming south to breed

The migration of the grey whale is one of the better known aspects of its behaviour. They spend the summer months in the far north, principally in the Bering Sea, where they live in mixed herds. As summer draws to a close they swim slowly southwards and come in close to the coast, particularly so when they approach California where they can be seen swimming only a mile or so offshore. Here the herds segregate; the females stay together and led by an older cow come really close into the bays and lagoons where they get shelter from the weather to give birth to the calves. These are usually born at about the end of January, measuring about 15 ft in length and weighing around 1 500 lb. Normally only a single calf is produced but twin births have been recorded, the calves suckling for about 9–10 months. As spring approaches the migration is reversed. The males, who have been waiting in deeper water, join the females with their newborn calves and the herds make their way back to the northern oceans to feed again in the colder waters where food is more abundant.

Chivalrous males

It has sometimes been noticed that grey whales show a one-sided faithfulness. If a female is injured or gets into difficulties one or more males may go to her aid, either to keep her at the surface where she can breathe, or to defend her from the attacks of killer whales. But if a male gets into similar difficulties, the females have been seen swimming away from the scene of trouble!

After man, killer whales are the greatest danger to grey whales. It is said that when a small school of grey whales are attacked by a large group of killers they may become so terrified by the attacks that they just float at the surface, belly uppermost, paralysed by fear and making themselves extremely vulnerable to further attack. The grey whales' habit of coming close inshore during the breeding season probably keeps them fairly clear of the attacks of killer whales who prefer deeper water. Sometimes grey whales have come so close inshore that they have practically run aground, and on one occasion a grey whale was seen playing about in the surf like a seal. They have also been found stranded at low tide, apparently without ill effect as they just floated off again at the next high water. This is most unusual since, for almost every other species of whale, stranding means death.

class	**Mammalia**
order	**Cetacea**
family	**Eschrichtidae**
genus & species	***Eschrichtius glaucus*** *grey whale*

Ground squirrel

Ground squirrels include two most remarkable animals, one living in hot desert, the other under arctic conditions.

There are 230 or more species of squirrels: flying, tree and ground squirrels. It is not easy in some instances to draw a line between the last two as some tree squirrels spend a lot of time on the ground and some ground squirrels often take to the trees. In some ground squirrels the tail is bushy but never so much as in tree squirrels, and it is usually not so long.

Ground squirrels are 8–31 in. long of which $\frac{1}{3}-\frac{1}{2}$ is tail, the proportions varying with the species. There are three kinds of colouring: almost uniformly yellowish grey, the same but with the back lightly spotted with light buff or yellowish white, and brownish grey with dark stripes, often with lines of yellowish spots. Their ears are small, their legs short and their feet bigger by comparison with tree squirrels such as the grey and the red. A few of the 32 species live in Africa, but most of the others live in North America, from Mexico to Alaska, where some of them are called gophers. There are seven species ranging from eastern Europe across northern and central Asia, which are usually called susliks and spermophiles.

Life among logs and rocks

Ground squirrels are active by day, alert to danger, sitting up on their haunches or standing on their hindlegs to watch for enemies, giving a twittering or whistling alarm call when an intruder is sighted. Some, like the rock ground squirrel, live in groups, others are more solitary. All either dig their own burrows or find shelter low in hollow trees, under logs, among rocks or in similar sheltered places.

In the southern parts of their range ground squirrels are active throughout the year except during bad weather. Farther north they hibernate, the length of the winter sleep being longer to the northward.

Cheek pouches for carrying food

The diet of ground squirrels includes seeds, nuts, roots, leaves, bulbs, fungi, insects, birds and eggs. They will also cat carrion. The amounts eaten of these foods vary from one species to another. The rock ground squirrel eats mainly plant food with some insects, Franklin's ground squirrel eats mainly plant food, with some insects but adds toads, frogs, mice and birds, and the thirteen-striped ground squirrel eats only a small proportion of vegetable matter. Food is hoarded, quantities of seeds, nuts and grain being carried away in cheek pouches to be stored.

Large families—not always welcome

There is usually one litter a year but there may be two in some species, and litters are usually large, up to 12 or more, born after a gestation of 23–28 days. The babies are born naked, blind and toothless. The eyes open 3–4 weeks later, and soon after this the young leave the nest. When food is plentiful high populations build up and together with other rodents, ground squirrels can cause widespread damage to crops. Some species are suspected of being carriers of diseases, including bubonic plague. The golden-mantled ground squirrel of the Rockies is one. It also tends to cause erosion by burrowing in the mountainside. The numbers of ground squirrels are controlled to a large degree by natural enemies: weasels, lynxes and hawks. In North America the coyote and bobcat are two of their main enemies. The golden-mantled ground squirrel readily forages on crops, where these are available. Elsewhere it visits houses and camp sites searching for food, and as a result is something of a favourite.

△ *Warm coat for Arctic ground squirrel.*
▷ △ *Who goes there? A South African ground squirrel* ***Geosciurus inauris*** *stands erect in typical posture on lookout for danger.*
▷ *White-tailed antelope squirrel gets its name from holding its tail high and exposing a white rump, like the pronghorn antelope.*

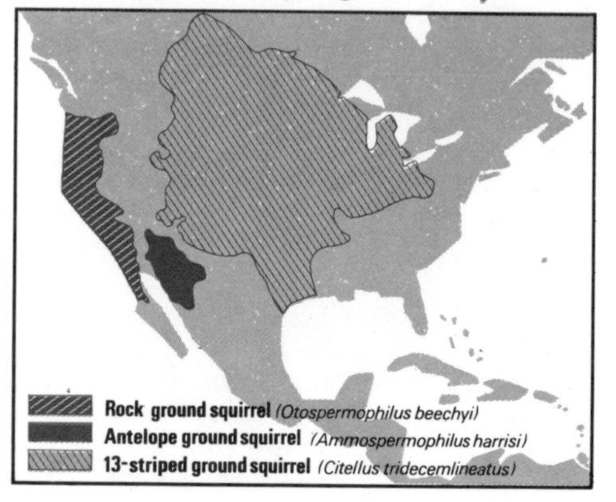

Extremes of temperature

A remarkable ground squirrel is that known as the antelope ground squirrel, living in the deserts of the southwestern United States. It gets its name from the way it carries its tail high exposing a white rump, like the pronghorn antelope. The antelope ground squirrel runs a fever every day yet is never ill from it. When the air temperature is 43°C/110°F or more and the sand beneath its feet is 66°C/150°F the squirrel dashes about from place to place. It shows no signs of discomfort even when its own body temperature rises above 110°F. It can stand exceptionally high temperatures because it loses little water in its natural functions. Its urine, for example, is almost solid. It does not sweat but loses heat by conduction, convection and radiation. To get rid of heat it retires to a shady spot and flattens itself on the ground or else it dives down into its burrow. Within 3 minutes its temperature drops from 42°C/107°F to 33°C/100°F. If things get bad it can lose heat by panting and it can, as an emergency measure, cool itself another way. It drools and then spreads the saliva over its head with its forepaws, as if washing itself. At a pinch the squirrel can endure several hours of the blazing desert heat by running around with its head soaking wet.

At the other extreme, in Alaska, at Point Barrow, the snow lasts for most of the year and the ground is permanently frozen to a depth of hundreds of feet. During the short summer the earth may thaw for a few inches, at most a few feet. There lives the Barrow ground squirrel, on 'islands' of sandy soil, sleeping 9 months of the year. In May this squirrel looks out, after waking, on a still snow-covered world and feeds on stores of leaves, stems, roots and seeds laid up the previous year. As the air warms slightly the ground squirrels mate, the young are born 25 days later, towards the end of June. Their eyes open at 3 weeks, and 2 days after this they begin foraging. In a little over a month they must grow, dig their own burrow and provision it before going to sleep for 9 months. For the adults things are little better because they have the burden of bringing up a family, feeding to recover their strength, lay in stores for the winter and re-furbish the burrow, all in 3 months. This is why the Barrow ground squirrel works for 17 hours each day, with restless urgency, whatever the weather, undeterred by rain and bitter winds.

class	**Mammalia**
order	**Rodentia**
family	**Sciuridae**
genera & species	***Citellus tridecemlineatus*** *13-striped ground squirrel* ***C. parryi*** *Barrow ground squirrel* ***Ammospermophilus harrisi*** *antelope ground squirrel* ***A. leucurus*** *white-tailed antelope ground squirrel* ***Otospermophilus beechyi*** *rock ground squirrel* *others*

Harp seal

The harp seal, sometimes known as the Greenland seal or saddleback, is found in the extreme northern parts of the Atlantic and nearby areas of the Arctic Ocean. Male and female are about the same size, up to 6 ft long and 400 lb weight. They are pale grey with very distinctive markings on face and back. These are generally black in the male and a dark grey in the female. Sometimes the markings in the female are spotted rather than a continuous band of colour. The darker markings consist of a dark area on the front of the face which reaches back as far as the eyes, and the characteristic 'harp' marking on the back and flanks. This is a horseshoe of darker colour which extends from the tail forwards along the flanks, sweeping upwards just behind the foreflippers to meet on the top of the back. It is this that gives the name of 'saddleback'.

There are about three main populations of harp seals. The smallest, about a million strong, live in the seas around Jan Mayen; the next, about 1½ million, is confined to Novaya Zemlya and the White Sea; and the third, of rather less than 2 million, lives around the coasts of Labrador and Newfoundland. Harp seals have been recorded very occasionally from Britain, usually Scotland and especially Shetland, but on one occasion a harp seal was seen in the Bristol Channel.

The Newfoundland population is best known scientifically as it has been intensively studied during the last 15 years, principally by Canadian scientists, and most of the following remarks apply to this population. But the situation is much the same in the other populations; only the dates of pupping and moulting differ and the migrations follow different routes.

Following the pack ice

The life history of the harp seal is very closely tied to movements of the pack ice and we can pick up the story with the adult harp seals moving south from the open sea between Baffin Land and West Greenland, where they have spent the summer months. They pass down the coast of Labrador around November and split into two distinct groups. One remains on the ice in the Labrador current to the north of Belle Isle. The other migrates around Newfoundland and onto the ice in the southern part of the Gulf of St Lawrence.

Adults lead the way

This migration consists almost entirely of adult seals, the younger animals following later. They haul out onto the pack ice and move away from the edge into the hummocked area, where they find greater protection from the elements and from predators. Pups are born between late January and early April, the bulk of them between February 20 and March 5. At this time most

of the adults on the ice are females, the males joining them soon after the pups are born. The actual timing of the births depends to a certain extent on ice conditions. Sometimes there is very little ice solid enough for breeding, so the seals move very close inshore and pup on the ice near the shores.

Whitecoats nursed for two weeks

At birth the pups are covered with a stiff white woolly coat, the 'whitecoat' of the sealers. They are 25–30 in. long and weigh around 10 lb.

Suckling takes place during the next fortnight and the pup builds up a reserve of fat and blubber to last it through the period of weaning. The females may occasionally leave the pups and enter the sea through ice holes which have been kept open in the 'leads' or gaps in the sea ice, just by the passage of seals arriving and the movements of the males. Mating takes place at about the time the pups finish feeding, each male mating with one female only. The pups lose their white coats at 3–4 weeks and are left with an attractive short-haired grey pelt speckled with darker grey and black markings. This, referred to as the 'beater', is particularly valued by the sealers. The pups, when abandoned, gradually make their way to the ice edge where they remain feeding on shrimps and other small bottom-living creatures, graduating to small fish such as immature capelin which are abundant in these waters at this time. As adults they will feed on herring, cod, haddock and flatfish.

Second wave of migration

After the pup has been abandoned and the seals have mated, they disperse briefly to the surrounding seas, and then haul out on the ice again to moult in April. At this time they are joined by the immature non-breeding seals who had followed them down from the summer feeding grounds. When moulting is finished the harp seals move again. This time they migrate to the north with the retreating ice and head again for the summer feeding grounds in the area centred around Baffin Island and Greenland.

This pattern of migration with the ice is also followed by the other two populations. Those in the White Sea area have their pups on the pack ice of the White Sea, then migrate towards Spitzbergen and Franz Josef Land where they spend the summer feeding, returning to the White Sea ice to have their pups in October. The Jan Mayen harp seals feed at sea in the area between Spitzbergen and Greenland and then in the spring move south to Jan Mayen to have their pups on the ice floes.

Tracing their movements

Scientists have followed the migrations of the harp seals in a number of ways. First they fastened tags to the tails or hind flippers of the pups; then by examining the seals as they were found on the ice floes they were able to trace their movements. One of the more unusual recoveries of a seal tag happened in 1955. A fisherman off Newfoundland found in a cod's stomach a seal tag that had been fastened to a harp seal pup some two months earlier. Presumably the pup had died and the tail, complete with tag, had been eaten by a hungry cod! Nowadays much of the tracking of the seals is done from the air.

Extensive use of aircraft, especially helicopters, has enabled scientists to follow seals on the ice and to plot their movements more accurately. Aircraft have also revolutionised the sealing industry: no longer do the sealers rely on experience and a little guesswork to find the seal pups, they make extensive use of light aircraft fitted with skis as well as helicopters to spot the breeding groups. Hunters are flown in, the seals are killed and often the skins are flown out by air as well. This is a highly dangerous operation, and each year many of the planes come to grief while trying to land on none-too-solid ice, or on a floe that is a few yards too short! The skies are full of activity as not only the sealers use aircraft, but so do government officials, checking that the sealing regulations are followed, and scientists who are studying the seals.

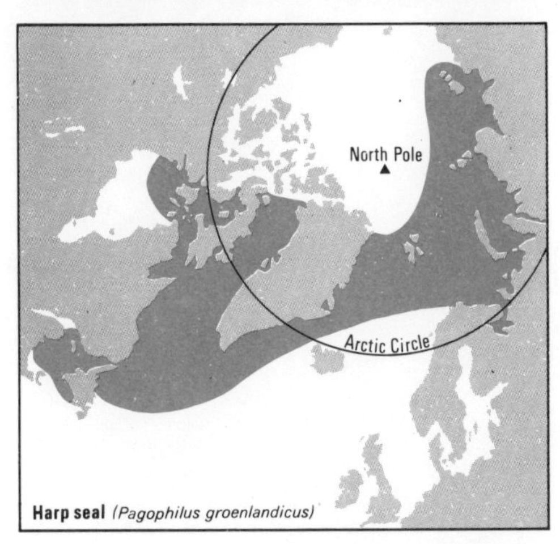

Harp seal *(Pagophilus groenlandicus)*

△△ Harp seal and her pup. The female has distinctive markings on her face and back. The sorrowful looking pup is covered with a stiff white woolly coat, the 'whitecoat' of the sealers. This white coat is shed after 3–4 weeks and the pup is left with an attractive short-haired grey pelt speckled with darker grey and black markings, the 'beater'.
△ There are three main populations of harp seal: in the seas around Jan Mayen, Novaya Zemlya and the White Sea and around the coasts of Labrador and Newfoundland.
▷ The pups are born between late January and early April. During this time most of the adults on the ice are females, the males coming onto the ice after the pups are born. The pups are suckled for a fortnight after birth. They build up a reserve of fat and blubber which will last them through the period of weaning.
◁ Surveying the ice. The harp seal's life history is very closely linked with the movements of the pack ice. Holes are made in the ice by the passage of seals arriving on the ice and by the movements of the males. A female may occasionally leave her pup and enter the sea through one of these holes.

Large-scale slaughter

Sealers take all classes of seal, the most valuable being the 'beaters', followed by the 'whitecoats'. The latter must, however, be taken before they are ten days old as after this time the hair starts to fall out, reducing the value of the skin. The immature seals are also taken. These are known to the sealers as 'bedlamers'. The history of sealing off Newfoundland goes back 200 years and the toll has always been heavy. In the 1920s the average annual kill of harp seals was around 300 000 but this total gradually dropped during the 1930s, and then built up again after the war years. The average catch in the past few years has again been of the order of 300 000 pelts each year, but this has severely reduced the populations. The Newfoundland herds have been reduced by almost half in the last 15 years. Recently the hunting in the Newfoundland area has been controlled by an agreement between Canada and Norway, which decides in advance when the hunting will start and when it should finish. During the spring hunt of 1969 the Canadian government proposed that, with international agreement, the Gulf of St Lawrence should be declared a sanctuary for seals, with hunting limited to international waters in the North Atlantic off Labrador and Newfoundland. Such a sanctuary could be developed as a tourist attraction, compensating for the loss of revenue from furs, and helping to conserve seal stocks for controlled hunting elsewhere. Harp seals are hunted by the Russians in the White Sea where the Norwegian sealers used to take seals in the 1930s but have not been allowed to do so since 1946. In the Jan Mayen area, however, both Russian and Norwegian sealers operate.

class	**Mammalia**
order	**Pinnipedia**
family	**Phocidae**
genus & species	***Pagophilus groenlandicus*** *harp seal*

A trio of hippos in single file moves ponderously through a mixed throng of cormorants, pelicans and gulls, with a hippo youngster in the lead.

Hippopotamus

Distantly related to the pigs, the hippopotamus rivals the great Indian rhinoceros as the second largest living land animal. Up to 14 ft long and 4 ft 10 in. at the shoulder it weighs up to 4 tons. The enormous body is supported on short pillar-like legs, each with four toes ending in hoof-like nails, placed well apart. A hippo trail in swamps shows as two deep ruts made by the feet with a dip in the middle made by the belly. The eyes are raised on top of the large flattish head, the ears are small and the nostrils slit-like and high up on the muzzle. The body is hairless except for sparse bristles on the muzzle, inside the ears and on the tip of the short tail. There is a thick layer of fat under the skin and there are pores in the skin which give out an oily pink fluid, known as pink sweat. This lubricates the skin. The mouth is armed with large canine tusks; these average 2½ ft long but may be over 5 ft long including the long root embedded in the gums.

Once numerous in rivers throughout Africa, the hippopotamus is now extinct north of Khartoum and south of the Zambezi river, except for some that are found in a few protected areas such as the Kruger National Park.

The pygmy hippopotamus, a separate species, lives in Liberia, Sierra Leone and parts of southern Nigeria in forest streams. It is 5 ft long, 2 ft 8 in. at the shoulder and weighs up to 600 lb. Its head is smaller in proportion to the body, and it lives singly or in pairs.

Rulebook of the river-horse

The name means literally river-horse and the hippopotamus spends most of its time in water, but comes on land to feed, mainly at night. It can remain submerged for up to 4½ minutes and spends the day basking lethargically on a sandbar, or lazing in the water with little more than ears, eyes and nostrils showing above water, at most with its back and upper part of the head exposed. Where heavily persecuted, hippopotamuses keep to reed beds. Each group, sometimes spoken of as a school, numbers around 20–100 and its territory is made up of a central crèche occupied by females and juveniles with separate areas, known as refuges, around its perimeter each occupied by an adult male. The crèche is on a sandbar in midstream or on a raised bank of the river or lake. Special paths lead from the males' refuges to the feeding grounds, each male marking his own path with his dung. The females have their own paths but are less exclusive.

The organisation of the territories is preserved by rules of behaviour which, in some of their aspects, resemble rules of committees. Outside the breeding season a female may pay a social call on a male and he may return this, but on the female's terms. He must enter the crèche with no sign of aggression and should one of the females rise on her feet he must lie down. Only when she lies down again may he rise. A male failing to observe these rules will be driven out by the adult females attacking him *en masse*.

Matriarch hippos

It was long thought that a hippopotamus school was led by the oldest male. It is in fact a matriarchy. For example, young males, on leaving the crèche, are forced to take up a refuge beyond the ring of refuges lying on the perimeter of the crèche. From there each must win his way to an inner refuge, which entitles him to mate with one of the females, by fighting. Should a young male be over-persecuted by the senior males he can re-enter the crèche for sanctuary, protected by the combined weight of the females.

The characteristic yawning has nothing to do with sleep. It is an aggressive gesture, a preliminary challenge to fight. Combats are vigorous, the two contestants rearing up out of the water, enormous mouths wide open, seeking to deliver slashing cuts with the long tusks. Frightful gashes are inflicted and a wounded hippo falling back into water screams with pain, but the wounds quickly heal. The aim of the fighting is for one hippo to break a foreleg of his opponent. This is fatal because the animal can no longer walk on land to feed.

Nightly wanderings

Hippos feed mostly at night, coming on land to eat mainly grass. During one night an individual may wander anything up to 20 miles but usually does not venture far from water. Hippos have been known to wander through the outskirts of large towns at times, and two surprised just before dawn by a motorist entering Nairobi showed him they could run at 30 mph.

Babies in nursery school

When in season the female goes out to choose her mate and he must treat her with deference as she enters his refuge. The baby is born 210–255 days later. It is 3 ft long, 1½ ft high and 60 lb weight. Birth may take place in water but normally it is on land, the mother preparing a bed of trampled reeds. The baby can walk, run or swim 5 minutes after birth. Outside the crèche

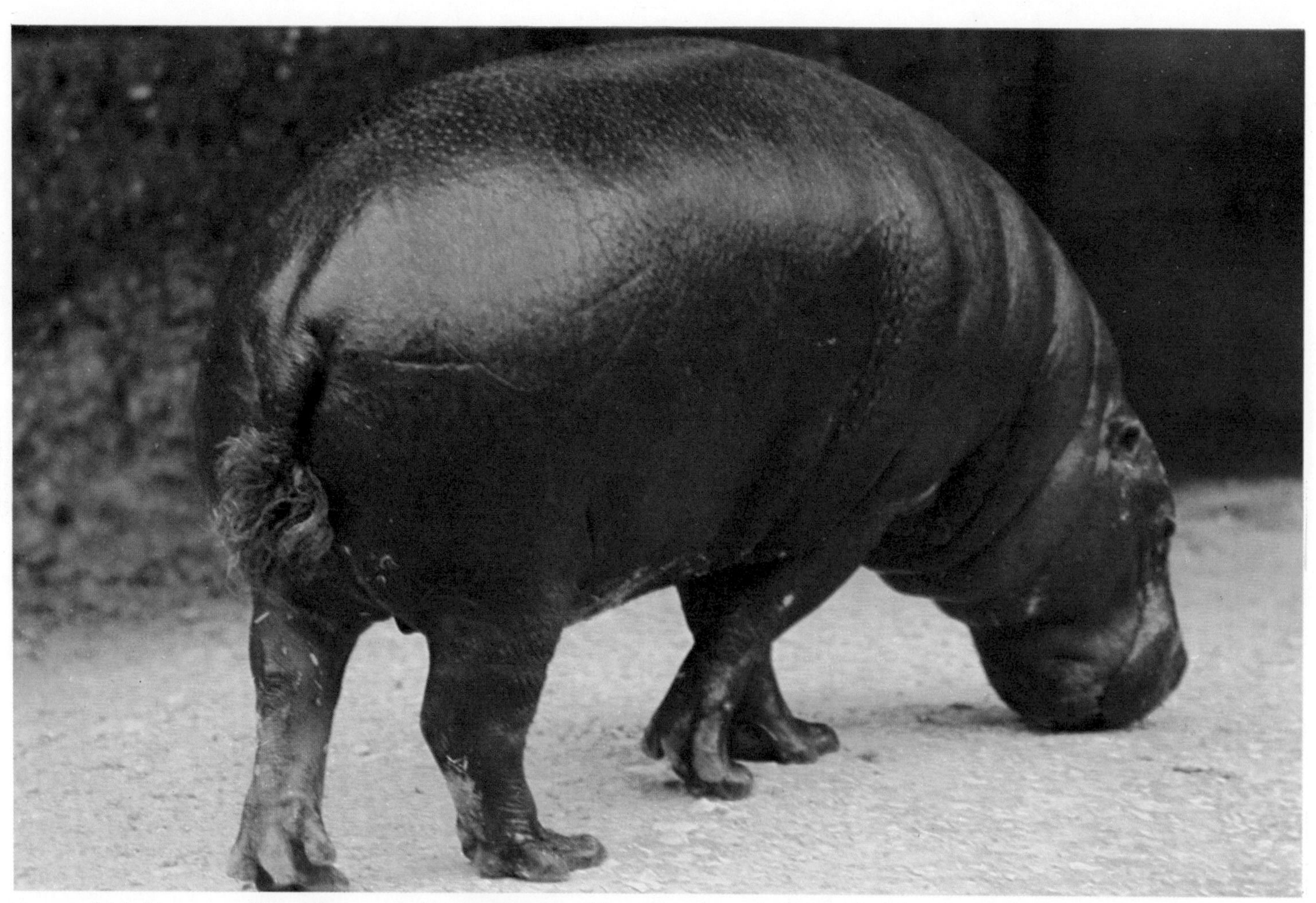

A mournful-looking pygmy hippo **Choeropsis liberiensis**, *from West Africa. Its well-oiled look is due to the secretion of a clear, viscous material through its skin pores. When frightened, the pygmy hippo prefers to head for the undergrowth, whereas the big hippos invariably seek safety in water.*

the organisation of the school is dependent on fighting and the females educate the young accordingly. This is one of the few instances of deliberate teaching in the animal kingdom. In a short while after its birth the baby hippo is taken on land for walks, not along the usual paths used when going to pasture but in a random promenade. The youngster must walk level with the mother's neck presumably so she can keep an eye on it. If the mother quickens her pace, the baby must do the same. If she stops, it must stop. In water the baby must swim level with her shoulder. On land the lighter female is more agile than the male, so she can defend her baby without difficulty. In the water the larger male, with his longer tusks, has the advantage, so the baby must be where the mother can quickly interpose her own body to protect her offspring from an aggressive male. Later, when she takes it to pasture, the baby must walk at heel, and if she has more than one youngster with her, which can happen because her offspring stay with her for several years, they walk behind her in order of precedence, the elder bringing up the rear.

Obedience, or else . . .

The youngsters must show strict obedience, and the penalty for failing to do so is punishment, the mother lashing the erring youngster with her head, often rolling it over and over. She may even slash it with her tusks. The punishment continues until the youngster cowers in submission, when the mother licks and caresses it.

Babysitting was not invented by the human race: hippos brought it to a fine art long ago. If a female leaves the crèche for feeding or mating she places her youngster in charge of another female, who may already have several others under her supervision. The way for this is made easy, for hippo mothers with young of similar age tend to keep together in the crèche.

The young hippos play with others of similar age, the young females together playing a form of hide-and-seek or rolling over in the water with stiff legs. The young males play together but they indulge in mock fights in addition to the other games.

Few enemies for the hippo

Hippos have few enemies apart from man, the most important being the lion which may occasionally spring on the back of a hippo on land, raking its hide with its claws. But even this is rare.

The wanderlust hippo

Many animals sometimes wander well inland for no obvious reason. Huberta was a famous hippopotamus that wandered a thousand miles. She left St Lucia Bay, in Zululand, in 1928 and wandered on and on until in 1931 she reached Cape Province. Each day she stopped to wallow in a river or lake, and her passage was noted in the local newspapers all along her route, so her journey is fully documented. Throughout that time she never came into contact with another hippo. Huberta became almost a pet of the people of South Africa and a law was passed to protect her. She was finally shot, however, by a trigger-happy person in April 1931, and was then found to be a male. So, it will never be known how much farther Hubert might have wandered.

class	**Mammalia**
order	**Artiodactyla**
family	**Hippopotamidae**
genera & species	***Hippopotamus amphibius*** ***Choeropsis liberiensis*** *pygmy hippo*

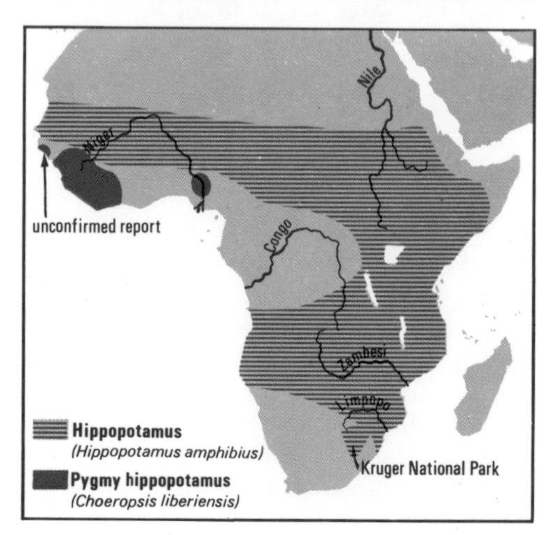

Horse

Wild horses, as distinct from asses and zebras which also belong to the genus **Equus**, *were widespread over Europe and Asia in prehistoric times. Early in the historic period their numbers were much reduced through being hunted for their flesh and for domestication. Only two remnants of what must have been big populations of wild horses persisted. One, the tarpan, survived in the Ukraine until 1851. The other is Przewalski's horse, also called the Mongolian wild horse, of Central Asia, which almost became extinct. The domesticated horse may have evolved from one or both of these, but when and how this happened is unknown. The best we can say is that the horse was domesticated at some time before 2 000 BC.*

Pastureland violence: a pair of horses locked in foot and tooth combat. This is probably a minor dominance battle, with few injuries.

The rough-tough tarpan

All that remains of the tarpan is two skulls, a drawing, some cave paintings and a description by Johann Friedrich Gmelin, a German naturalist who visited the Ukraine in 1769 expressly to find the tarpan. He described it as small, a very shy and very swift animal with a long, mouse-coloured coat, dark feet, pointed ears, fiery eye, erect mane and tail covered with hair. The loss of the tarpan was partly due to hunting and partly through dilution – that is, by hybridizing with domesticated horses. Gmelin told how these wild horses were a nuisance through coming into the fields to eat the hay stooked for carting, two of them eating a stook in one night. He also told how the wild stallions lured domestic mares away after defeating the domestic stallions in bloody fights that often ended in the death of the tame stallions. Small horses painted on the walls of Palaeolithic caves at Lascaux, in France, are also believed to be tarpans.

Mongolian wild horses

Przewalski's horse was discovered by Nicolai Przewalski, Russian explorer, in 1881, in Central Asia. At that time they seemed to have been numerous but numbers decreased in the disturbed times following the Revolution of 1918. It is believed that 40 of them survived this but experts on this horse suggest that even these had been diluted by breeding with domestic horses. Even so, efforts are now being made to protect the surviving stock in the wild and preserve this horse in zoos. Przewalski's horse is stocky, $4\frac{1}{2}$ ft at the shoulder, sandy-orange brown with a black erect mane, no forelock and a longhaired black tail. The summer coat has a dark stripe down the middle of the back. There is an indistinct shoulder stripe and the legs show bars up to the knees.

Semi-wild horses

Horses readily go feral. This has happened in so many places that it helps us to understand how easily the wild horses surviving after domestication should have become diluted by crossing with domestic stock. Well known examples are the half-wild mustangs of the United States and the brumbies of Australia. In Britain the New Forest, Exmoor and Dartmoor ponies are probably in the same class but have been semi-wild for so long that their ancestry is open to argument.

Smiling with anger

Wild horses live in large herds on steppes and grassy plains. A herd consists of mares, foals and colts led by a stallion. As the male colts reach maturity they are driven to the fringes of the herd by the boss stallion. Feral and domestic horses tend towards the same formations but in smaller numbers. Being herd animals they have the usual need for communication between individuals by the use of facial expression and posture. Aggression is shown by laying the ears back and opening the mouth to show the teeth, the corners of the mouth being drawn up. Friendliness or greeting is similar but with the ears erect and the corners of the mouth not drawn up although the mouth is open to show the teeth. Appeasement or submissiveness, as when a young stallion approaches the boss stallion without intent to challenge his authority, is shown by holding the head low with the ears depressed sideways to almost the horizontal and opening the mouth, this time to make nibbling movements with the front teeth. A more marked sign of friendliness is to nibble the skin of the dominant individual near the root of his tail. This is enough to turn hostility to harmony and the dominant will tend thereafter to present his hindquarters for the tail root to be nibbled. A mare in season also shows her teeth to the stallion but with ears erect. At the same time she raises her hindquarters and turns her tail to one side.

Fighting between stallions is with the hoofs of the forefeet, the contestants rear-

ing up on hindlegs to do so. At a later stage the teeth are used for savage biting, especially at the neck. For defence against natural enemies, such as wolves, kicking with the hindhoofs is probably the first line of defence when a single individual is attacked. Since feral horses sometimes form into a line abreast, when alarmed, we may suppose this represents the formation of a phalanx to protect the herd as a whole, and especially as a screen to protect the young.

The foal is born about 11 months after mating. A young male is called a colt until 4 years old, while a female is called a filly until that age. Horses commonly live to 20 years; ages of 30–40 years are not infrequent, the record being 62.

Horses in battle

Domestication of the horse has made a tremendous impact on the history of mankind. From the many antiquities on which the horse is portrayed it seems that horses were first used for chariots, more than for single riders. In time, however, they were used as pack animals, for heavier transport—and especially for cavalry. The tendency has always been for the best horses to be reserved for monarchs and the nobility. They were, and still are, a status symbol. In the Near East, especially, a horse meant everything to its rider, valued as highly as, or even more highly than, his own children. In war, the conquests of Genghis Khan and of the Mohammedans would have been impossible without horses. The latter swept across North Africa and up through Spain. They were stopped at Poitiers in 732 AD when they met the Franks, who were mounted on heavier horses. The mounted horsemen of the Spanish *conquistadores* were decisive in the conquest of South America. In North America the aboriginal Indians quickly learned to use the horse, while the pioneers were dependent on it not only to fight the Indians but to round up cattle.

Domestic breeds

The monarch of the horse world is the Arab, a lightly built and delicate breed noted for its bulging forehead and concave profile, slender muzzle, arched neck, short and straight back and long graceful legs. As with almost everything else connected with the history of horses it is a matter for argument how the Arab and the other breeds came into being. It seems likely there were several subspecies of wild horse across Europe and Asia and that the earliest breeds of domesticated horses reflect the differences between them. It is possible also that domestication arose independently in different places. What seems certain is that there were eventually the Arabs and the northern horses. The latter were mainly of two types, a heavy horse and a lighter Celtic horse, of which the Iceland pony is a typical form. In the last thousand years or so, and increasingly as time went on, Arab stallions were imported into Europe to improve the strains of other breeds.

In time the heavy northern horses came to carry the heavily-armoured knights of the age of chivalry. When armour went out of fashion the heavy cavalry persisted but the breeds became more especially diverted to menial tasks, drawing heavy wagons and drays. Yet the horses retained their magnificence in the shire, Clydesdale and Suffolk, in the Belgian as well as in the Percheron, a somewhat lighter breed. Others of the 60 or so breeds are the Hackney, the high-stepping horse of the days of carriages, before the internal combustion engine; the Welsh pony and, at the opposite end of the scale from the shire and the Clydesdale, the Shetland pony. A famous breed on the other side of the Atlantic is the Morgan horse which sprang from a single progenitor. In

Sunset scene in the Camargue, southern France, a region famous for its herds of semi-wild horses. After a roundup, a group of them is herded along the skyline under the watchful eye of one of the skilled herdsmen.

1795 a small bay stallion was foaled in Vermont which sired a breed named after its owner Justin Morgan. The breed became the favourite carriage horse in the United States and was used also as a cavalry and police mount. It probably had thoroughbred blood but was heavier than the thoroughbred, the breed with Arab blood, famous in horse racing.

Horse evolution

In 1838 William Colchester, a brickmaker of Kingston in Suffolk, was digging out clay for bricks when he unearthed a fossil tooth. The next year, at Studd Hill in Kent, William Richardson was collecting fossils when he found what looked like the skull of a hare. As time passed more fossils of this kind were found of an animal to which the name *Hyracotherium* was given. It has also been called *Eohippus,* the 'dawn-horse' which lived 70 million years ago. Although its teeth are not like those of a horse the skull is, but its forefeet had four toes and the hindfeet had three. During the century following those first two discoveries in England many fossils have been dug out of the ground and put together. When arranged in a chronological series they give us an almost complete picture of the geological history of horses. As we pass along the row we see the animal growing in size. At the same time the skull twists slightly on its long axis to turn it from what seems to be almost a hare's skull to the heavy skull of the modern horse. The cheek teeth at first are small with cusps, not unlike our own molars, and these gradually change to the grinding teeth of the modern horse, with ridges on the upper surface, well suited to grinding grasses, the basic food of present-day horses. And while these changes are going on the legs are getting longer and one by one the toes are dwindling in size and dropping out altogether, except for the middle toe bones which are growing longer and stouter. In the end, in the modern horse, the bones of the lower leg are made up of the bones of the middle toe with the nail enormously enlarged to form a hoof.

There are few animals for which we have so complete a series of ancestral bones. Yet there are those, especially people fond of horses, who find it hard to accept the idea that this noble beast started as an animal the size of a hare. If we stretch a point and say that man has been selectively breeding horses for 3 500 years we find he has managed to produce such contrasts as the Shetland pony at most 42 in. high—Shetlands are always measured in inches—and the shire horse 17 hands high—a hand is 4 in.—capable of pulling a load of 5 tons. The changes from *Hyracotherium* to the wild horse are somewhat greater, it is true, but the time during which they took place is 20 thousand times longer than the $3\frac{1}{2}$ thousand years needed to produce the Shetland and the Clydesdale.

class	**Mammalia**
order	**Perissodactyla**
family	**Equidae**
genus & species	***Equus caballus*** ***E. caballus gmelini*** *tarpan* ***E. caballus przewalskii*** *Przewalski's horse*

▽ *Wet stampede: a herd of Camargue horses shatters the calm of a shallow stream.*

Hare of the plain and prairie: the jack rabbit of the western United States has two obvious adaptations for grassland life—very long ears, useful for detecting predators at a distance, and long hind legs with which it runs up to 45 mph in a series of bounding leaps.

Jack rabbit

The jack rabbits of the western United States are hares belonging to the genus **Lepus**—*they are close relatives of the brown hare, the varying hare and the snowshoe rabbit. The white-tailed jack rabbit, also known as the plains or prairie hare, has a brownish coat in the summer which changes to white in the winter. Only the 6in. black-tipped ears and 4in. white tail remain unchanged all the year round. This jack rabbit, which weighs up to 10 lb, lives in the prairies of the northwest, but to the south lives the smaller black-tailed or jackass hare. The latter name is derived from the 8in. black-tipped ears. The coat is sandy except for the black upper surface of the tail. It does not turn white in winter. This species lives in the arid country from Oregon to Mexico and eastwards to Texas. There is also a small population in Florida which has come from imported jack rabbits, used in training greyhounds, that have gone wild.*

The remaining jack rabbits, the two species of antelope or white-sided jack rabbits, live in restricted areas of Arizona and New Mexico.

Safety in bounding leaps

Like all hares, jack rabbits live on the surface of the ground and do not burrow. The exception is the white-tailed jack rabbit which in winter burrows under the snow for warmth and also gains protection against predators such as owls. Otherwise jack rabbits escape detection by crouching among the sparse vegetation of the prairies and semi-desert countryside. They lie up in shade during the day and come out in the evening. Each jack rabbit has several forms, hollows in the ground shaded and concealed by plants, within its home range. If flushed, jack rabbits will run extremely fast, sometimes reaching 45 mph in a series of 20ft springing bounds like animated rubber balls. Every so often they leap up 4 or 5 ft to clear the surrounding vegetation and look out for enemies.

Water from cacti

Jack rabbits feed mainly on grass and plants such as sagebrush or snakeweed, and often become serious pests where their numbers build up. To protect crops and to save the grazing for domestic stock, hunts are organised or poisoned bait put down. In the arid parts of their range, when the grass has dried up, jack rabbits survive on mesquite and cacti. They can get all the water they need from cacti providing they do not lose too much moisture in keeping cool. To eat a prickly cactus a jack rabbit carefully chews around a spiny area and pulls out the loosened section. Then it puts its head into the hole and eats the moist, fleshy pulp which it finds inside.

Born in the open

The length of the breeding season varies according to the range of the jack rabbit, being shorter in the north. At the onset of breeding jack rabbits indulge in the typical mad antics of hares. The males chase to and fro and fight each other. They rear up, sometimes growling, and batter each other with their forepaws. They also bite each other, tearing out tufts of fur or even flesh and occasionally violent kicks are delivered with the hindlegs. A carefully-aimed kick can wound the recipient severely; otherwise the fight continues until one of the combatants turns tail and flees.

The baby jack rabbits are born in open nests concealed by brush or grass and lined with fur which the female pulls from her body. The litters are usually of three or four young but there may be as few as one or as many as eight. The babies weigh 2–6 oz and can stand and walk a few steps immediately after birth, but they do not leave the nest for about 4 weeks.

Precarious heat balance

Large ears are a characteristic of desert animals, such as bat-eared and fennec foxes, and it is usually supposed that as well as improving the animal's hearing they act as radiators for keeping the body cool. There is, however, a drawback to this idea. If heat can be lost from the ears it can also be absorbed. The problem has now been resolved because it has been realised that a clear sky has a low radiant temperature and acts as a heat sink. In the semi-arid home of the black-tailed jack rabbit a clear, blue sky may have a temperature of 10–15°C/50–59°F to which heat can be radiated from the jack rabbit's ears that have a temperature of 38°C/100°F. Only a slight difference in temperature is needed for radiation to take place and the large difference between ears and sky allows efficient heat transfer.

Jack rabbits rely on radiation to keep them cool, for, as we have seen, they do not get enough water to be able to use evaporation as a means of cooling. In hot weather jack rabbits make use of every bit of shade and in their forms the ground temperature is lower than the air or body temperature and so acts as another heat sink.

The heat balance of a jack rabbit is, however, very precarious. On a hot day it is possible for two men easily to run down a jack rabbit. By continually flushing it and keeping it in the open the jack rabbit soon collapses from heat exhaustion and is soon ready for the pot!

class	**Mammalia**
order	**Lagomorpha**
family	**Leporidae**
genus & species	***Lepus californicus*** *black-tailed jack rabbit* ***L. townsendi*** *white-tailed jack rabbit* *others*

◁▽ *On the look-out. White-tailed jack rabbit crouches by sparse vegetation of the prairies.*
▽ *Black-tailed portrait. Named jackass hare, after its 8in. black-tipped ears, this jack rabbit does not change to white in winter as does the white-tailed jack rabbit.*

Jaguar

This is the largest of the American cats and is known as **el tigre** *in South America. The jaguar is no longer than a leopard but is more heavily built; head and body are 5–6 ft long and the tail is about 3 ft. It may weigh up to 300 lb. The ground colour of the coat is yellow, becoming paler underneath. All over the body is a pattern of black spots up to 1 in. diameter. The jaguar's coat is usually easy to distinguish from the spotted coat of a leopard because the spots on the jaguar skin are arranged in a rosette of 4–5 around a central spot. These rosettes are not so marked on the legs or head where the spots are more tightly packed. Black and albino jaguars are known.*

Jaguars range from the southern United States to Patagonia. In places it is still quite common but elsewhere it has been shot as a cattle-stealer and for its beautiful coat and is now rare or missing altogether.

△ *'Jaguara'—the South American Indian name for the jaguar—is said to mean the 'carnivore that overcomes its prey in a single bound'.*

Little-known cat

Although they are found on the plains of Patagonia and the mountains of the Andes, the main home of jaguars is dense forests, especially near water. They come into open country when food is plentiful there. They are good climbers, rivalling the leopard in their ability to prowl through trees, often hunting their prey along branches. In some areas that are flooded for part of the year, jaguars are confined to trees—except when they take to water for they are extremely good swimmers as well. Jaguars are difficult to find in their forest haunts because their spotted coats blend in so well with their surroundings. As a result less is known about jaguars than about other big cats. Our knowledge is based mainly on the stories of South American Indians and the accounts of explorers and hunters.

It is likely that jaguars have territories or ranges which they defend against other jaguars. This explains reports by travellers in the South American forests of being followed by a jaguar for many miles, then suddenly being abandoned. Apparently the jaguars had been seeing the men off their territories.

Man-killers

Jaguars take a wide variety of food. Their main prey is capybaras and peccaries and they also capture large animals such as tapirs and domestic cattle as well as sloths, anteaters, monkeys and deer. They fight, kill and devour alligators and wait for freshwater turtles to come ashore to lay their eggs. The turtles are tipped onto their backs and the jaguars rip open their shells. Their eggs are also dug up.

Another favourite food is fish, which are caught by the jaguar lying in wait on a rock or a low branch overhanging the water and flipping the fish out with a quick strike of the paw. It is sometimes said that jaguars flip their tails in the water to act as a lure. Their tails may hang into the water and attract fish accidentally, and as cats often flick their tails when excited this could be especially attractive to fish. Having done this by chance and met with success it would not be beyond the mental powers of a jaguar to learn to do it deliberately.

Inevitably there should be stories of man-eating connected with the jaguar, as there are with any big carnivore. According to some accounts it is more dangerous than the lion or leopard, partly because of its habit of defending a territory and partly because it becomes possessive about its prey. When the prey is domestic stock men living and working nearby are likely to be attacked. In these cases, however, the jaguars are man-killing rather than man-eating. Another cause of attacks is the wounding of jaguars by hunters. Jaguars are hunted with dogs that bring them to bay after tiring them, and an enraged jaguar may charge, killing dogs or hunter. In exoneration of the jaguar, the famous explorer and naturalist Humboldt tells of two Indian children playing in a forest clearing. They were joined by a jaguar who bounded around them until it accidentally scratched the face of one child. The other seized a stick and smote the jaguar on the face; the wild beast slunk back

Prowling around—jaguar cubs. The spotted coat has groups of black rosettes similar to the leopard's, but in jaguars each rosette has a central spot.

into the jungle. How much truth lies in the story is impossible to say, but it is not the first account of a large beast playing with children.

Year-round breeding

Jaguars breed at any time of the year and have cubs every other year. Gestation lasts about 100 days and 2–4 cubs are born in each litter. Virtually nothing is known of courtship or care of the young in the wild because of the difficulty of making observations. Jaguars have lived in captivity for 22 years.

Peccary foes

It is thought jaguars 'take possession' of their prey and defend it. Indians in Guyana have told of jaguars attaching themselves to herds of peccaries and following them about, preying on those that get separated from their fellows. It is also said that the peccaries, which can be very savage, will attempt to rescue their companions and send the jaguar running to the nearest tree. This is supported by the account of an English explorer in Brazil who, hearing a great commotion one evening, found a jaguar perched on an anthill surrounded by about 50 furious peccaries. Inadvertently the jaguar lowered its tail which was immediately grabbed by the peccaries, who pulled the jaguar down and tore it apart. This may be a tall story—but at the moment much of our knowledge of jaguars has to rest on such stories.

class	**Mammalia**
order	**Carnivora**
family	**Felidae**
genus & species	***Panthera onca*** *jaguar*

Kangaroo

The best-known of the five kangaroos are the great grey and the red. The great grey or forester is up to 6 ft high, exceptionally 7 ft, with a weight of up to 200 lb. Its head is small with large ears, its forelimbs are very small by comparison with the powerful hindlimbs and the strong tail is 4 ft long. The colour is variable but is mainly grey with whitish underparts and white on the legs and underside of the tail. The muzzle is hairy between the nostrils. The male is known as a boomer, the female as a flyer and the young as a Joey. The great grey lives in open forest browsing the vegetation. The red kangaroo is similar to the great grey in size and build but the male has a reddish coat, the adult female is smoky blue, and the muzzle is less hairy. Unlike the great grey kangaroo it lives on open plains, is more a grazer than a browser, and lives more in herds or mobs, usually of a dozen animals.

*The 55 species of kangaroo, wallaby and wallaroo make up the family Macropodidae (**macropus** = big foot). Only two are called kangaroos but there are 10 rat kangaroos and two tree kangaroos which, with the wallabies, will be dealt with later. A third species is known as the rock kangaroo or wallaroo. There is no brief way of describing the difference between a kangaroo and a wallaby except to say that the first is larger than the second. An arbitrary rule is that a kangaroo has hindfeet more than 10 in. long.*

The red is found all over Australia. The great grey lives mainly in eastern Australia but there are three races of it, formerly regarded as species: the grey kangaroo or western forester of the southwest; that on Kangaroo Island off Yorke Peninsula, South Australia; and the Tasmanian kangaroo or forester. The wallaroo or euro lives among rocks especially in coastal areas. It has shorter and more stockily built hindlegs than the red or the great grey.

Leaps and bounds

When feeding, and so moving slowly, kangaroos balance themselves on their small forelegs and strong tail and swing the large hindlegs forward. They then bring their arms and tail up to complete the second stage of the movement. When travelling fast, only the two hindfeet are used with the tail held almost horizontally as a balancer. They clear obstacles in the same way, with leaps of up to 26 ft long. Usually the leap does not carry them more than 5 ft off the ground but there are reports of these large kangaroos clearing fences up to 9 ft. Their top speed is always a matter for dispute. They seem to be capable of 25 mph over a 300yd stretch but some people claim a higher speed for them.

Eating down the grass

Kangaroos feed mainly by night resting during the heat of the day. The red kangaroo, because it eats grass, has become a serious competitor with sheep, important in Australia's economy. By creating grasslands man has helped the kangaroo increase in numbers. In turn the kangaroo tends to outgraze the sheep, for which the pastures were grown, not only through its increased numbers but by its manner of feeding. Sheep have lower teeth in only the front of the mouth, with a dental pad in the upper jaw. Kangaroos have front teeth in both lower and upper jaw which means they crop grass more closely than sheep. At times, it is reported, they also dig out the grass roots. They can go without water for long periods, which suggests they were originally animals of desert or semi-desert, but where water is supplied for sheep kangaroos will, if not kept out, take the greater share.

Kangaroos set a problem

Enemies of the larger kangaroos are few now that the Tasmanian wolf has been banished. The introduced dingo still claims its victims but that is shot at sight. The loss of natural enemies, the creation of wide areas of grassland and the kangaroo being

A place in the sun: a red kangaroo group whiles away a lazy sociable afternoon. The powerful hindlegs and long tails can be clearly seen.

able to breed throughout most of the year, has created a problem, especially for sheep graziers, in Australia. Fencing in the pastures, often thousands of acres in extent, is costly – about £200 a mile – and kangaroos have a trick of squeezing under the fence at any weak spot. So kangaroos are shot. In one year, on nine sheep properties totalling 1 540 000 acres, 140 000 kangaroos were shot and it would have needed double this number of kills to keep the properties clear of them. Another problem is that kangaroos often bound across roads at night and collide with cars causing costly damage and endangering those in the cars.

Bean-sized baby

The manner in which baby kangaroos are born and reach the pouch had been in dispute for well over a century. In 1959-60 all doubts were set at rest when the birth process of the red kangaroo was filmed at Adelaide University. About 33 days after mating the female red kangaroo begins to clean her pouch, holding it open with the forepaws and licking the inside. She then takes up the 'birth position' sitting on the base of her tail with the hindlegs extended forwards and her tail passed forward between them. She then licks the opening of her birth canal or cloaca. The newborn kangaroo, $\frac{3}{4}$in. long, appears headfirst and grasps its mother's fur with the claws on its forefeet. Its hindlegs are at this time very small. In 3 minutes it has dragged itself to the pouch, entered it and seized one of the four teats in its mouth. The birth is the same for the great grey except that the female stands, with her tail straight out behind her. The baby kangaroo, born at an early stage of development, weighs $\frac{1}{35}$ oz at birth. It remains in the pouch for 8 months, by which time it weighs nearly 10 lb. It continues to be suckled for nearly 6 months after it has left the pouch and can run about, putting its head in to grasp a teat. Meanwhile, another baby has probably been born and is in the pouch. The red kangaroo has lived for 16 years in captivity.

Overlooking the obvious

The truth about kangaroo birth took a long time to be established. In 1629 Francois Pelsaert, a Dutch sea captain, wrecked on the Abrolhos Islands off southwest Australia, was the first to discover the baby in the pouch of a female wallaby. He thought it was born in the pouch. This is what the Aborigines also believed. In 1830 Alexander Collie, a ship's surgeon on a sloop lying in Cockburn Sound, Western Australia, investigated the birth and showed that the baby was born in the usual manner and made its way unaided into the pouch. From then on various suggestions were put forward: that the mother lifted the newborn baby with her forepaws or her lips and placed it in the pouch, or that the baby was budded off from the teat. In 1883 Sir Richard Owen, distinguished anatomist, came down heavily on the side of those who said the mother placed the baby in her pouch holding it in her lips, yet in 1882 the Hon L Hope had shown Collie to be correct. In 1913 Mr A Goerling wrote a letter to the Perth *Western Mail* describing how he had watched the baby make its way to the pouch with no help from the mother. It was not until 1923, however, that this view was generally accepted, when Dr WT Hornaday, Director of the New York Zoological Gardens, watched and described the birth. Finally, in 1959-60, the whole process of birth was filmed by GB Sharman, JC Merchant, Phyllis Pilton and Meredith Clark, at Adelaide University, setting the matter at rest for all time. It seems so obvious to us now!

class	**Mammalia**
order	**Marsupialia**
family	**Macropodidae**
genus & species	***Macropus giganteus*** *great grey kangaroo* ***M. robustus*** *rock kangaroo or wallaroo* ***Megaleia rufa*** *red kangaroo*

Full steam ahead: a shallow water sprint shows the versatility of bounding movement.

Killer whale

The killer whale is closely related to the false killer whale and also the pilot whale. It has a very bad reputation for ferocity which is probably unjustified. Killer whales are small for whales, the females growing up to a maximum of about 15 ft, but an old male may be as long as 30 ft. They are one of the few whales in which there is a marked difference in size between the sexes, the sperm whale being another example. The colour is very striking and distinctive, both sexes having similar markings, which are black on the back and white on the underside. Occasionally the white is somewhat yellowish. The chin is white and there is a characteristic white oval patch just above and behind the eye. There is a small whitish patch just behind the dorsal fin which varies quite considerably in shape and hue in different animals. The white on the underside sweeps up towards the tail and the flanks are white between the dorsal fin and the tail. The flippers, which are broad and rounded, are black all over, but the underside of the tail flukes are white. The dorsal fin is very conspicuous, usually about 2 ft high, but in the old males it may be 6 ft. The oldest males also have very long flippers, up to $\frac{1}{5}$ the animal's total length, the average length of the flipper in juvenile males and adult females being $\frac{1}{9}$ only.

Killer whales are found in all seas but are particularly numerous in the Arctic and Antarctic where there is abundant food to satisfy their voracious appetite. They are not uncommon around the British Isles, where a number have been stranded, mainly on the north and east coasts. These strandings take place in most months of the year. A larger number than usual were stranded on British coasts during the last war, mostly on the North Sea coast, probably due in part at least to anti-submarine activities.

Killer whale showing off its strength and beauty. Despite their reputation for ferocity, killer whales kept in oceanaria have been unaggressive and many are hand-tame.

△ *Running at the surface with blowhole open, a killer whale in relaxed mood.*

▽ *Affectionate play between a pair of killers. Sensory pits can be seen on the head.*

Living in packs

Killer whales hunt together in packs made up of both sexes. They are inquisitive and appear to take a close interest in anything likely to be edible. Nothing is known about their movements in the oceans or how much, if at all, the populations in different oceans mix. In the Antarctic they are often seen around whaling factory ships and probably they tend to follow the ships around as they offer an easy source of food. Otherwise very little is known about their habits.

Ruthless hunters

The killer whale is a voracious feeder and will take anything that swims in the sea. Included in its diet are whales, dolphins, seals, penguins, fish and squid. It will attack even the larger blue whales and quite often killers will hunt in packs numbering from two or three up to as many as 40 or more. When attacking a large whale they are said to work as a team. First one or two will seize the tail flukes to stop the whale thrashing about and slow it down, then others will attack the head and try to bite the lips. Gradually the whale becomes exhausted and its tongue lolls from its mouth—to be immediately seized by the killers. At this point all is over for the whale: the tongue is rapidly removed and the killers take their fill, seeming to favour a meal from around the head of their monster victim.

Apart from attacking fully-grown and healthy whales, killers have earned the hate of whalers because they often take the tongues from whales that have been harpooned and are lying alongside the factory ship waiting to be processed. They will even take the tongue from whales being towed by the catcher boat, and in an effort to stop this looting a man may be posted with a rifle to deter the killers. If he should injure a killer all the others in the pack turn on it and it very soon becomes their next meal.

Killer whales also eat seals and porpoises, and there are a number of records of complete seals found in a killer's stomach. The greatest number recorded is the remains of 13 porpoises and 14 seals that were taken from the stomach of one killer whale, while another contained 32 full-grown seals. Off the Pribilof Islands in the Bering Sea, killer whales are often seen lying in wait for the young fur seal pups swimming out into the open sea for the first time. The number of seals actually taken by killers is not certain but it is likely that large numbers of pups must meet their end in this way before they reach the age of one year.

In the Antarctic, penguins form an important part of the killer whale diet. On many occasions killer whales have been seen swimming underneath ice floes, either singly or sometimes several at a time, and then coming up quickly under the floe either to tip it or break it up, thereby causing the penguins to fall into the water and into the waiting jaws of the killers.

Once killer whales were seen cruising close to an island where there was a colony of grey seals. As the killers came close in the seals hurried ashore in spite of a couple of people standing nearby. The certain danger from killer whales was more important to the seals than possible danger from man. It is said that when killer whales attack grey whales, these become so terrified that they just float on their backs unable to make any effort to escape.

Seven-footer calves

Very little is known about the breeding habits of the whale. They are thought to produce their young towards the end of the year, in November and December after a 16-month gestation. This is supported by examination of some of the stranded whales washed up on the beach and found to be pregnant. The calf at birth is about 7 ft long. The females suckle the young in the same way as other whales, but how long this lasts is not known.

No enemies

The killer whale probably has no real enemies. A few are killed by man, usually irate whalers. They are not a very valuable catch to a whaler although some Russian whaling fleets do catch a few, usually if there is nothing else worth shooting.

◁ *Flukes aloft, a killer sounds with a minimum of splash—a tribute to its streamlining.* △ *A killer pack surges round the edges of encroaching ice.* ▽ *Killer curiosity.*

Chased by killers

The most famous story of killer whales is that told by Herbert Ponting who was the official photographer to the British *Terra Nova* Antarctic expedition led by Captain Scott in 1911. While the ship's cargo was being unloaded onto the ice some killer whales appeared nearby. Ponting went to take some photographs carrying the bulky photographic apparatus of those days over the floes. As he went across the ice the killers thrust up alongside and then followed him as he crossed the floes, tipping them from beneath. Ponting just managed to get to the safety of the fast ice in front of the killers—a lucky escape.

Ponting's experience must have been terrifying, yet it is often found that a reputation for ferocity is unfounded. Divers who have met killer whales have not been molested and several killer whales have been kept in oceanaria. All have been unaggressive or even hand-tame. One story tells of a fisherman of Long Island, New York, who threw a harpoon at a killer whale. The whale pulled free and followed the boat and its terrified occupants to shallow water, but it made no attempt to harm them despite such severe provocation.

class	**Mammalia**
order	**Cetacea**
family	**Delphinidae**
genus & species	***Orcinus orca*** *killer whale*

Koala

The koala is probably Australia's favourite animal. It is known affectionately as the Australian teddy bear although there are a dozen names to choose from. At various times it has been called bangaroo, koolewong, narnagoon, buidelbeer, native bear, karbor, cullawine, colo, koala wombat and New Holland sloth! The last two have an especial interest. For a long time it was believed the koala was most nearly related to the wombat and was placed in a family on its own, the Phascolarctidae, near that of the wombat. Now it is placed in the Phalangeridae with the opossums. In habits the koala recalls the slow loris and the sloth, two very different animals which also move in a lethargic way.

The koala is like a small bear, 2 ft high, up to 33 lb weight, with tufted ears, small eyes with a vertical slit pupil and a prominent beak-like snout. Tailless except for a very short rounded stump, it has a thick ash-grey fur with a tinge of brown on the upper parts, yellowish white on the hindquarters and white on the under parts. It has cheek pouches for storing food and the brood pouch of the female opens backwards. All four feet are grasping. On the front feet the first two of the five toes are opposed to the rest and the first toe on the hindfoot is opposed. Also on the hindfoot the second and third toes are joined in a common skin.

Ace tree-climbers

The koala is essentially tree-living, only occasionally descending to lick earth—apparently to aid digestion—or to shuffle slowly to another tree. If forced to the ground its main concern is to reach another tree and climb it, scrambling up even smooth trunks to the swaying topmost branches where it clings with the powerful grip of all four feet. Although its legs are short they are strong and there are sharp claws on the toes. When climbing a trunk its forelegs reach out at an angle of 45° while the hindlegs are directly under the body. It climbs in a series of jumps of 4–5 in. at a time. During the day it sleeps curled up in a tree-fork. It never enters hollows in trees. Koalas are inoffensive although they have harsh grating voices, said to be like a handsaw cutting through a thin board; it has been

▽ *A koala squatting up a telegraph pole on Phillip Island, off eastern Australia.*

▷ *Year-old koala who will soon leave mother.*

claimed that they have the loudest Australian voice, other than the flying phalanger.

Fussy feeders

At night the koala climbs to the topmost branches to find its only food: the tender shoots of eucalyptus, 12 species of which are eaten. A koala is said to smell strongly of eucalyptus. Bernhard Grzimek, well-known German zoologist and ethologist, has spoken of koalas as smelling like cough lozenges. Their feeding is, however, more restricted than this. Different races of koala eat only certain species of gum tree. Koalas on the east coast of Australia feed only on the spotted gum and the tallow wood, in Victoria only the red gum. Even then they cannot use all the leaves on a chosen gum. At certain times the older leaves, sometimes the young leaves at the tips of the branches, release prussic acid – a deadly poison – when chewed. So, as more and more gum trees have been felled, koalas have become increasingly hemmed in, prisoners of their specialised diet. One of the difficulties of saving the koala by having special reserves is to supply enough trees for them of the right kind. Koalas are said to eat mistletoe and box leaves as well, and a koala in captivity was persuaded to eat bread and milk, but without gum leaves they cannot survive.

Get off my back!

Another drawback to preserving the koala is that it is a slow breeder. Usually the animal is solitary or lives in small groups. At breeding time a boss male forms a small harem which he guards. The gestation period is 25–35 days and there is normally only one young at a birth, $\frac{3}{4}$ in. long and $\frac{1}{5}$ oz weight. It is fully furred at 6 months but continues to stay with the mother for another 6 months after leaving the pouch, riding pick-a-back on her, which has led to many endearing photographs. On weaning it obtains nourishment by eating partially digested food that has passed through the mother's digestive tract. The young koala is sexually mature at 4 years, and the longest lived koala was 20 years old when it died.

Pitiless persecution

Until less than a century ago there were millions of koalas, especially in eastern Australia. Now they are numbered in thousands. In 1887–89 and again in 1900–1903 epidemics swept through them, killing large numbers. This was at a time when it was a favourite 'sport' to shoot these sitting targets, often taking several shots to finish one animal which meanwhile cried piteously, like a human baby, a fact that caused Australian naturalists to condemn the sport as the most callous. At all times koalas are a prey to forest fires as well as to land clearance for human settlement. Moreover a market was developed for their pelts, their fur being thick and able to withstand hard usage. In 1908 nearly 58 000 koala pelts were marketed in Sydney alone. In 1920–21 a total of 205 679 were marketed and in 1924 over two million were exported. By this time public opinion was being aroused and before long efforts were being made to protect the surviving populations and to establish sanctuaries for them and ensure their future.

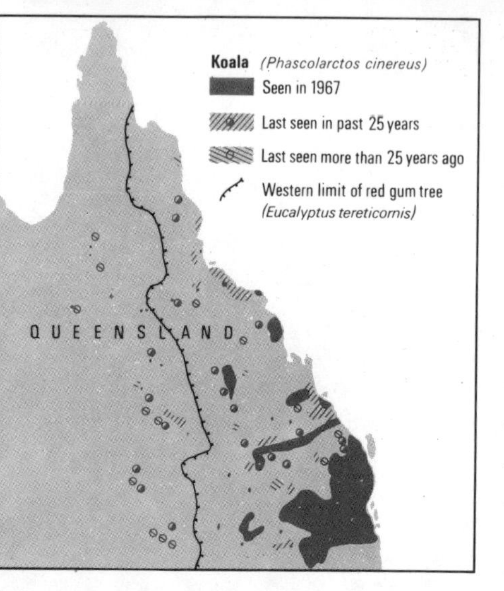

Curious cuddly: favourite of millions, the koala is the Australian teddy bear. It spends most of its time shuffling about its eucalyptus tree-top home. The baby above has climbed onto its mother's back from a downward opening pouch. At a year it will be ready to leave its mother and find its own gum tree. Numbers have seriously decreased in the last 100 years mainly due to fires destroying their gum trees and from persecution by man. From a 1967 survey in Queensland the present-day distribution was established in that state (left).

class	**Mammalia**
order	**Marsupialia**
family	**Phalangeridae**
genus & species	***Phascolarctos cinereus*** *koala*

Lemming

This is a rodent linked in our minds with mass suicides and one of the main aims here must be to put this longstanding story into perspective. The mass migration story is usually told about the Norwegian lemming, which is only one of 12 species of these rodents, living in the northern hemisphere. These include five species of collared lemmings in Arctic Canada, Siberia and European Russia, two bog lemmings of North America, one wood lemming from Norway to Siberia and four species of true lemmings of northern Europe, Asia and America. All are stout bodied 4–6 in. long with 1 in. or less of tail, with thick fur, blunt muzzles, small eyes, and ears small and hidden in the fur. Here we concentrate on the Norwegian lemming, which also ranges across Sweden, Finland, and northwest Russia.

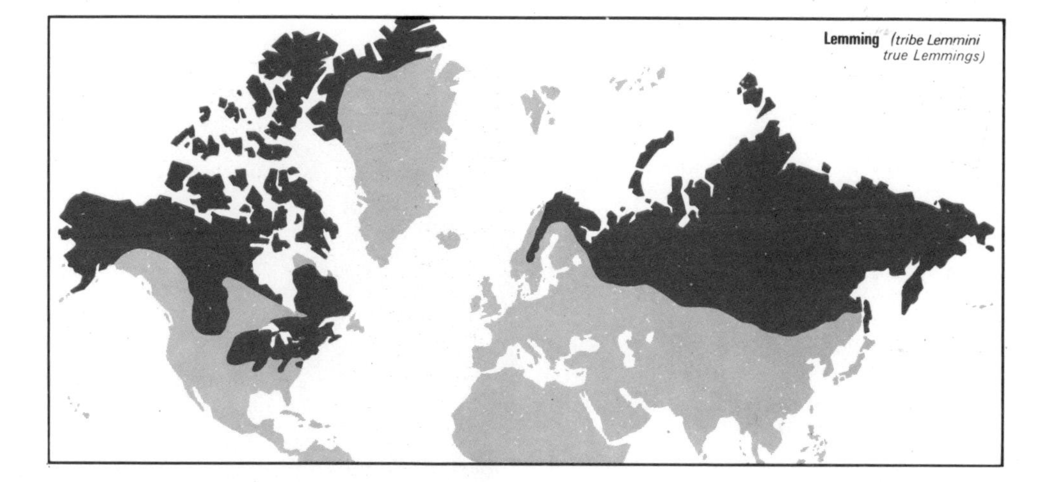

Safe under snow

The Norwegian lemming lives at 2 500–3 300 ft above sea level, above the line of willows. In summer lemmings occupy moist stony ground partly covered by sedges, willow shrubs and dwarf birch. They make paths through the carpet of lichens and rest in natural hollows or cavities in the vegetation. In autumn they move into drier areas, at or about the same level as the summer quarters. In winter they usually live under the snow, protected from cold and from enemies, building rounded nests of grass that are sometimes left hanging on twigs when the snow has melted. They make extensive tunnels under the snow. Because their food is lichens, mosses and grasses wintry conditions do not interrupt their feeding and in an ordinary winter they continue to breed.

Several litters in a year

Several litters are produced by each female in a year. The gestation period is 20–22 days and each litter has 3–9 young, born in a spherical nest of shredded fibres, moss and lichens, made under cover of a rock or in a burrow.

Their enemies are ermine (stoats), weasels, rough-legged buzzards, ravens, longtailed skuas, the various members of the crow family and snowy owls. In winter the lemmings are safe under their covering of snow from all but ermine and weasels, and even they are present in fewer numbers. So there is a marked difference in predation between summer and winter.

Mass migrations

Lemmings, like many small rodents, are subject to fluctuations in numbers from year to year. They also share with voles, to which they are closely related, cyclic rises and falls in numbers. The populations build up over a period of years to abnormal numbers and then comes a crash fall and numbers are reduced to normal. The interest, so far as lemmings are concerned, lies in the causes of these rises and falls and in what actually happens when their numbers are abnormally high. The scientific explanation in the past has been that in years of abnormal numbers the lemmings migrate down the mountainsides into the fertile valleys in search of food. This is near the truth. The popular stories, aided by artists' impressions in the form of pictures, based on local hearsay, is of columns of lemmings in headlong dash down to the sea, where they are drowned in a sort of mass suicide. In the last ten years this has been particularly studied and several informed accounts have been published, including one by Kai Curry-Lindahl, in *Natural History* for August–September 1963.

A matter of climate

Curry-Lindahl gives three main reasons for the population explosions. First, an early spring and a late autumn produce favourable climatic conditions that not only yield an abundance of food but give the lemmings a longer period in which to take advantage of it. Secondly, mild winters with thaws and also severe winters are damaging to winter breeding. In winters that are between these extremes there is a high rate of breeding and of survival among the young. The third is that because of the lack of enemies during winter there is no brake on the increase in numbers from the first two.

In the years 1960-61 there was an explosive eruption of lemmings, in three waves, one in May, the second in June and a third in August. They were noticeable mainly in places where there were obstacles to their spreading out evenly, for example, a long lake or where two rivers meet making a kind of funnel. Then there comes an accumulation of lemmings followed by a kind of panic in which they march recklessly but not in any special direction. They may go up the mountains as well as down into valleys. They may go to any point of the compass. They may go over glaciers or swim across rivers or lakes or, as in Norway where the mountains run down to the sea, into the sea. Lemmings swim well, with the body and head well out of water. Their fur is waterproof, so they take no harm from a wetting, and if the water is calm they can cross a river or lake. If it is choppy, as in the sea, or on lakes in windy weather, many are drowned.

During 1960 there were abundant lemmings but no crash in numbers. This came in 1961, and the explanation is probably supplied by an investigation of the collared lemming in America by WB Quay. Briefly, he found that under warm conditions especially when there was stress or tension the lemmings suffered an upset of their internal balance. One symptom was abnormal deposits in the blood vessels of the brain. The result was severe exhaustion and finally death.

△ Norwegian lemming—the biggest and best known of all lemmings with its characteristic markings. It is thought to be the only existing mammal indigenous to Scandinavia.

△ In peak years the Norwegian lemming is completely unafraid of predators including man. Their main enemies are ermine or stoats, weasels and many birds of prey.

No massed columns

Another study of Norwegian lemmings has shown that although the animals live solitary lives on the mountain tops, and are probably intolerant of each other's company, when they move down into the valleys they become gregarious. For a while, therefore, they crowd together, feed amicably side by side and share burrows. This may be what happened in 1960. As their numbers rise, so the stress mounts, with a high death rate, as in 1961. The important thing to remember is that there are no massed columns of lemmings all of which can be seen flowing in one direction until they reach the cliffs by the sea.

Rains of lemmings

The first published account of the suicidal tendencies of the Norwegian lemmings was written by Zeigler, a geographer of Strasbourg, in 1532. It was based on information given him in Rome by two bishops from Norway. He told how in stormy weather lemmings fell from the sky, that their bite was poisonous and that they died in thousands when the spring grass began to sprout. In 1718, Joran Norberg, writing of the march of Charles XII's army over the mountains in Norway, said 'People maintain that clouds passing over the mountains leave behind them a vermin called mountain mice or lemmings'. Eskimos in Arctic America have similar beliefs about rains of lemmings, and their name for one species in Alaska—anticipating by many years modern ideas about UFOs—is the 'creature from space'.

class	**Mammalia**
order	**Rodentia**
family	**Cricetidae**
genera & species	***Lemmus lemmus*** *Norwegian lemming* ***Dicrostonyx hudsonius*** *collared lemming, others*

▽ *Lemmings usually migrate in search of food over the Norway mountains at a fixed altitude, as shown by arrows. But when there is a population explosion panic follows and the lemmings move haphazardly in all directions. They do not rush to water to commit suicide, in fact they swim quite well, but many die of stress leading to exhaustion.*

△ *Alaskan varying lemming or collared lemming in its white winter coat. It is the only true rodent that becomes white in winter, being brown or grey in the summer. Like all lemmings it is small and fat and has a short tail. Its small ears are completely covered by thick fluffy hair. It can burrow under snow and wander over it even in the dead of the Arctic winter.*

◁ *These footprints in the Norwegian snow beside the ski tracks show that the lemming does not hop but walks.*

◁ *Snow claws of the Alaskan varying lemming. In winter the third and fourth claws of the forefeet become enlarged by the growth of a thick horny shield under the permanent claws. The purpose of this shield is not known but probably has something to do with its snow-shovelling activities.*

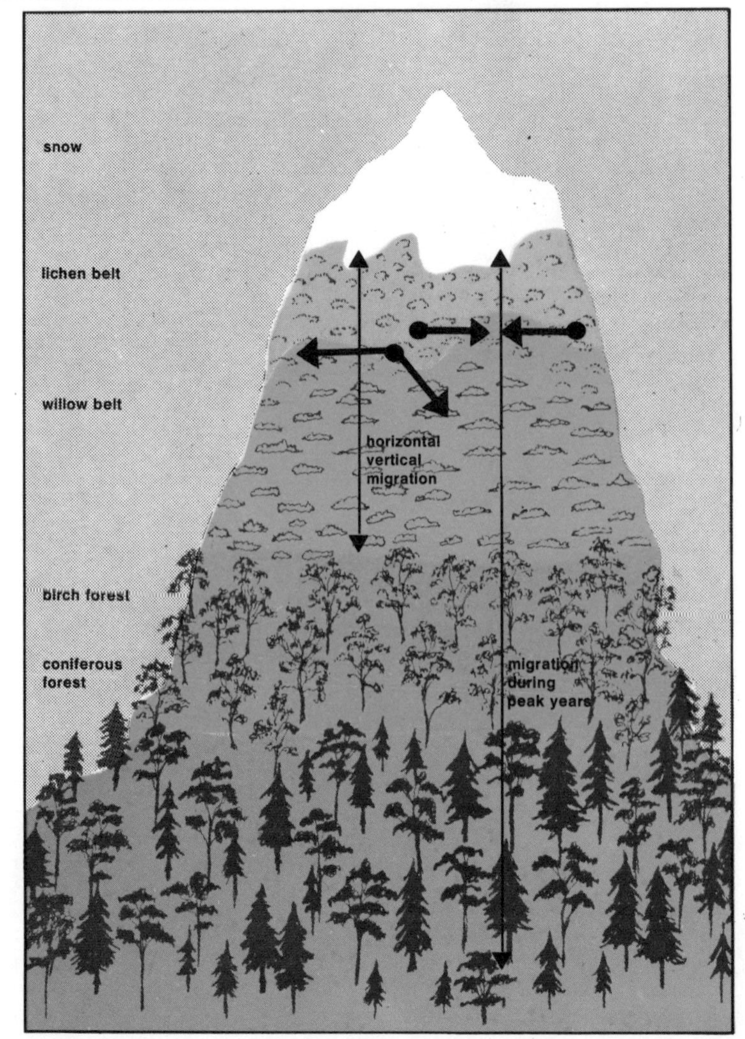

Leopard

The size of the leopard, one of the big cats, varies from one part of its range to another, the range being southern Asia and much of Africa. The males may be up to 8 ft long, including 3 ft of tail, and weigh up to 150 lb. The females are smaller. The colour and length of the fur also vary with the locality and with the climate. Its ground colour is a tawny yellow, whitish on the underparts, with many small black spots which on most of the body are arranged in rosettes. The panther is an alternative name for the leopard used especially for the black individuals that occur sporadically but are particularly common in south-east Asia.

▷ *With prey in the offing, a leopard starts silent routine—each paw fastidiously placed and carefully lifted, with body low and tense for the final mighty bound. If the quarry gets restless the cat will freeze, its dappled body merging perfectly with the surroundings.*

▷▽ *Big cat, big stretch: a leopard limbers up.*

▽ *Going up: a leopard scrabbles its way to the branches. Once among them, it is nimble and well-camouflaged. Older ones often lurk aloft until they can drop on an unfortunate passer-by, and are strong enough to climb with the carcase.*

Powerful cat

Always shy and wary, with keen senses, the leopard's ability to hide makes it harder to track down than a lion or tiger. Leopards will live wherever cover is available: in forest, bush, scrub or on rocky hillsides. They are mainly solitary except at the breeding season. When living in areas where they are hunted they are nocturnal. Elsewhere they are active in the early morning and again in the late afternoon then continuing into the night. During the day, where they are nocturnal, or in the heat of the day, leopards lie up in 'day-beds' in thick undergrowth. Leopards are very powerful for their size and can carry a large kill up into

◁ *Too late: the prey's first inkling of danger was this leopard's charge, leaving time for only a futile bound before death.*

▽ *Treetop larder: hunger brings a leopard back to a carcase stored for future reference.*

a tree. They climb trees well and often take a kill up into a fork to cache it. They are powerful leapers and one which leapt over a railway truck 7 ft high had made a leap 20 ft long. The leopard's voice is a grunting or harsh coughing, or a sawing roar.

Strong food preferences

A leopard will eat almost anything that moves, from dung beetles to antelopes larger than itself. The wide range of prey includes impala, steenbuck, bushbuck, reedbuck, nyala, klipspringer, waterbuck, zebra foals, wildebeest calves, warthog, dassies, as well as cane rats, hares and ground birds. The smaller prey are taken more especially by the young and the old. A favourite trick of leopards in their prime is to lie along a branch waiting for prey to pass beneath, then to drop onto it, seizing it by the throat or, at times, sinking the teeth into its skull.

It has several times been found that individual leopards have a taste for one kind of prey. One fed largely on impala, another ate only bushpig and would travel 2 miles each night from its day-bed to hunt, never molesting the game around its day-bed. The famous leopard of Kariba seemed to eat nothing but the fish *Tilapia,* lying at the water's edge until the fish came to the surface, then catching them with its paw. Another leopard is said to have caught frogs. This tendency to specialise may explain the man-killers. Once they have killed a human they develop a taste for the flesh. It may be the same with leopards that kill livestock because there are known cases of leopards living near farms that never molested the cattle. Some leopards seem to be unusually partial to dogs as food. There are records of leopards eating leopard. It seems that when males fight, over territory or in the breeding season, one may be killed and the other eats it.

Cubs are born blind

There are no firm records for the times of breeding but it seems likely that it may take place at all times of the year. Gestation is 90–105 days and the number of cubs at a birth is usually 3 but may be 1–6, born blind. The cubs stay with the parents until sub-adult.

Leopards in pest control

Because of the attacks of individual leopards on human beings, domestic stock and dogs, they have been mercilessly hunted in parts of their range. It is now realized that leopards can confer a direct benefit where they prey on antelopes, bushpigs and baboons that ravage crops. One leopard is known to have lived almost entirely on cane rats that are a scourge of several crops, such as grass and sugar cane.

Even more intensive hunting for the sake of their skins has further reduced their numbers as compared with former times. At a meeting of the Survival Service Commission of the International Union for the Conservation of Nature at Nairobi in 1963 it was reported that nearly 50 000 leopard skins were being poached in East Africa each year, as a result of women's fashions. Most of these were smuggled out through Somalia and Ethiopia because elsewhere leopards were protected by law. A full-sized leopard coat takes 5–7 skins. The poachers were catching the animals by nooses, springjaw traps, poison and other agonising methods. The eventual price of the skin in London or New York was 100 times more than the poachers received.

Strong and agile

Mr S Downey, honorary game ranger for the former Tanganyika Territory, reported in 1953 seeing a leopard with the carcase of a Thomson's gazelle in the lower branches of a tree. A lioness trotted over, looked at the kill and leapt into the tree. The leopard, without hesitation, took the carcase to the topmost branches of the tree, moving the carcase this way and that, getting it nicely balanced in a fork. The lioness seemed to decide she was at a disadvantage. She jumped down from the tree, trotted over to a shady spot 50 yards away, lay down and went to sleep. The leopard seemed now to believe it safe to remove its kill. It descended with it to the ground and made for a larger tree nearby. Suddenly the lioness was on her feet and the leopard raced back to the first tree and up into the branches, only just ahead of the lioness, but sufficient to take the gazelle out of the lioness's reach to the top of the tree once more. Altogether, it is a first-class illustration of the strength and agility of the leopard.

class	**Mammalia**
order	**Carnivora**
family	**Felidae**
genus & species	***Panthera pardus*** *leopard*

Lion

Lions were once common throughout southern Europe and southern Asia eastwards to northern and central India and over the whole of Africa. The last lion died in Europe between 80–100 AD. By 1884 the only lions left in India were in the Gir forest where only a dozen were left, and they were probably extinct elsewhere in southern Asia, for example, in Iran and Iraq, soon after that date. Since the beginning of this century the Gir lions have been protected and a few years ago they were estimated to number 300. A census taken in 1968, however, puts the figure at about 170. Lions have been wiped out in northern Africa, and in southern Africa, outside the Kruger Park.

The total length of a lion may be up to 9 ft of which 3 ft is tail, the height at the shoulder is 3½ ft and the weight up to 550 lb. The lioness is smaller. The coat is tawny; the mane of the male is tawny to black, dense or thin, and maneless lions occur in some districts. The mane grows on the head, neck and shoulders and may extend to the belly.

▷Shady business: lioness evades the heat.
▷▷An aspiring lion claims a higher position.
▷▷▷Romantic yearnings? Lion in the dusk.
▽Pride of the bush: lionesses with their cubs.

Prides and the hunting urge

Lions live in open country with scrub, spreading trees or reedbeds. The only sociable member of the cat family, they live in groups known as prides of up to 20, exceptionally 30, made up of one or more mature lions and a number of lionesses with juveniles or cubs. Members of a pride will co-operate in hunting, to stalk or ambush prey, and they combine for defence. The roar is usually not used when hunting although lions have been heard to roar and to give a grunting roar to keep in touch when stalking. A lion is capable of speeds of up to 40 mph but only in short bursts. It can make standing jumps of up to 12 ft high and leaps of 40 ft. Lions will not normally climb trees but lionesses may jump onto low branches to sun themselves and they as well as lions will sometimes climb trees to reach a kill cached in a fork by a leopard. There is one record of a lioness chasing a leopard, apparently with the intent to kill it, into a tree, but she was foiled by the leopard going into the slender top branches which failed to bear the lioness's weight.

Not wholly carnivorous

Although strongly carnivorous lions take fallen fruit at times. Normally, in addition to the protein, fat, carbohydrate and mineral salts, lions get their vitamins from the entrails of the herbivores they kill. Typically, lions first eat the entrails and hindquarters working forward to the head. In captivity, lions flourish best and breed successfully when vitamins are added to a raw meat diet. Although lionesses often make the kill the lions eat first (hence, 'the lion's share') the lionesses coming next and the cubs last. In general, antelopes and zebra form the bulk of lions' kills but almost anything animal will be taken, from cane rats to elephant, hippopotamus, giraffe, buffalo and even ostrich.

A survey in the Kruger Park showed that in order of numbers killed the prey-species were: wildebeest, impala, zebra, waterbuck, kudu, giraffe, buffalo. A later survey showed a preference descending through waterbuck, wildebeest, kudu, giraffe, sable, tsessebe, zebra, buffalo, reed-buck, impala. When age or injury prevents a lion catching agile prey it may turn to porcupines and smaller rodents, to sheep

▽ Violence afoot: hampered by the water around them, a pair of paddling lions make the opening moves of a soggy trial of strength.

and goats, or turn man-killer, taking children and women more particularly. Man-eating can become a habit, however; once a small group of lions at Tsavo held up the building of the Uganda railway through their attacks on the labourers. Dogs may be killed but not eaten.

Exaggerated story of strength

A favourite story is of a lion entering a compound, killing a cow and jumping with it over the stockade. R Hewitt Ivy argues in *African Wild Life* for June 1960 that this is impossible. His explanation is that lions visiting a cattle compound do not all go inside. Possibly one leaps the fence, makes a kill and drags it under the fence to those waiting outside. Should the cattle panic and one leap the fence it will be pulled down by the rest of the pride outside and there eaten.

The lion hunts in silence and it is the lioness that most often kills the prey. The usual method of killing is to leap at the prey and b eak the neck with the front paws. Alternatively a lion may seize it by the throat with its teeth or throttle it with the forepaws, on the throat or nostrils. Another method is to leap at the hindquarters and pull the prey down. A lion will kill a hippopotamus by scoring its flesh with the claws in a running battle. Lions will kill and eat a crocodile and will also eat carrion, especially if it is fresh, and lion will eat dead lion. An old story tells of the lion's habit of lashing itself into a fury with a spur on the end of its tail, in order to drive itself to attack. Some lions do have what appears to be a claw at the tip of the tail. But this is only the last one or two vertebrae in the tail out of place, due to injury.

Natural control of populations

Lions begin to breed at 2 years but reach their prime at 5 years. The males are polygamous. There is a good deal of roaring before and during mating, and fights with intruding males may take place. Gestation is 105–112 days, the number of cubs in a litter is 2–5, born blind and with a spotted coat. The eyes open at 6 days, weaning is at 3 months after which the lioness teaches the cubs to hunt, which they can do for themselves at a year old. There is a high death rate among cubs because they feed last, so suffering from a diet deficiency, especially of vitamins. This serves as a natural check

▽ *On firmer ground. The skirmish begins as one lion lumbers up onto his rear legs and lunges at his equally cumbersome opponent.*

on numbers. Should numbers fall unduly in a district—as when lions are hunted by man or culled in national parks—prey is more easily killed and there is more food to spare. Lionesses will then kill for their cubs and then the cubs eat first. This richer diet makes for a high survival rate among the cubs, so restoring the balance in the population number.

Dangers for the King of Beasts

There are no natural enemies as such, apart from man, but lions are prone to casualties, especially the young and inexperienced. A zebra stallion may lash out and kick a lion in the teeth, after which the lion may have to hunt small game. The sable antelope is more than a match for a single lion and other antelopes have sometimes impaled lions on their horns. A herd of buffalo may trample a lion or toss it from one set of horns to another until it is dead, although two lions will overcome one large buffalo. One female giraffe attacked a lioness trying to kill her calf. Using hoofs of fore and hindlegs, as well as beating the lioness with her neck, she severely mauled her—and chased the lioness away over a distance of 100 yards. This is a better performance than a rhinoceros can manage. A lion will kill rhino up to three-quarters grown.

class	**Mammalia**
order	**Carnivora**
family	**Felidae**
genus & species	***Panthera leo*** *lion*

△▷ *Plan of campaign: lion's strategy observed in Kruger Park. Detecting wildbeest, 16 lions deliberated and split into three parties.*
▷ *Drink up! One lioness remains on watch.*
▽▷ *Lioness casually accepts a mother's duty.*
▽ *An affectionate nudge from mother.*

Llama

If the name of this South American humpless camel were given its correct pronunciation 'yama' we could avoid confusion with the spiritual head of the Tibetans. The llama has long been domesticated. Its wild ancestor was probably the guañaco, the other wild camel of South America being the vicuña. The alpaca was probably also derived from the same wild ancestor.

The guañaco ranges from North Peru to the southern plains of Patagonia. It is nearly 6 ft in head and body length, including a fairly long neck, and its tail is nearly 10 in. long. It stands nearly 4 ft at the shoulder, 5 ft to the top of the head, and it weighs up to 210 lb. Its woolly coat is tawny to brown and the head is grey. It lives in herds of a few females with one male, but there are leaderless herds of males up to 200 strong. For defence guañacos rely on speed, up to 40 mph, and they readily take to water. Mating takes place in August and September the single young being born 10–11 months later. It runs swiftly soon after birth and is weaned at 6–12 weeks.

The vicuña is slightly smaller and weighs only 100 lb. It ranges from north Peru to the northern parts of Chile, keeping more to the mountains, usually above 14 000 ft. It is up to 3 ft at the shoulder, with a light brown coat and yellowish-red bib.

The chief interest in what is sometimes called the llamoid group—the llama, alpaca, guañaco and vicuña—lies in the history of their domestication and the relationships between the four animals, which are still unsolved. When in 1531 the Spanish conquistadores overran the Inca Empire in the high Andes they found the llama and alpaca both numerous as domesticated animals. The llama is larger than the wild guañaco, weighing up to 300 lb. Its coat is of a long, dense fine wool. It was and still is used mainly as a beast of burden, with an uncertain temper, a greater tendency to spit than a camel and to bite. The alpaca, smaller than the llama, was selectively bred for its coat which makes a finer quality wool than that of any other animal. It is much longer also and was formerly woven into robes used by Inca royalty.

△ *Old fashioned? Life in the Andes and for the llamas has changed little since the Incas.*

Which ancestor?

Bones agreeing with those of both the llama and the guañaco have been unearthed in human settlements of a date which suggests that their domestication goes back 4 500 years. It may have been even earlier than this, so we can only guess at their wild ancestors and in this there is a sharp difference of opinion among the few zoologists who have made the study. The first to do so was O Antonius, the Austrian palaeontologist who, in 1922, argued that the llama was derived from the guañaco and the alpaca from the vicuña. Thirty years later, several others re-examined the arguments and decided Antonius was wrong, and that a study of the skulls more especially showed that the two domesticated forms were both derived from one wild ancestor, the guañaco. Both schools of thought have their followers today, and the matter cannot be settled by breeding. The llama and the alpaca interbreed to produce fertile offspring, which suggests strongly they belong to one species. In captivity both the guañaco and the vicuña interbreed to give fertile offspring, which suggest they too are one species. There is, however, one marked difference between the two wild forms. The vicuña is less easy to tame because of its shyness than the guañaco, although the Jesuits in South America showed that it could be domesticated. There are other views on this matter. One is that the alpaca may have been derived from hybrids between the wild vicuña and the domesticated llama. The other is that possibly both llama and vicuña were bred from a wild species that long ago became extinct.

Many uses

Although today the motorised vehicle tends to oust the llama as a pack animal it is still of prime importance to the people living high in the Andes. It and the other llamoids are fitted for life at these heights because their haemoglobin can take in more oxygen than that of other mammals and their red corpuscles have a longer lifespan—235 days as against the 100 days of human blood corpuscles. Apart from their use for transport—and a llama usually refuses a load of more than 50 lb—llamas provide meat, wool for clothing, hides for sandals and fat for candles. Its wool when braided is used for ropes and its dung is dried and used for fuel. Once a llama is dead little is wasted from its carcase.

Persecution for tourism

Besides making use of the domesticated llamoids the wild species have long been corraled, sheared and then released. In the modern world, with the development of tourism, there has grown up a big demand for products from these animals. One result is a heavy persecution of the wild species so they are much more rare than formerly and fears are being expressed that they may become extinct. As an example of wasteful manufacture a vicuña bedcover displayed for sale to tourists was made up of the skins from at least 40 animals.

Scientists kept out

It is a little surprising that we should have so little information on the biology of animals that have been domesticated for over 4 000 years, or of their wild relations. It is much the same story for the South American fauna as a whole and it is explainable, at least partly, by the early history of the Spanish Conquest. Francisco Pizarro, who defeated the Incas, is said to have sent back false reports of this new-found land to prevent the Spanish crown from realising the full extent of its wealth. In any case, the Spanish authorities had their own reasons for not making public the full potential of their colonies. As a result permission to enter the interior of South America was strictly withheld until towards the end of the 18th century, when the first naturalists were allowed in—and these 200 lost years have not yet been overtaken.

class	**Mammalia**
order	**Artiodactyla**
family	**Camelidae**
genus & species	***Lama pacos*** *alpaca* ***L. huanacos*** *guañaco* ***L. peruana*** *llama* ***L. vicugna*** *vicuña*

Lynx

Lynxes are bobtailed members of the cat family and one, the bobcat, which stands somewhat apart from the others, has been separately dealt with (page 27).

The original animal to be given this name, now distinguished as the European lynx, is up to 3½ ft long in the head and body with an 8in. tail, weighing up to 40 lb. It has a relatively short body, tufted ears and cheek ruffs, powerful limbs and very broad feet. Its fur varies from a pale sandy-grey to rusty red and white on the underparts. Its summer coat is thin and poor, with black spots, its winter coat being dense and soft and usually lacking the spots. It ranges through the wooded parts of Europe, except south of a line from the Pyrenees and Alps. In Asia it extends eastwards to the Pacific coast of Siberia and southwards to the Himalayas.

Closely related to it is the Spanish lynx, which is smaller, with shorter and more heavily spotted fur. The Canada lynx is larger, has longer hair and is almost without spots.

The latest view of many leading zoologists is that lynx types all constitute one species ***Lynx lynx,*** *but for the moment we are following the accepted pattern.*

Champion walkers

Lynxes live in forests, especially of pine, which they seldom leave. They are solitary beasts, hunting by night, using sight and smell. Their keen sight is proverbial and is summed up in the way we describe anyone with keen eyes or keen powers of observation as lynx-eyed. Ancient writers credited the lynx with being able to see through a stone wall. Lynxes run very little but are tireless walkers, following scent trails relentlessly for miles to pursue their prey. Alternatively, as they climb trees well, lynxes will lie out on branches to drop on to passing prey, or will lie in ambush. They swim well, and their broad feet carry them over soft ground or over snow. The voice is a caterwauling similar to that of a domestic tomcat but louder, and like a cat a lynx uses claws and teeth in a fight. Within its home range a lynx buries its urine and faeces but near the boundary of this range both are deposited on prominent places, such as hillocks. These serve as boundary marks to be recognized by the occupant and by neighbours.

Instant death

Lynxes used to be numerous in Europe where today, in most parts, they are scarce, having been wiped out because of their alleged raids on sheep, goats and other livestock. Some zoologists claim that lynxes are less interested in farmstock than in wild game. The natural food of the European

Feline ferocity: the Canadian lynx hunts at night in pine forests, using sight and smell to track down prey such as the snowshoe rabbit.

Lagomorph meal: death is quick as the lynx bites into the shoulders and then the nape.

lynx includes hares, rabbits, ground birds and small deer. The latter are killed by a bite at the nape of the neck which severs the spinal cord, or the lynx may use a two-way bite, into the shoulders and then into the nape. Death is instantaneous with both methods. Lynxes also kill squirrels, foxes, badgers, fish, beetles, especially the wood-boring species, and many small rodents. They tend to kill small game such as rodents in summer, turning to larger game such as deer in winter. The prey of the Canada lynx is similar. The snowshoe rabbit, one of the North American hares, is its main prey.

Helpless kittens

Mating takes place in March, the young being born after a gestation of 63 days in the European lynx and after 60 in the Canada lynx. The litter is usually 2–4 kittens born well furred but blind. The eyes open at 10 days and the kittens are weaned at 2 months but remain with the mother for another 7 months. Although the kittens are somewhat advanced at birth they are slow in developing. Even at 8 months or more they still have milk teeth and their claws are still feeble. The young lynx must therefore feed on small rodents or on food killed by the mother. Should it become separated from its mother when the first winter snows fall its chances of survival are slim, with the small rodents living under 2 ft or more of snow. The females mature at a year old and the recorded life span is up to 20 years.

One enemy

Man is the only enemy of the lynx but his hand has been heavy. Lynx fur has long been valued for garments and trimmings. A gown of crimson damask furred with lynx is listed in an inventory of the belongings of the Duke of Richmond, taken in 1527. In Canada, lynx fur was prominent in the transactions of the Hudson Bay Company. More than hunting, the destruction of forests in Europe and in Canada, or, as in Europe, the husbanding of forests, has deprived the lynx of its best and most natural habitat. In Sweden, for example, hunting and changes in the forest drove the lynx northwards during the 19th century until what had been the northern limit of its range became the southern limit. This meant less food and, more important, less chance of survival for the kittens with the longer and more rigorous winters in the higher latitudes. Finally, in 1928, the Swedish lynx was given legal protection. Its numbers have since begun to rise and the lynxes have moved farther south.

class	**Mammalia**
order	**Carnivora**
family	**Felidae**
genus & species	***Lynx canadensis*** *Canada lynx* ***L. lynx*** *European lynx* ***L. pardellus*** *Spanish lynx*

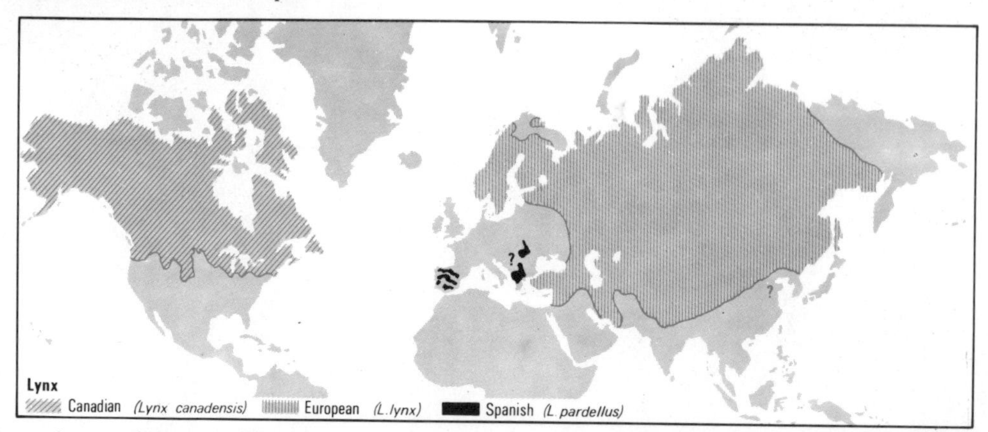

▷*Mandrill magnificence: the exposed parts of the mandrill's skin, face and hindquarters, are the most highly coloured of any mammal.*

Mandrill

Nobody who has been to a zoo where the mandrill is kept can have failed to have noticed it or to have been impressed by its colourful, even repulsive, appearance. The mandrill is a forest-living baboon and, with its close relative the drill, differs so strongly in some respects from other baboons that the two are singled out here for special attention.

The mandrill's body is thick-set, up to 33 in. long, with a stumpy tail only 4 in. long. The muzzle is long and deep and has long thick ridges on either side of the nose. The nostrils are broad and round and the male has long sabre-like canine teeth. The male mandrill is brightly coloured and most people who have seen it remember the colours on its face and the colours it presents as it turns and walks away. The general colour of its coat is dark brown with white cheek whiskers, yellow beard and a crest on the crown of the head. The nose is bright red and the ridges on the muzzle are an equally bright blue. His genitalia have a similar colouring and there is also a blue patch on the rump, on either side of the ischial callosities, or sitting-pads, which are pink. The female is altogether more drab; her face is grey black and her genitalia are not brightly coloured.

The drill is slightly smaller than the mandrill, being up to 28 in. long and its general colour is olive-brown, the face is a deep black and there are no grooves on the muzzle ridges. The drill has, however, a white fringe round the face.

The mandrill is restricted to Cameroun, Rio Muni, Gabon and the Congo (Brazzaville). The drill is found in Gabon and Cameroun, to the Cross River in Nigeria, also on Fernando Póo, but strangely it does not occur in Rio Muni. Both live in tropical rain forest, the drill to quite high altitudes on Mount Cameroun.

Mandrill manners: a yawn means aggression rather than tiredness in primate language.

Secret vegetarians

Mandrills and drills are said to live in small family groups in the wild, with one male and one or more females with their offspring making up each group. They walk on their toes and fingers, the soles and the palms not touching the ground. Although they seem to spend most of their time on the forest floor they climb to the middle layer of the trees to feed and to sleep. Their food in the wild is a matter for speculation. Probably, like other baboons living on the savannah, they eat mainly plant food together with a certain amount of insects and other animal food. It is likely, however, that their diet is more vegetarian than that of other baboons.

Monthly breeding cycle

In England a baby mandrill was born at the Chester Zoo in 1937, and one at the London Zoo in 1953. In the United States, however, births in zoos have been more common. Mandrills usually bear one baby at a time, but twins were born in the Zoological Gardens of Baltimore in 1961. The female is in season every 33 days and this is shown externally by the swelling around her genitalia, which begins after menstruation and reaches a peak at ovulation. There seems to be no special season for breeding. The development of the young is similar to that of other baboons and that of monkeys generally. In zoos mandrills are long-lived; one reached 26½ years and another 46.

Why the colours?

Mandrills and drills are among the largest of the baboons and they are the most fearsome to look at, as well as being the most powerful. Leopards may take the babies from time to time but they certainly would not attack a full-grown male. The fearsome appearance is to some extent due to the colours, and zoologists, especially those who study animal behaviour, have been hard put to it to suggest what purpose the colouring serves. Wolfgang Wickler has suggested that the mandrill being highly coloured at both ends tends to baffle his enemies. He puts it this way, that a flash of colour from the mandrill's rear end would make a potential enemy quake and run away, thinking that what was in fact a mandrill going away from him was a mandrill coming towards him. Few zoologists feel tempted to accept this theory, at least in its present form. Ramona and Desmond Morris, in their book *Men and Apes*, have offered an alternative theory, which is perhaps slightly more plausible. We know that in some monkeys, such as the vervets, the males have brightly coloured genitalia and two males will show these colours to each other as a form of threat display. The Morris's theory is that the mandrill uses its colours in much the same way, but having a coloured face as well there is a double impact of threat towards an opponent. One drawback to this theory is that the drill, so closely related to the mandrill and so like it in almost every other way, has only a black face.

The problem of the use of colours by the mandrill is one that probably can be solved only by close study in the wild, and that seems to be a long way off.

Gesner's ape-wolf

The Swiss naturalist Konrad von Gesner, who lived in the 16th century, has left us valuable records of the contemporary knowledge of natural history, some based on travellers' tales, others on fact. One such tale concerned an animal that he called an ape-wolf or bear-wolf. Gesner wrote: 'This animal was brought to Augsburg with great wonder and was shown in the year 1551. It is found in the great wilderness of the Indian land but is very rare. On its feet it has fingers like a man, and when anyone points at it, it turns its buttocks to him. It is by nature frolicsome especially towards women, to whom it displays its frolicsomeness.' It is now believed that Gesner's ape-wolf was the mandrill.

class	**Mammalia**
order	**Primates**
family	**Cercopithecidae**
genus & species	***Mandrillus sphinx*** *mandrill* ***M. leucophaeus*** *drill*

Maned wolf

The maned wolf is a South American fox. It looks like a large red fox on stilts because of its long legs. Head and body together measure 51 in., and the tail 16 in. It stands $2\frac{1}{2}$ ft at the shoulder and weighs up to 51 lb. Apart from the northern wolf it is the largest member of the dog family. Its coat is shaggy and yellowish red and the legs are black for most of their length. There is some black about the mouth, and sometimes on the back and tail. The tip of the tail, chin and throat are sometimes white. Its name is from a mane on the back of the neck and shoulders which is erected in moments of excitement.

It ranges from the southern parts of Brazil to the Argentine.

Long legs puzzle

The maned wolf is solitary, mainly nocturnal and speedy. Like other foxes, it is shy and wary of man. Not surprisingly, therefore, little is known of its habits. It lives where small areas of forest are interspersed with open country, and it can be assumed that in foraging it ranges widely over the countryside. Otherwise its extraordinarily long legs have no meaning unless they are an adaptation for seeing above the long grass while running through it. This seems unlikely since the maned wolf lives on small prey that could not be seen any the better from a height of 2 ft. It is said that maned wolves are agile when going uphill but clumsy when descending, because their hindlegs are slightly longer than their forelegs.

Banana addict

The prey is said to be pacas and agoutis, both of them large and fast-running rodents, also insects, reptiles and birds. The maned wolf eats fruit, sugar cane and other plant foods. In fact, maned wolves have been successfully kept alive in the Antwerp Zoo on 2 pigeons and $4\frac{1}{2}$ lb of bananas a day. Other foods were tried but lean meat was brought up, and eggs and milk were refused. Three maned wolves in the San Diego Zoo took only bananas at first but later turned from these and were fed cubed meat. In the wild, the wolves dig out snails from soft ground and also small rodents. Surprisingly they dig with their teeth, and are said never to use the claws for digging, which if true must be unique for a carnivore.

Pups like fox cubs

Courtship is similar to that of domestic dogs. The number of pups is usually two but there were three litters of three each born in the San Diego Zoo in 1953-4. The baby coats are dark brown, nearly black, and the tails have white tips. Unfortunately none of these survived so nothing is known of the later growth stages. The record life span in captivity is $10\frac{1}{2}$ years.

Fastest dog

There are no natural enemies, yet there is a real fear that maned wolves may soon become extinct. Having no thick underfur, their pelts are useless. The animals themselves are not common in zoos. The reason why they are becoming rare is that they are killed because of alleged attacks on lambs, calves and foals. One way they are hunted is on horseback, with a lasso. The wolf is vulnerable to this because, having no enemies, it merely runs for a while then stops and looks round. The idea of continuous flight seems to be absent from its makeup, so it fails to make the most use of its speed, which is said to be greater than that of any other member of the dog family, and comparable with that of the cheetah in the cat family.

Unusually sweet-toothed

In many ways the so-called South American foxes are fox-like, yet they resemble dogs in other ways. Their English common names reflect this. Besides the maned wolf and the bushdog there are the crab-eating fox, the small-eared dog and others, of which Azara's dog is one. It looks like a short-legged fox but in its anatomy is much nearer the wolf and the jackal. Its home is in the forests of Paraguay and northern Argentina where it feeds much as the maned wolf does, on insects, lizards, small mammals and ground birds. Like the maned wolf it eats seeds and fruits. It has also developed a liking for sugar cane and will leave the shelter of its forest home to create havoc among the cane crop. More than this, it seems to have an obsession for the sweet sap and will slash and chew the canes as if its life depended on it.

class	**Mammalia**
order	**Carnivora**
family	**Canidae**
genera & species	***Chrysocyon brachyurus*** *maned wolf* ***Dusicyon azarae*** *Azara's dog*

With pointed muzzle, very large erect ears and body covered with long, reddish-brown hair the handsome maned wolf reclines peacefully.

Marmoset

Marmosets, the smallest of all monkeys, are South American. They differ from the more familiar monkeys in having claws, and in lacking wisdom teeth, the third molars. These, with its brain, are thought to be primitive features. For these and other reasons they are put, together with the tamarins, in a family separate from other South American monkeys.

Most marmosets are about 8 in. long, with a tail about 12 in., but the pygmy marmoset is under 6 in. long, with an 8in. tail, and is the smallest monkey in the world. The lower incisors of marmosets are almost as long as the canines and are used in grooming. The common marmoset, from the Brazil coast, has silky fur, marbled with black and grey, a black head and long white tufts of hair around the ears. There are three or four closely related species from the Brazil coast, as far south as the Paraguay border, and the Amazon River basin. The pygmy marmoset, of the Upper Amazon, has no ear-tufts and the hair of the head is swept backwards over the ears. It is brown, marbled with tawny.

Males fight over territory

Marmosets are very active animals, bounding along branches, scurrying like a squirrel, with very jerky movements and helped by their claws. Family groups, made up of a mated pair and their offspring, live in the upper canopy of the trees, feeding on insects, fruit and leaves. Whether the pairs are territorial is not known but Gisele Epple, to whom most of our knowledge of marmoset behaviour—albeit of captive animals—is due, says that two members of the same sex will fight if caged together. This suggests that they are to some extent territorial. When two male marmosets meet, they threaten each other with rapid flattening and erection of the conspicuous white ear-tufts. They walk near each other with backs arched, pulling back the corners of the mouth, and flattening the ear-tufts.

Father carries the burden

Breeding may take place at any time of the year. Unlike many other higher primates, marmosets have a courtship display. The male walks with his body arched, smacking his lips and pushing his tongue in and out. The two lick each other's fur, and groom each other using their long lower incisors as a comb. As in other monkeys the female menstruates at approximately monthly intervals. When she is in season the male is very active, marking objects by pressing the glands on his scrotum against them. Gestation is only 20 weeks. Twins are the rule rather than the exception among marmosets; in two-thirds of the births of common marmosets, and in 90% of those in pygmy marmosets, there are twins. Triplet births are commoner than single births. The male usually carries the young about; they cling to his back. The female may carry them but as a rule she has them only at feeding time. This caused a tragedy the first time marmosets were bred in captivity. The laboratory workers decided to take the male out 'to be on the safe side', so they removed the animal that was not carrying the young, that is, the female. The youngsters starved to death. Young marmosets are completely independent at 5 months.

Lurking enemies

There are several small cats in the Brazilian jungle that might kill marmosets as well as some birds of prey. A tame marmoset showed no fear of snakes when presented with either a live snake or a rubber model. By contrast, a tame marmoset fell to the ground in a state of seeming terror at the sight of a polished tortoise shell. This sheds no light on the marmoset's enemies except in comparison with certain African monkeys, which have been seen to fall out of trees in a dead faint at the sight of a leopard's eyes shining through the foliage.

With immense plumed topknot of pure white fur that looks like ostrich-feather headdresses worn by African chieftains, the cotton-topped marmoset ***Oedipomidas oedipus*** *has the most exaggerated headgear.*

Mid-day 'hibernation'

Some tree-living animals, such as squirrels, have a group of long bristles on each wrist. These are tactile hairs (organs of touch) and marmosets are the only members of the higher primates to have them. This is one more primitive character to add to those already mentioned at the beginning. Perhaps the most primitive feature is the body temperature, which may vary by as much as 4 Centigrade degrees, from one part of the day to another, and is lowest about midday. This suggests that the marmosets have a period of torpor at the time when most other mammals rest because the sun is then at its hottest. A related South American monkey, the douroucouli, has a very even body temperature which does not vary by more than 1 Centigrade degree throughout the day. Moreover, the douroucouli maintains a steady body temperature when the

△ *Pygmy marmoset: smallest of all living monkeys, its adult body length is only 4 in. For many years it was thought to be a juvenile form of some other marmoset species. When it was definitely established as a distinct type the dealers were unscrupulous, offering young common marmosets as the rare pygmy species.*
▷ *The most brightly coloured of all living mammals, the golden lion marmoset* **Leontideus rosalia** *sets a problem as no full explanation can be given why his coat is such an intense shimmering golden yellow. The most exotic of the marmoset species, the first living specimen seen in Europe was apparently owned by Madame Pompadour.*

△◁ *White-eared marmoset* ***Callithrix aurita*** *is one of the plumed marmosets. The small face has a triangular light blaze on the forehead and largish ears partly concealed by plume-like hairs that sprout from the cheeks.*
△ *Silvery marmoset* ***Mico argentata*** *with zoo bred youngster—quite a catching colour with the glistening hair of the adult's black face and the baby's bright pink face and ears. As in other marmosets the male assists at birth, receiving and washing the newly born young. The father transfers the young to the mother at feeding time then accepts them from the mother again after feeding, often wearing them like a scarf round his neck.*
◁ *White-faced marmosets* ***Callithrix jacchus geoffroyi*** *grooming, an important part of social life.*

surrounding air drops to as much as 8°C/46°F at night. A widely fluctuating body temperature is a feature of the lower vertebrates, such as reptiles, and when we find it in a mammal this almost certainly indicates that that animal is primitive.

class	**Mammalia**
order	**Primates**
family	**Callithricidae**
genera & species	***Callithrix jacchus*** *common marmoset* ***Cebuella pygmaea*** *pygmy marmoset, others*

Mink

There are two species of these valuable fur-bearers: the American mink and the European mink. They are close relatives of the ermine (stoat) and weasel and belong to the family Mustelidae which includes the badger. The mink is very similar in appearance to the ermine. Males are 17–26 in. from the snout to the tip of the tail, which is 5–9 in. long. Females are about half the size of the males. The ears are short and set close together, the tail is bushy, and the toes are partly webbed. Wild mink are light to dark brown with a white patch on the lower lip and chin and a few white spots on the belly, but a number of colour varieties, from white to almost black, have been bred by artificial selection on mink farms.

The American mink is found in most parts of North America, from the Arctic Circle southwards to Mexico. It is absent from the southern half of Florida and parts of the southwest United States. The European mink is much rarer than it was formerly and it is difficult to tell exactly what is its present range because the introduced American mink is so similar. It is still found in Russia, parts of eastern Europe, Finland, possibly France, and within the last century it has spread across the Urals into the Siberian plains.

Riverbank dweller

Mink hunt on land and, like weasels, kill more than they need, but they also hunt in the water, like otters. Mink live along the banks of wooded streams, rivers, marshes and lakes. Being nocturnal, they are rarely seen, but their droppings, footprints and remains of prey that are found along the bank are sure evidence of mink.

Male mink wander long distances while females have small home ranges. Each mink has a den among rocks, in holes in trees or in burrows of other animals. In North America they take over woodchuck or musk-rat holes and in Britain they enlarge the burrows of water voles.

▽ *The silent watcher. Unlike their relatives, mink are equally at home in water and on land.*

Surplus food stored

Mink prey on both land and water animals. In water, they feed on crayfish and frogs as well as fish, including salmon and trout. On land they catch many kinds of small mammals including muskrats and water voles, and birds such as moorhens, ducks and other birds that nest on the ground. They also raid poultry runs. Surplus food is stored in the den; one in the United States contained 13 muskrats, 2 mallards and a coot.

Playful kits

Mating, which takes place from February to March, is the only time when minks make much noise; both sexes purr. Each male may mate with several females as he passes through their territories but he usually stays with the last one. As with all members of the mustelid family, there is delayed implantation of the embryos and the kits are not born until 45–50 days later, although it may be any time between 39 and 76 days. There are usually 5 or 6 kits in a litter.

The kits stay in the nest of fur and feathers for 6–8 weeks, although they are weaned at 5 weeks. They are very playful, sliding down banks like otters and indulging in noisy, rough-and-tumble mock fights. Until autumn they accompany their parents on hunting trips, then the family breaks up and each leads a solitary life until the next mating season.

Introduced vermin

Mink is one of the most valuable fur bearers and has been bred commercially since 1866 either on large farms or ranches or in backyard cages as a source of extra income. In 1951 the pelts of two million ranch-reared mink were sold on the United States market. Farms have been set up elsewhere from Iceland to the Falkland Islands, and in Russia American mink have been introduced to replace the vanishing European mink, which, anyway, has an inferior pelt.

In many places mink have escaped from farms and colonised the surrounding country. In Iceland they have attacked nesting ducks and waders, with severe effects. In Scandinavia they have turned their attentions to salmon. The first mink farm in the British Isles was established in 1929 and soon feral mink were being recorded. Sometimes there were reports of pine martens in southern England, but whenever it was possible to follow up reports they were found to be mink. Wild mink were first recorded in one area when a fur dealer was sent a mink pelt in mistake for an otter pelt.

It was originally thought that mink would never become established in Britain but in 1956 they were found breeding in Devon.

Signs of breeding were later found in other parts of England as well as Wales and Scotland and mink have spread rapidly along many river systems. They may become a major nuisance as there are reports of raids on poultry, and of moorhens and water voles being wiped out from ponds and stretches of river.

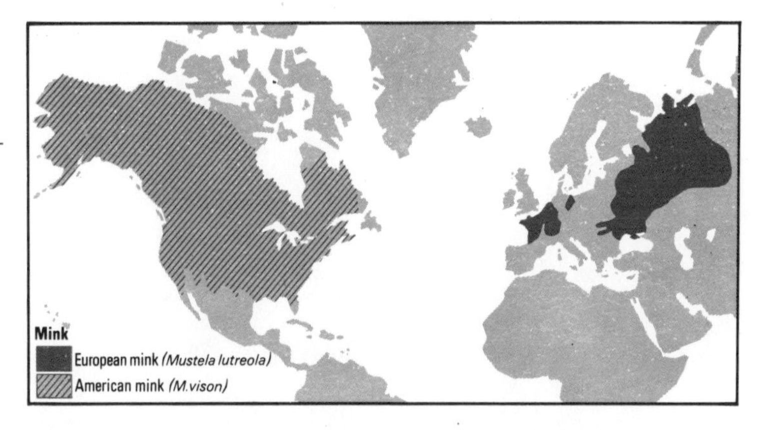

class	**Mammalia**
order	**Carnivora**
family	**Mustelidae**
genus & species	***Mustela lutreola*** *European mink* ***M. vison*** *American mink*

◁*Cutting a figure. Profile of a mink as it slinks through the snow.*
◁▽ *Platinum mink, a variety of American mink from selective breeding, is bred on farms for the colour of its fur. It does not occur in the wild.*
▽ *Inquisitive bystander. Eyes glued and all senses alert, an American mink pulls itself up onto its front legs and cranes its neck for a better view of the world around.*

Moose

The largest living deer, moose are up to $9\frac{1}{2}$ ft long and weigh up to 1 800 lb. They have long legs, standing $7\frac{3}{4}$ ft at the shoulder, with the rump being noticeably lower. The antlers, worn by males only, span up to 78 in. and have flattened surfaces with many tines or snags. The summer coat is greyish or reddish brown to black above, being lighter on the underparts and legs. The winter coat is greyer than the summer coat. The head is long with a broad overhanging upper lip, large ears and there is a tassel of hair-covered skin which hangs from the throat and is known as the 'bell'.

The moose lives in wooded areas of Alaska and Canada and along the region of the Rockies in northwest United States. Known as the elk in the Old World, the species lives in parts of Norway and Sweden and eastwards through European Russia and Siberia to Mongolia and Manchuria in northern China.

In summer moose spend a lot of time in water. A cow (above) wades out to look for food while a magnificent bull (below) stands at the water's edge.

△ *A proud mother watches her twin calves, waiting patiently for them to finish their succulent meal. Moose calves stay with their mother for about two years by which time they are sexually mature.*
▷ *A bull munches away at a tasty meal of young green shoots. The palmate antlers of this moose are still quite small, nevertheless it is a powerful fighting animal and will look for an adversary once its velvet is shed.*

Diets for winter and summer

Moose tend to be solitary but in winter a number of them may combine to form a 'yard'; an area of trampled snow in a sheltered spot surrounded by tall pines and with plenty of brushwood for feeding. They stay there until the food is used up and then move on to make another yard. Moose are most at home in well-watered woods and forest with willow and scrub, and ponds, lakes or marshes. Much of their time in summer is spent wading into lakes and rivers to feed on water-lilies and other water plants. By doing this they also escape to some extent the swarms of mosquitoes and flies. They will submerge completely to get at the roots and stems of water plants. In winter they use their great height to browse the shoots, leaves and branches of saplings.

A moose will also straddle a sapling, bending it over to reach the tender shoots. Bark, particularly of poplar, is also eaten. The Algonquin Indian name *musee* (now moose) means wood-eater.

Growth of antlers

The bulls shed their antlers in December. In April to May new antlers begin to sprout and by August they are full grown and shedding their velvet. At first the exposed antlers are white but after a bull has rubbed them well on bushes and low branches they become polished and brown. A yearling bull has spikes 6–8 in. long; in a two-year-old they are forked, and by the time he has reached 3 years he has narrow hand-shaped antlers with 3–4 points.

Calves remain with mother

The breeding season, September to October, is marked by fighting between the bulls, who spar with each other with their antlers, normally doing little damage. They are polygamous, a mature bull mating with several cows.

In fighting a rival a bull moose tries to knock his opponent sideways with his antlers. If successful he then tries to follow through with a second thrust, and this can result in the prongs of the outer edge of the antler passing between the ribs and inflicting serious damage. As a rule, the weaker of the two gives up the fight after the first blow, or at least before serious damage has occurred. It is when the two are more evenly matched that gashes and broken antlers result. Seldom do the antlers become interlocked.

The bulls bellow for females and, on hearing their answers, smash their way noisily through the thick brushwood to find them. Indian hunters used to call up the bull moose by using a birch bark trumpet to imitate the cow's call. The bull may stay with the chosen cow until her calf is 10 days old. The gestation is 240–270 days, after which 1–3 calves, normally 2, are born. The first time a cow gives birth she has a single calf, but after this twins seem to be the rule, with triplets rarely. The calf is a uniform reddish brown. The calves run with the mother at about 10 days old. For the first 3 days the calf is unable to walk much and the cow keeps close beside it, often squatting low or lying down for the calf to reach her udder, the calf calling to bring the mother to it when it is hungry. Calves remain with the mother for 2 years, by which time they are sexually mature. Moose are long-lived animals, sometimes reaching 20 years of age.

Formidable hoofs

The main enemies are bears and wolves, and to a lesser extent pumas, coyotes and wolverines, which prey on the young. An adult moose is a match for any of these, defending itself not with its antlers but by striking downwards with its large hoofs and then trampling on its opponent.

Learning the life-style

An outstanding feature of the life of any moose is that for part of the year it is solitary, and for part of the year it is gregarious. There is, therefore, a big change in its mentality; at one time it shuns its fellows, at another time it joins them. The secret is in its early training: the mother stays close to her calf when it is first born, as we have seen. As the weeks pass the area over which the mother may wander is gradually extended but all the time the calf keeps close to the mother, following at heel. If it does not, and shows signs of wandering away, she brings it back. Mother and child may be said to cling to each other, and the mother drives everything out of her territory, not only other moose but other deer as well as horses, and even people.

Left to itself, however, a moose calf will follow any deer, horse or man that comes by. So it seems that by nature a moose calf is sociable but by training it becomes antisocial. When the rutting season begins, however, the mother is less inclined to drive moose or anything else out of her territory, so very soon the calf has the company of its mother, a bull and one or more cows that the bull brings in. The calf has company now, and in fact plays a part in the courtship. If the bull is rude to it, its mother will leave him, taking her calf with her. Otherwise all is well and the calf stays with this small group.

The following spring, when the new calf or calves arrive, the older calf stays with the mother but now she makes it keep to the perimeter of her territory. At the next breeding season, in September or October, the calf, its brothers or sisters and its mother, will be joined by a bull and one or more cows, possibly with their calves. If the calf, now in its second year, is a male, it will have to keep its distance or the bull courting its mother will treat it as a rival. It will creep back to its mother any time the bull goes off to drive away a full-grown rival, but it will have to retreat again when he returns. If a calf is a female its own mother will treat it as a rival and drive it away, making it keep its distance whenever the bull, her mate, is near. So each calf learns to be solitary but also learns that it can enjoy the company of others at one time of the year, so it is readily able to live in a group for winter feeding, yet be solitary for the rest of the year apart from the rut.

◁ *Evidence of moose—a barked aspen tree. During the winter when food is scarce the moose out of necessity eats the bark of trees, twigs and any small plants in the snow.*
▷ *The breathtaking view is not even noticed; more important for these two young moose is to satisfy their appetites. Moose are the largest of all deer, they have quite pronounced humped shoulders and heavy manes.*

class	**Mammalia**
order	**Artiodactyla**
family	**Cervidae**
genus & species	***Alces alces***

Muskrat

The muskrat is a large vole with a head and body length of 9–12 in. and a tail of 7–11 in. The name is derived from the secretions of two glands at the base of the tail. The fur colour ranges from silvery brown to almost black. Under the long, coarse guard hairs there is a short, dense layer of very soft hair. The short undercoat is very hardwearing and is sold under the name of musquash, the name given to the animal by Canadian Indians.

Muskrats are adapted to an aquatic life. The scaly, almost hairless tail is flattened from side to side so acting as a rudder. The hindfeet are partly webbed and fringed with short, stiff hairs.

The two species are native to North America. The Newfoundland muskrat is restricted to that island, while the other species ranges widely over the mainland from Alaska and Labrador southwards to Baja California and South Carolina. Muskrats have been introduced to several parts of Europe.

Two kinds of nests

The presence of muskrats in an area is shown by well-defined channels through the vegetation, slides down banks and by networks of tunnels. They are usually found away from water only when they are searching for new feeding grounds. They live in marshes, lakes and rivers, preferably where there is dense vegetation that provides them with both food and cover. Muskrats make two types of nest: in open swamps a pile of water vegetation is made, perhaps 4 ft high and 5 ft across. The walls are cemented with mud and a nest of finely shredded leaves is made in the middle. Several tunnels connect the nest with underwater exits. The other type of nest is built along the edges of ponds and rivers, where muskrats dig elaborate tunnels to a nest above the high water mark. The entrances are either underwater, below the level of the thickest ice, or above water, when they are camouflaged by piles of vegetation.

Muskrats can stay underwater for up to 12 minutes. Normally they dive for shorter periods, only staying submerged when danger threatens. In the winter they keep open breathing holes in the ice by continually breaking the ice as it forms. As the ice around the hole thickens it is plugged with plants, and these provide the muskrat with shelter from the sight of enemies when it surfaces to breathe.

Water plants feeder

The food of muskrats consists mainly of roots and leaves of water plants such as water lilies, wild rice, cattails and arrowheads. In winter they do not feed on stored food but on the roots of water plants which they grub out from the mud. Crops such as maize, alfalfa, clover and peanuts are plundered and also some animal food is eaten, such as crayfish, mussels, and occasionally fish.

Rapid breeding

In the northern parts of their range, muskrats breed from spring to autumn and several families are raised in that time. In the south breeding continues all the year round, with a peak of births from November to April. Gestation lasts only 22–30 days and the young muskrats are only 2–3 in. long when born. There are usually 5–7 in a litter. They are blind, naked and helpless at birth, and are weaned when one month old, although they will have started nibbling plants some time previously. When they are weaned the mother drives them out of the nest because she has mated shortly after they were born and will soon give birth to another litter of rats.

Floods are dangerous

Muskrats are particularly vulnerable during floods, and are more common in lakes and rivers that are not susceptible to flooding. As the muskrats are driven from their nests they seek refuge on rafts of floating vegetation and on trees where bark and twigs form an emergency food. Fighting is common under these conditions and many muskrats die of drowning, disease and cold.

The lack of cover and the weakened condition of the muskrats also makes them an easy target for their many predators, which include red-tailed hawks, great horned owls, bald eagles, foxes, coyotes and raccoons. In the water they are attacked by alligators, snapping turtles, pike and water snakes such as the water moccasin. Perhaps their worst enemy is the mink, but it probably only kills the excess population, that is the young and sick. The American naturalist Paul Errington in his book *On Predation and Life,* has described vividly the limitations on muskrat overpopulation: 'Life among the muskrats proved to be a most hectic succession of fights, evictions, trespasses, abandonments of litters, and other troubles. Young animals died of a skin disease. . . They were preyed upon by minks. They were eaten by cannibalistic members of their own kind.'

◁ *Water home, a muskrat builds a winter retreat.*
▷ *At home in the water, the muskrat is well adapted for swimming. The hindfeet are partially webbed and the tail which is flattened is used as a rudder to guide the animal.*
▽ *Bankside contentment, the muskrat sits happily in the shallow water. This large vole has a coat of long coarse guard hairs and a short dense layer of soft hairs. This underfur is sold as musquash.*

Friend or foe?

Muskrat fur is very valuable. Each year pelts worth tens of millions of dollars are traded in North America. For this reason alone the muskrat is a very useful animal, but they are also welcomed in many places as they keep waterways open by eating water plants. They are sometimes deliberately imported for this purpose. Elsewhere, however, muskrats are a pest because their tunnelling breaches banks and dykes. In Europe muskrats are raised on ranches for their fur but many escape and go wild. They may cause serious damage as there are few large marshes where they can live without coming into conflict with man. They are a severe pest in Holland and were only just prevented from becoming established in Britain. Their rapid rate of breeding and the absence of natural enemies in Europe allowed them to spread rapidly. In 1927 five females and four males escaped from a farm in Scotland. Three years later nearly 900 of their descendants had been trapped. One point in their favour though is that they are very tasty, and are marketed in the United States as 'marsh rabbit' or 'Chesapeake terrapin'.

class	**Mammalia**
order	**Rodentia**
family	**Cricetidae**
genus & species	*Ondatra zibethicus* *O. obscura* *Newfoundland muskrat*

Wide-eyed beauty—the murine opossum, a pretty creature about the size of a large mouse, is a tree-living animal. It is quick-witted and completely fearless, opening its mouth at anything that threatens, and sometimes forgetting to close it for half an hour or more!

Opossum

The opossums are the only marsupials which live outside Australasia, except for the little known caenolestids or 'rat' opossums of South America. The best known opossum is the Virginia opossum, common in many parts of the United States and ranging into South America. The Virginia opossum is often called the 'possum', but this name is also, confusingly, given to Australian marsupials of the family Phalangeridae such as the brush-tailed opossum. Virginia opossums are the size of a small dog, with a head and body length of 12–20 in., but their appearance is rat-like with short legs and a pointed muzzle. The tail is almost as long as the body and is naked for most of its length. The ears are also hairless. The rough fur varies from black to brown or white. The hindfeet are rather like human hands. The first toe is clawless and opposable in exactly the same way as our thumb.

The other opossums which live in Central and South America are similar to the Virginia opossum in appearance. Some have a bushy tail and the water opossum or yapok, has webbed hindfeet and a waterproof pouch. Most of them are not known at all well, except the murine or mouse opossum that sometimes damages banana and mango crops and is occasionally found in consignments of bananas.

Resilient marsupial

The spread, in historic times, of the Virginia opossum from the southeast United States north to Ontario in Canada is remarkable for an animal whose original home is in tropical and subtropical climates. It is even more remarkable when one remembers that it is a marsupial, a group that in most parts of the world has become extinct in face of competition from the placental mammals. The opossum's spread may be due to a decrease in the number of its predators as these have been killed off by man. It is surprising that opossums have survived in the northern parts of their new range as opposums are sometimes found with parts of their ears or tail lost through frostbite. Although they do not hibernate, they become inactive during very cold spells, subsisting on fat stored during the autumn.

Opossums generally live in wooded country where they forage on the ground, climbing trees only to escape enemies such as dogs and sometimes to find food. They can, however, climb well gripping with their opposable toes, while their tail is nearly prehensile. A young opossum can hang from a branch with its tail but adults can use them only as brakes or as a fifth hand for extra support.

Each opossum has a home range of usually 6–7 acres, although it is sometimes twice as large. They feed mainly at night and spend the day in a nest in hollow tree trunks, abandoned burrows, or under piles of dead brushwood. The nest is made of dead leaves that are carried in a most unusual way: they are picked up in the mouth and passed between the front legs to be held between the belly and the tail which is folded under the body.

Varied diet

Opossums have sometimes been described as scavengers, mainly because they are often found feeding on rubbish around human habitations. This merely shows their adaptability for they will eat a very wide variety of foods. Small ground animals such as earthworms, grasshoppers, beetles, ants, snails and toads are taken in large numbers in the summer and autumn, together with voles, mice, snakes and small birds. Poultry runs are sometimes raided. Plants are eaten especially during late autumn and winter when animal food is becoming scarce and an easy way of finding opossums is to search for them when they are feeding among pokeberries and persimmon at night.

Carrying the babies

After a gestation of 12–13 days, the tiny young, numbering 8–18, emerge from the mother's body in quick succession and crawl into her pouch to grasp her nipples. As she usually has 13 nipples, some of the litter may be doomed to perish from the start. The young remain in the pouch for 10 weeks, and after this they sleep huddled together in the nest. When the mother goes out foraging she carries the youngsters, now the size of rats, clinging to her back, and if she has a large litter, she may find it difficult to walk. The young are weaned shortly afterwards and they become independent when about 14 weeks old. Females breed before they are 1 year old, producing one litter in the northern part of their range but two or three in the south. The usual life span in the wild is about two years.

Good to eat?

Opossums are eaten by many animals such as bobcats, coyotes, foxes, hawks and owls. They are trapped for their fur which is not of very good quality but is used to make simulated beaver or nutria. In some places they are eaten and 'possum and sweet taters' is highly esteemed by some.

Playing possum

The phrase 'playing possum' has come from the amazing habit of opossums of feigning

Opossum litter in the pouch. The young have partly-developed limbs and rudimentary eyes and ears. They remain in the pouch for 10 weeks, and after this they sleep huddled together in the nest.

death when frightened. The habit is not confined to opossums and has been recorded in foxes, African ground squirrels and the hog-nosed and grass snakes. When an opossum is confronted by a predator such as a dog, and cannot escape quickly, it turns at bay, hissing and growling, and trying to attack. If the dog succeeds in grabbing and shaking the opossum, it suddenly goes limp, rolling over with eyes shut and tongue lolling out as if dead. Strangely enough, the dog then loses interest and presumably this is the case with natural predators that do not eat carrion, that is dead animals. A few minutes later the opossum gradually recovers and runs off.

We can only presume that this strange trick is effective in persuading predators to leave opossums, otherwise the opossums would only be playing into the predators hands. It is also a complete mystery as to how the trick is worked. It has been suggested that paralysing substances are released into the brain to produce the death-like state and these gradually diffuse away so the opossums recover. More recently experiments have been made with an electro-encephalogram, a machine which records the patterns of minute electric currents in the brain, showing differences between waking and sleeping states for instance. Recordings made of opossums feigning death showed that they are, in fact, in a 'normal, waking, highly alert behavioural state'. They are not in a trance or catalepsy, but wide awake, so they are really living up to their name and 'playing possum'.

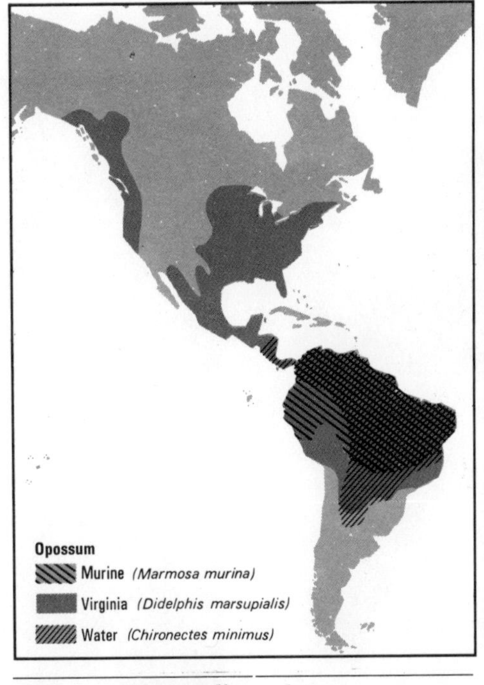

class	**Mammalia**
order	**Marsupialia**
family	**Didelphidae**
genera & species	***Chironectes minimus*** *water opossum* ***Didelphis marsupialis*** *Virginia opossum* ***Marmosa murina*** *murine opossum, others*

△ *Life-line—the Virginia opossum looks stranded but don't be deceived for it can hang from a branch by its naked, scaly, prehensile tail for considerable periods.*

▽ *Good-natured mother allows her young to scramble over her. This opossum,* ***Marmosa*** *sp, does not have a pouch. It can breed three times a year in the cooler parts of its range, and all the year round in tropical areas.*

Growing old. Youngsters (above) stay with their mother for 5 years but they are 10 or 12 before they are mature (opposite page) and they may live to 40.

Orang utan

The orang utan is one of our more interesting relatives. It occupies an intermediate place within the Hominoidea, the superfamily that comprises the apes and man, as it is less closely related to man than the gorilla or chimpanzee, but more man-like than the gibbon.

The big male orang stands 4½ ft high when upright, and may weigh as much as a man. Females stand only 3 ft 10 in. at the most, and weigh half as much as the male. The arms are 1½ times as long as the legs, both hands and feet are long and narrow and suited for grasping, and the thumb and great toe are very short since they would only 'get in the way' of the hook-like function of the hand. The skin is coarse and dark grey, and the hair, which is reddish, is sparse, so the skin can be seen through it in many places. The male develops large cheek-flanges of unknown function, and grows a beard or moustache, the rest of the face being virtually hairless. There is a great deal of variation in facial appearance; orangs are as individual and instantly recognisable as human beings. Both sexes have a laryngeal pouch, which in the male can be quite large, giving it a flabby appearance on the neck and chest. The forehead is high and rounded, and the jaws are prominent. Youngsters have a blue tinge to the face.

Orang utans are found on Borneo and Sumatra. There are slight differences between the two races, and these are more marked in the male. The Borneo race is maroon-tinted, and the male looks really grotesque, with enormous cheek-flanges and great dewlaps formed by the laryngeal sac. The Sumatran race is slimmer and lighter-coloured, and males can look quite startlingly human, with only small flanges and sac, a long narrow face, and a long gingery moustache.

Old man of the woods

The orang utan is strictly a tropical forest animal. It generally lives in low-lying, even swampy forests, but is also found at 6 000 ft on mountains in Borneo. Here, at any rate, most individuals are entirely arboreal. They swing from branch to branch by their arms, though they may use their feet as well, or walk upright along a branch, steadying themselves with their hands round the branch above. It is reported by the Dyaks of Borneo that big old males become too heavy to live in the trees, so they spend most of their time on the ground.

When they are on the ground, orangs move quadrupedally, with the feet bent inwards and clenched, and the hands either clenched or flat on the ground. This contrasts with the gorilla and chimpanzee, which live mainly on the ground and 'knuckle-walk', with their feet flat on the ground and their hands supported on their knuckles. In captivity, orangs easily learn—or discover for themselves—how to walk erect, but because the leg muscles are insufficiently developed to do this easily, the knee is kept locked and the leg straight.

Anti-social 'burping'

At night the orang utan makes a nest, between 30 and 70 ft above the ground. There is often a kind of sheltering roof over this nest, to protect the orang from the rain—a structure which is not found in nests made by chimps or gorillas. The nest is otherwise much more sketchily made than that of chimps or gorillas. It takes only 5 minutes to make and the orang usually moves on and makes a new nest at its next night's stopping place. Sometimes the same one is used again and the previous night's nest may be used for a daytime nap.

Unlike gorillas and chimpanzees, orang utans seem to have no large social groupings. A female with her infant often travels with other such females for a while, forming something like a smaller version of the chimpanzee's 'nursery group'. A male may join this group, but adult males live alone most of the time. Adolescents of both sexes tend to travel around in groups of twos or threes. It is possible that male orangs, like gibbon families, may be territorial, spacing themselves vocally. The laryngeal sac is filled with air, making the animal swell up terrifyingly, and the air is then released to produce what has been described as a 'loud, two-tone booming burp'. They communicate within a group by making a smacking sound with their lips every few seconds. The most terrifying sound which an orang makes is a roar. This begins on a high note and the tone gets deeper and deeper as the laryngeal sac fills with air. Roaring is heard at night and before dawn, and orangs are said

▽ *An aggressive orang utan burps defiantly.*

to make the same noise when wounded. The Dyaks report that male orangs fight and scars are quite common.

There is no special birth season, food being available all the year round in the Indonesian rain forest. Gestation lasts 9 months. The young orang weighs only $3\frac{1}{2}$–4 lb at birth, and is sparsely covered with hairs on the back and head. At first it clings to its mother's fur, usually slung on her hip, but when it is a little older, it wanders about on its own, sometimes walking along the branch behind its mother, clinging to her rump hairs. At about 5 years or so, orangs seem to leave their mothers and form adolescent bands.

Source of contention

Man is the principal enemy of the orang utan. Orangs love the juicy, evil-smelling durian fruit, and so do human beings, so this is often a source of contention. An orang will react to a human intruder by making a great deal of the smacking sound, and breaking off branches, keeping up a continuous shower of them which is often annoying enough to drive the humans away. A Dyak recently reported that he was attacked for no reason by a huge male orang that he came upon unexpectedly on the ground. It has few other enemies. There are no tigers in Borneo—Dyaks claim to have exterminated them about 1 000 years ago—and in Sumatra there are only a few. Leopards are unknown on both islands.

Zoos are a danger

The orang's distribution has been steadily declining. Its ancestors' remains have been found in 14 million-year-old deposits in the Siwalik Hills, Punjab, India. In the Pleistocene, 500 000 years ago, the orang was found as far north as China, and as far south as Java. Today it occurs all over Borneo—the largest and least populated of the East Indian islands—and in the north of Sumatra. It seems that deforestation and heavy human populations have affected its distribution very adversely and there are now fears that it may become extinct altogether in the wild. One reason for its decline is its slow breeding rate. A female breeds every fourth year or so, and usually not until the previous young has left her. It is possible that the average female may bear only three or four young in her life.

The biggest threat, however, to the orang's survival is, sad to say, the zoo trade. Every zoo wants a young ape to display to its visitors, and orangs are the easiest to obtain. Many unscrupulous private zoos, especially in the United States, have paid high prices for baby orangs, and there has been quite a lucrative trade in them in Southeast Asia. Baby orangs are obtained by shooting their mothers. The dealer does not make much effort to ensure the captive's welfare as he probably bought it from the hunter at a low fee, so many youngsters die. For every one orang that reaches a zoo alive, ten orangs have probably died. It is now illegal in Singapore to possess orangs, and smugglers are penalised, but other ports in Southeast Asia are still open for this trade. There is now a list of animals in danger of extinction which, under an international convention, cannot normally be imported into the countries, including the United States, which signed the convention. They can only be imported under special licence, usually for research purposes. This may have some effect on the situation. The deforestation problem, however, remains.

In 1963 Barbara Harrison, working with her husband, Tom, then Government Ethnologist and Curator of Sarawak Museum, estimated that only 2 000 wild

orangs remained in Sabah, 1 000 in Kalimantan (Indonesian Borneo), 700 in Sarawak and 1 000 in Sumatra. Of these, only the Sabah population seems to be anything like adequately protected. In 1964 another estimate put the Sumatra population at only 100. Tom and Barbara Harrison undertook a programme in Sarawak of reintroducing into the wild, young orangs which had been illegally bought by people. This has met with a certain amount of success. There are about 300 in zoos all over the world and breeding has been achieved several times. Most zoos that breed them now keep the Bornean and Sumatran races separate, which will help to save the Sumatran race.

class	**Mammalia**
order	**Primates**
family	**Pongidae**
genus & species	***Pongo pygmaeus pygmaeus*** *Bornean orang utan* ***P. p. abeli*** *Sumatran orang utan*

◁ *A young orang secure in its mother's arms.*

△ *Brotherly love.* ▽ *The old man of the woods. A male orang with his large cheek flanges.*

Otter

The various species of otter are all much alike in appearance and habits. They are long-bodied, short-legged mammals, with a stout tail thickened at the root and tapering towards the tip. There is a pair of scent-glands under the tail. The head is flattened with a broad muzzle and numerous bristling whiskers. The ears are small and almost hidden in the fur. The sleek, dark brown fur consists of a close fawn underfur which is waterproof and an outer layer of long stiff guard hairs, which are grey at their bases and brown at their tips. The throat is whitish and the underparts pale brown. Each foot has five toes, bearing claws in most species, the forefeet are small, the hindfeet large and webbed.

The common or European otter ranges across Europe and parts of Asia, to Japan and the Kurile Islands. It is 4 ft long, including the tail, but may reach $5\frac{1}{2}$ ft, and weighs up to 25 lb. The bitch is smaller than the dog otter. The Canadian otter, of Canada and the United States, is very similar to the European but has an average larger size. It is sometimes spoken of as the river otter, to distinguish it from the sea otter, a markedly different animal. The small-clawed otter, of India and southeast Asia, is much smaller than the European species but the clawless otter of western and southern Africa is larger and is a marsh dweller, feeding on frogs and molluscs. The giant Brazilian otter is the largest of all the otters. It reaches $6\frac{1}{2}$ ft in length, and has a tail that is flattened from side to side.

Solitary and elusive

Except during the mating season otters are solitary, extremely elusive and secretive, and always alert for any sign of disturbance. They will submerge in a flash, leaving few ripples or, when on land, they will disappear among vegetation. Their ability to merge into their background on land is helped by the 'boneless' contortions of the body and the changing shades of colour in the coat which is aided by the movements and changes in the guard hairs. For example, the coat can readily pass from looking sleek and smooth to looking, when damp, spiny and almost porcupine-like.

Otters do not hibernate. They will fish under ice with periodic visits to a breathing hole. It has been said that otters will use a trick known in aquatic insects; that is, to come up under ice and breathe out, allowing the 'bubble' to take in oxygen from the air trapped in the ice and lose carbon dioxide to the ice and water, then inhale the re-vitalised 'bubble'. This has not yet been proved, however.

▽ *Prenuptial affectionate play. Usually solitary, otters are sociable in the mating season.*

Master-swimmers

At the surface an otter swims characteristically showing three humps each separated by 5–8 in. of water. The humps are the head, the humped back and the end of the tail curved above the water line. When drifting with the current only the head may be in view. Occasionally an otter may swim with the forelegs held against the flanks, the hindlegs moving so rapidly as to be a blur. When this is done at the surface there is a small area of foam around the hindquarters, with a wake rising in a series of hump-like waves. It will also use this method when submerged, although more commonly it swims with all four legs drawn into the body which, with the tail, is wriggled sinuously, as in an eel. Leaping from the water and plunging in again, in the manner of a dolphin, is another way in which an otter can gain speed in pursuit of a large fish. Underwater it will often progress in a similar, but smoother undulating manner.

An otter shows its skill better in its ability to manoeuvre. It will roll at the surface, or when submerged, pivotting on its long axis, using flicks of the tail to give momentum. It can turn at speed in half its own length, using tail and hindquarters as a rudder, or it may swim round and round in tight circles, creating a vortex that brings mud up from the bottom. This last tactic is used to drag small fishes up that have taken refuge under an overhanging bank.

When an otter surfaces it stretches its neck and turns its flattened, almost reptilian head from side-to-side reconnoitring before swimming at the surface or coming out on land.

Otters are nomads, fishing a river or lake then moving on to take their next meal elsewhere. They are said at times to cover up to 16 miles overland in a night. Certainly the European and Canadian otters are met at times far from the nearest water. Overland they move by humping the back. A favourite trick is to take a couple of bounds then slide on the belly for 4–5 ft. On a steep slope the glide may take them 40–50 ft. On a muddy or snow-covered slope the slide becomes tobogganning, otters often retracing their steps to slide repeatedly down the slope in a form of play.

Otters live in rivers and lakes, especially small rivers running to the sea or to large lakes. They particularly like those free of weed and undisturbed by human beings. In times of scarcity otters will move to the coast and are then spoken of as 'sea-otters', not to be confused with the real sea otters.

Eels and crayfish favoured

The European otter has a varied diet of fish, small invertebrates, particularly crayfish and freshwater mussels, birds, small mammals, frogs and some vegetable matter. The main fish food seems to be eels and slow-moving fishes but salmon and trout are also eaten.

Otter families play sea-serpents

Mating takes place in water, at any time of the year, with a peak in spring and early summer. After a gestation of about 61 days 2 or 3 cubs, exceptionally 4 or 5, are born, blind and toothless, with a silky coat of dark hair. There is uncertainty about when the eyes open, the only reliable record being 35 days after birth. The cubs stay in the nest for the first 8 weeks and do not leave their mother until just before she mates again.

△ *A backflip from a European otter. This photo caught an otter leaping playfully into the water for the pure joy of living.*

△ *The African clawless otter has only a small connecting web at the base of the toes.*

Young otters swim naturally, as is shown by cubs hand-reared in isolation. The indications are, however, that the mother must coax them, or push them, into the water for their first swim. In the early days of taking to water a cub will sometimes climb onto the mother's back, but normally the cubs swim behind their mother. On rare occasions two or more family parties will swim one behind the other. When this does happen a line of humps is seen, and as the leading otter periodically raises her head to take a look around the procession resembles the traditional picture of the sea-serpent.

Otters as lake monsters

It has been said that any schoolboy knows an otter when he sees one. This is so only as long as the otter runs true to form, but otters are quick-change artists and highly deceptive. Sir Herbert Maxwell has recorded how, at the turn of the century, four gentlemen crossing Loch Arkaig in a steam pinnace saw a 'monster' rise from the depths almost under the bows of their boat, create a tremendous flurry of water at the surface, then dive again out of sight. All were puzzled as to its identity, but when the stalker, a Highlander, present with them in the boat, was questioned later, he was in no doubt that the 'monster' was an otter.

The monster of Loch Morar, near Loch Arkaig, is traditionally 'like an overturned boat towing three overturned dinghies', which could serve as a reasonable description of a bitch otter followed by her three cubs. The ogo-pogo of Canada is believed to be founded on otters swimming in line, and at least one lake monster in Kenya was proved to be a line of otters.

When President Theodore Roosevelt was big game hunting in 1911 he was out in a boat on Lake Naivasha, in Kenya, when the three humps of the local monster appeared. Roosevelt fired once, two humps disappeared, the third stayed on the surface. The skin of the otter was sent to the American Museum of Natural History in New York.

class	**Mammalia**
order	**Carnivora**
family	**Mustelidae**
genera & species	***Amblonyx cinerea*** *Indian small-clawed otter* ***Aonyx capensis*** *clawless otter* ***Lutra canadensis*** *Canadian otter* ***L. lutra*** *European otter* ***Pteronura brasiliensis*** *giant Brazilian otter* *others*

△ *The many-roomed mansion of a pack rat.*
◁ *Pack rat portrait shows typical vole features including muzzle and small eyes.*

▽ *Caught in the open: a dusky-footed wood rat. These shy creatures are rarely seen by day and when they do emerge they keep to shady places.*

Pack rat

Although called rats the pack rats, with their blunt muzzles, smallish eyes, fairly large ears and hairy tails, belong to the vole family. A packman was a pedlar who carried goods in a pack for sale. Pack rats have a similar reputation of 'bartering': when they take something, they always leave something in exchange. Their other names are trade rat and wood rat.

Pack rats are about $1\frac{1}{2}$ ft long of which about half is tail and they weigh $\frac{3}{4}$ lb. They are dark brown to buff grey with white or buff underparts. In some species the hairs are sparse on the tail, but in others the tail is almost bushy.

The 22 species are found from British Columbia southwards to Nicaragua and throughout most of the United States. A typical species is the eastern or Florida pack rat. The bushy-tailed wood rat, 2 ft long, of which 8 in. is tail, ranges from British Columbia to California and eastward to the Dakotas. West of the Rockies is the dusky-footed wood rat which is the one that has been most studied.

Thief in the night

These nocturnal rats seem to have an insatiable habit of picking things up and hoarding them. If on the way to its nest a rat sees something more attractive it drops the object it is carrying and takes the new one. They collect and hoard all sorts of objects, especially bright or coloured objects, or those made of metal. Whatever they take, they tend to replace it with something else. Two of the more amusing examples concern miners' camps. In one was a box of metal nuts. When the time came to use the nuts the box was found filled with stones. In the other the rats stole worthless trinkets from a prospector's cabin and left gold nuggets in their place—a good deal for the miners!

House builders

Some of the pack rats build houses of sticks with several 'rooms': some for storing food, others for sleeping, and one or more for their rubbish. The pack rats living in desert areas in the southwest United States, build their nests of pieces of cactus. They collect fragments of all kinds, including dead or living cactus and pile them against a living cactus, to a height of 2 ft and several yards across. Inside the heap is the nest, of dried grass and other soft materials, and leading to it is a maze of passages protected by the hard sharp cactus spines, with just enough room for a rodent to squeeze through without impaling itself.

Cactus reservoirs

Pack rats live in woodlands, swamps, on rocky ground or in deserts. When near lakes or streams, although they do not enter water, they may build their nests over the water on or in fallen or leaning trees, and in mangrove swamps they build in the trees. They are mainly vegetarian, eating nuts, berries, leaves and roots, and a few small invertebrates. They drink little, getting their water from juicy plants. In deserts they get it almost entirely from the fleshy stems of cacti. Sometimes they live near farmhouses and then they can be a pest, not by their 'bartering' but by taking crops.

Handing over house to the children

Some pack rats are said to be monogamous, the male remaining with the female even when she has young. The male courts the female by drumming with his feet, this behaviour also being used as an alarm signal. In the northern parts of their range there may be one litter a year but in the southern parts there are two litters, and breeding

△ *House bound. A dusky-footed wood rat asleep for the day. Its large house, with several rooms in the wild, is usually so placed as to be quite inaccessible to other animals.*

may occur throughout the year. Gestation is 33–39 days and there are 2–6 in a litter. The naked babies weigh ½ oz at birth. They begin to grow fur at 4 days and are fully furred in 2 weeks. They are weaned at 3 weeks, when the mother leaves her family to occupy the nest while she goes off to find or build a new one. Pack rats have the habit in common with a number of other rodents of licking saliva. The young lick the mother's saliva and possibly adults lick saliva as well. For the young it may be a way of transferring antibodies from mother to infant. In the adults it may have social significance, in promoting friendship and harmony. When two rats meet they stop a short distance from each other, sniff at each other's noses, then come close to lick each other's lips, mouth, face and head.

Another point of view

The species most extensively studied, the dusky-footed wood rat, does not seem to be monogamous. He mates with the female in the nearest nest but he is also a wanderer, going from one nest to another. The nest of this species is made of sticks. It has walls and a roof, usually two storeys, with passages and chambers. To an extent the nests are owned communally, in that the rats tend to swap houses much as we do. When a male mates he continues to live with the female for a while. Once the female is pregnant, however, she drives the male out. Should he be reluctant to go she lunges at him furiously with her forefeet, striking forward and down, biting his ears, face, legs or breast—and he does not retaliate. The attack may prove fatal through suppuration of the wounds. Should a female attack a male near his own nest, however, it is quite a different story.

Ten-year experiment

The differing opinions on whether or not the female tolerates the male reflects how little is known about animals, especially those small ones that move about at night. In 1951 two American zoologists, JM Linsdale and LP Tevis Jnr published their book, *The Dusky-footed Wood Rat.* Its 664 pages contained the results of 10 years' study of this one species. In spite of the information it contains the authors said that they still could not give a complete account of the rat's behaviour. One of their difficulties was that the rats usually came out only at night. If one came out during the day it sought the shady places. Even on moonlit nights the rats would stay at home, and an electric torch always made them scamper for shelter. The two zoologists painstakingly live-trapped the rats, marked them, released them, and watched which house they went into. Sometimes, by good luck, a trapped female would give birth, and Linsdale and Tevis could see how she behaved towards her young ones and what they did. They put little numbered labels in front of each house to identify it. For the rest they could only watch and listen and take note of every tiny detail. There is probably no other rodent in the world that has been observed for so long and so continuously in one study.

class	**Mammalia**
order	**Rodentia**
family	**Cricetidae**
genus & species	***Neotoma cinerea*** *bushy-tailed wood rat* ***N. floridana*** *eastern or Florida pack rat* ***N. fuscipes*** *dusky-footed wood rat* *others*

Pangolin

Pangolins have sometimes been called animated pine cones because the hair on their backs has been converted into large overlapping brown scales covering the head, back, tail and legs. The underside of the body is, however, soft and hairy. The pangolin's body is long, with a long tail.

Its snout is pointed, with a small mouth at the end and with toothless jaws. Its long tongue can be thrust out for nearly a foot. The pangolin has small eyes and hidden ears. Its legs are short and the five toes on each foot have stout claws used in digging. In Africa there are four species of pangolin, or scaly anteater as it is sometimes called, and three in southern Asia.

The large African pangolin of equatorial Africa is 5 ft or more long as is the giant pangolin. Other African species are the black-bellied or long-tailed pangolin and the small-scaled tree pangolin, from West Africa to Uganda, both 3 ft total length. The largest Asiatic species is the Indian pangolin, 3½ ft long. The Chinese pangolin, of Nepal, southern China, Hainan and Formosa, and the Malayan pangolin are both under 3 ft.

Ground dwellers and tree climbers

Most of these strange scaly beasts climb trees, using their sharp claws and their tail, either wrapping the tail around a branch and sometimes hanging by it, or using it as a support by pressing it against the trunk of a tree. The giant and Indian pangolins both live on the ground, however, the latter sometimes climbing trees for safety when chased. All pangolins are active mainly at night, the ground-living forms resting in burrows dug by other animals, the tree dwellers resting in cavities in the trunks. When on the ground they walk on the sides of their forefeet, or on their knuckles, with their long claws turned inwards. They will sometimes walk on their hindlegs with their body raised semi-erect, their tail raised above the ground as a counterpoise.

Hot meals

This attitude, with the tail supporting the erect or semi-erect body, is also used when a pangolin is tearing open a termites' nest with its long front claws and exploring the galleries of the nests with its long tongue. The tongue is sticky and is flicked in and out to carry the termites into the mouth. Ants are also eaten: adults, pupae, larvae, and eggs. The tough skin of the head protects the pangolin from attacks by soldier termites or the stings of ants. The nostrils and ear openings can be closed and the eyes are protected by thick lids. Ants crawling onto the body are shaken off, and those swallowed are soon ground by the thick muscular walls of the stomach and by the small pebbles that the pangolin swallows. Tree-climbing pangolins eat mainly tree ants. A pangolin drinks by rapidly darting its tongue out and in.

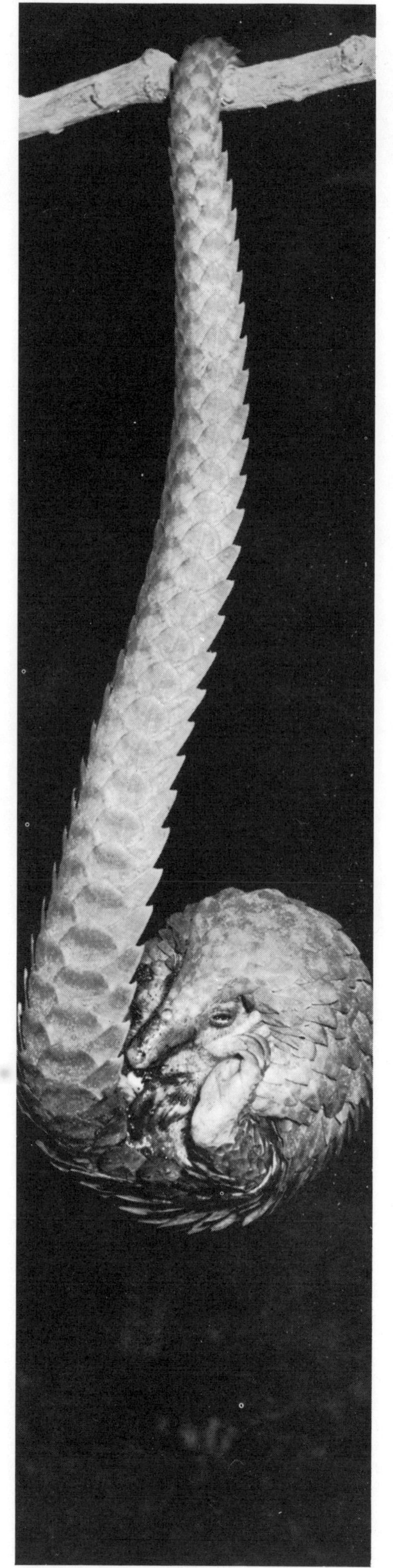

Pangolins do not usually survive long in captivity, a few weeks at most, and post mortem examinations have shown their digestive organs to be heavily parasitized. One lived over 4 years in the New York Zoological Park on finely ground raw beef, cooked cereal, evaporated milk, ant's eggs, with occasional raw egg, cod-liver oil and vitamin concentrate. But it seems likely that termites and ants are essential to them.

Babies ride pick-a-back

Very little is known about the breeding habits of pangolins since they fail to breed in captivity. They have one young, rarely two, in the wild, probably every year. The scales do not harden until the second day after birth. Later the baby rides on the mother clinging to her tail.

Ant-bathing

The main enemy of pangolins is probably man. Animals, such as leopards, sometimes examine them but are, it seems, put off by their scales. They are killed locally for their flesh and their scales are used for ornaments and charms, as well as for their supposed medicinal value. In Africa boys are sent into the burrow to put a rope round a pangolin's tail, to drag it out. Its defence is to roll up but even a light touch on a pangolin's body makes it snap its sharp-edged scales flat and this may act as a deterrent. Some pangolins, possibly all, can give off an obnoxious fluid from glands under their tail. They are said sometimes to hiss when molested.

There is a story that pangolins allow ants to crawl under their scales, then snap the scales down to kill them, afterwards eating the dead ants. The probable explanation is that the scales are snapped down because the observer touches the pangolin, or makes a movement that alarms it. That they do take an ant bath seems likely. There are local beliefs that a pangolin will lie in an ants' nest, allowing the insects to crawl over it, and under its scales onto the soft skin beneath. There are reports about a variety of animals taking ant baths. It is presumed they get satisfaction from the formic acid stimulating the skin. Cecil S Webb, an animal collector, was of the opinion that a pangolin's skin absorbed this acid and it was essential to the animal's health. He suggested this was one reason why they failed to survive in captivity.

An animal puzzle

Pangolin skins brought back from Africa and Asia were known to the Romans and also to the scientists of the 16th century and later. All were puzzled by them, as were the peoples in whose countries they lived. Arabs called the pangolin *abu-khirfa*, 'father of cattle', the Indians named it *bajur-kit*, 'jungle fish', the Chinese name was

◁ *It's easy when you have the equipment! A small-scaled pangolin hangs by its prehensile tail.*

▷△ *Small-scaled youngster grips to mother's tail while she is curled up asleep.*

▷ *Pangolins are quite widespread but not very numerous in any part of their range.*

▷▷ *The pangolin gets its name from its habit of rolling itself into a ball.*

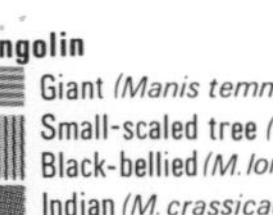
Pangolin
Giant (Manis temminckii)
Small-scaled tree (M. tricuspis)
Black-bellied (M. longicaudata)
Indian (M. crassicaudata)
Chinese (M. pentadactyla)
Malayan (M. javanica)

lungli, 'dragon carp', and the Romans called it an earth-crocodile. The name 'pangolin' is from the Malay *peng-goling*, the roller, from its habit of rolling into a ball. Pangolin skins puzzled the scholars of Europe until early 19th century when Baron Cuvier decided it was a mammal.

class	**Mammalia**
order	**Pholidota**
family	**Manidae**
genus & species	***Manis crassicaudata*** *Indian* ***M. javanica*** *Malayan* ***M. longicaudata*** *black-bellied* ***M. pentadactyla*** *Chinese* ***M. temminckii*** *giant* ***M. tricuspis*** *small-scaled tree*

◁ *A black-bellied pangolin illustrates its common description—'animated pine cone'.*
◁ ▽ *Ants swarm over the scales of a giant pangolin intent on eating all it can.*

▽ *A long elastic tongue protruded from the toothless mouth of a large African pangolin seeks out ants and termites on which it feeds.*

Pig

Domesticated pigs, or hogs, are derived from two wild species, the European wild boar and the Chinese wild pig. The Indian or crested wild boar may have contributed, but this is doubtful. It is not certain whether these three animals represent three different species or whether they are one species ranging across Europe and Asia as far as the East Indian islands, as well as North Africa. The tendency today is to accept the latter idea. One reason for the uncertainty is that the wild pigs in question show a great deal of variety. One that probably must be separated from the rest is the pygmy hog of Nepal which is only 1 ft high at the shoulder.

The Eurasian wild boar—the Chinese and Indian being grouped with it—usually grows to 4 ft head and body length, sometimes up to 6 ft. Its tail may be up to a foot long, and height at the shoulder up to 3 ft. The boar's weight may be up to 420 lb, the sow's up to 330 lb. Its tusks may be a foot in total length, including the continually growing root. The Eurasian wild boar is pale grey to brown or black in colour, the body hairs being sparse bristles with some finer hairs; the tail has only short hairs. Some individuals have longer hairs on the cheeks or a slight mane, or both.

The family party

The social unit of the pig is usually the family party but in the autumn family groups come together to form bands of up to 50 females and youngsters, the old males mainly remaining solitary. Pigs live mainly in open woodlands, especially where there are mud wallows in which they will spend many hours at a time if undisturbed. They also make crude shelters by cutting long grass then crawling under it to lift it so that it becomes entangled with the tall herbage around to form canopies. Quick footed and good swimmers, pigs normally avoid combat but will act vigorously when provoked, slashing with their tusks.

Nocturnal rootings

Wild pigs may travel far in a night rooting for anything edible. They will eat acorns and beechmast, roots of various kinds, fallen fruits, even the roots of ferns which few other animals eat. They have a natural tendency to dig for the potato-shaped fungi known as truffles. They will also eat insects, lizards, eggs, leverets, fawns, mice, voles, carrion, and any birds they can seize, in fact a very mixed diet. If allowed, they play havoc among cereal crops and root crops such as beet or turnip, and among potatoes. For this reason they have been hunted for centuries, although they have also been hunted for their flesh and for the sport they give.

△ *Two wild boar emerge into a forest clearing. The snow makes it harder for them to get food.*

▽ *Two inhabitants of the forest, a boar and a fox, seek food and water on a winter evening.*

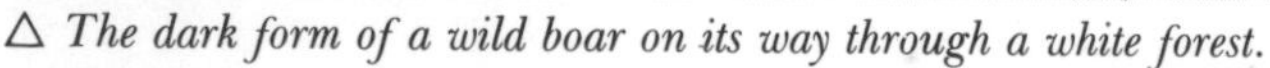

△ *The dark form of a wild boar on its way through a white forest.*

△ *A wild boar kicks up its heels and runs. They are not often aggressive.*

Striped young

The sow is in season every 3 weeks and produces litters of 3–12 after a gestation of 112–115 days. She has 8–14 teats and as each piglet takes a teat at feeding time the weaklings in a large litter will die. They suckle for about 12 weeks before being completely weaned onto food which they find while rooting around, never very far from the protection of their mother. The boar takes no part in caring for the young. The young are striped at first, becoming sexually mature at 18 months, and reach full size at 5—6 years of age. The pig has a life-span of up to 27 years.

Early domestication

Wild pigs cannot be readily herded but they take well to life in sties or in houses, so we can be fairly sure that their domestication must have come about when men ceased to be hunters and settled down to agriculture. Another clue is in the taboo on eating pig flesh, which seems to have originated in the nomads' contempt for the agrarian communities, expressed in a supposed disgust at the pigs they kept. This was probably reinforced by the disease trichinosis, that could be contracted from eating insufficiently cooked pork. The earliest domestication date is uncertain but it is unlikely to have been before the Neolithic period, with its agricultural revolution.

The European wild boar is larger than the Chinese pig. They represent extremes of a range in size and the evidence is that pigs were domesticated from local races, so giving domestic pigs of various sizes. There is evidence also that in prehistoric times there was importation of breeds from one part of Eurasia to another as well as some selective breeding. So there was a mixture before modern selective breeding began, making it difficult to trace the lineage of present-day domestic pigs.

The pig has been bred almost solely for its flesh and for its fat. Its bristles have been used to some extent for making brushes and the hide for making sandals and other fancy leather goods. Even the bones may be ground up for bone meal fertilizer. Nevertheless, the domestic pig as a living animal has been put to a variety of uses. It has been used for sacrifices; and for a curious custom among the armies of the Roman period, of swearing an oath on a pig or a piglet. Pigs have at various times been used for pulling carts. They have also been trained to detect truffles, which the owner then dug up. In Ancient Egypt pigs were used for treading in corn, their sharp hoofs making holes of the correct depth for the seed to germinate. Most surprising of all, they were trained in mediaeval England as pointers and retrievers for illicit hunting, in the New Forest, for example. There commoners were forbidden to keep any but the smallest dogs, capable of passing through King Rufus' Stirrup, an iron stirrup $10\frac{1}{2}$ in. high by $7\frac{1}{2}$ in. across. In southern India, where there lives a primitive tribe whose buffaloes wander into the marshes, an old woman was seen to speak to a pig which 'at once trotted into the marsh, rounded up the buffaloes and herded them to her, like a well-trained collie'.

class	**Mammalia**
order	**Artiodactyla**
family	**Suidae**
genus & species	***Sus cristatus*** *Indian wild boar* ***S. salvanius*** *pygmy hog* ***S. scrofa*** *European wild boar* ***S. vittatus*** *Chinese pig*

▽ *The good life? A fat, pink, domesticated sow feeds her hungry offspring.*

Platypus

Today the platypus is accepted as an unusual animal of quaint appearance, but it is not difficult to imagine its impact on the scientific world when it was first discovered. So strange did the creature appear that one scientist named it **paradoxus,** *and a paradox it was with duck-like bill, furry mammalian coat and webbed feet.*

Known as the duckbill, watermole or duckmole, the platypus is one of Australia's two egg-laying mammals, the other being the spiny anteater. The platypus is about 2 ft long including a 6in. beaver-like tail and weighs about $4\frac{1}{2}$ lb, the males being slightly larger than the females. The 'bill' is a sensitive elongated snout and is soft, like doeskin, not horny as is popularly supposed.

Although bizarre in appearance, the platypus is well adapted to its semi-aquatic life. The legs are short with strong claws on the toes and the feet are webbed. The webbing on the forefeet extends well beyond the toes, but can be turned back when on land, leaving the claws free for walking and digging. The eye and the opening to the inner ear lie on each side of the head in a furrow which can be closed when the platypus submerges. There are no external ears, thus the platypus is blind and deaf when under water. Young have teeth, but these are replaced in the adult by horny ridges.

Thick loose skin makes the barrel-shaped body of the platypus appear larger than it is. The pelt consists of a dense woolly undercoat and long shiny guard hairs. The colour varies from sepia brown to almost black above and is silver, tinged with pink or yellow underneath; females can be identified by the more pronounced reddish tint of their fur. Adult males have hollow spurs, connected to venom glands, on the ankle of each hind limb. The poison from them can be quite harmful to a man, although not fatal.

The platypus was not discovered until 1796, nearly 200 years after the first wallaby, for instance, had been seen by a European. This is not as strange as might appear at first sight, for aquatic animals tend to be elusive particularly if, like the platypus, they are nocturnal.

Its range can be seen on the map. The western limits are the Leichhardt River in North Queensland, and the Murray, Onkaparinga and Glenelg rivers, just within the border of South Australia. It is found in all fresh water, from clear icy streams at 5 000 ft to lakes and warm coastal rivers.

▷ *Out of its front door and into the river. When underwater the platypus is blind and deaf so it relies mainly on its sense of touch, highly developed in the soft rubbery bill.*

Hearty appetite

Like many small energetic animals the platypus has a voracious appetite, and probably needs more food, relative to its weight, than any other mammal. It feeds mainly in the early morning and late evening, on crayfish, worms and other small water animals. It probes for these with its bill and at the same time takes in mud and sand, which are apparently necessary for breaking up the food. During the day the platypus rests in burrows dug out of the banks, coming out at night to forage for food in the mud of the river-bottom.

Egg-laying mammal

The breeding season is from August to November and mating takes place in the water, after an elaborate and unusual courtship. Among other manoeuvres, the male will grasp the female's tail and the two will then swim slowly in circles. The female digs a winding, intricate burrow in a bank 25–35 ft, sometimes as much as 60 ft, long, 12–15 in. below the surface of the ground. At the end, a nesting chamber is excavated and lined with wet grass and leaves. The female carries these by wrapping her tail around a bundle. Usually two soft-shelled white eggs are laid, each ½ in. diameter. They often stick together, which prevents them rolling, and the wet leaves and grass keep them from drying out. Before retiring to lay her eggs, 2 weeks after mating, the female blocks the tunnel at intervals with earth, up to 8 in. thick, which she tamps into position with her tail. During the incubation period of 7–10 days she rarely leaves the nest but each time she does so these earth blocks are rebuilt. Presumably this is a defensive measure, but in fact today the platypus has virtually no natural enemies, although a carpet-snake or goanna may occasionally catch one. The inference is that in past ages natural enemies did exist in some numbers and the earth-block defences were very necessary. This is an example of what is known as 'fossil behaviour' and the platypus itself is a living fossil.

Blind for 11 weeks

The young platypus is naked and blind, and its eyes do not open for 11 weeks. It is weaned when nearly 4 months old, at which age it takes to the water. The mother has no teats; milk merely oozes through slits on her abdomen where it is licked up by the babies. A platypus matures at about 2½ years and has a life span of 10 years or more.

Competing with rabbits

Formerly hunted ruthlessly for its beaver-like pelt, the platypus is now rigidly protected. Too often, however, it falls foul of wire cages set under water for fish. Should the platypus enter one it cannot escape and will drown, as it is not able to stay under water for much more than 5 minutes. The introduced rabbit of Australia threatens the platypus in a different way. Where rabbits have driven too many tunnels the platypus cannot breed: it needs undisturbed soil for its breeding burrows. Fortunately, although reduced in numbers, it is now well protected by the Australian authorities and it is in no danger of extinction.

Creature of contrast

Fortunately for the sanity of naturalists, the paradoxical facts that the platypus, a mammal, laid eggs and suckled its young were not known when it was first discovered. In 1884, WH Caldwell, who had gone to Australia specially to study the platypus, dissected a female which had already laid one egg and was ready to lay another. Thrilled by this discovery he electrified members of the British Association for the Advancement of Science, then meeting in Montreal, with his laconic telegram—'Monotremes oviparus, ovum meroblastic' (monotremes egg-laying, egg only partially divides). Delegates stood and cheered, for controversy over this point had raged in the scientific world for some years.

Long before this, in 1799, the first dried skin reached London and came into the hands of Dr Shaw, then assistant-keeper in the Natural History section of the British Museum. When Dr Shaw saw the skin he literally could not believe what he saw. At that time visitors to the Far East were bringing back fakes such as the 'eastern-mermaid', made from the skin of a monkey skilfully sewn to the tail of a fish. It is not surprising, therefore, that Dr Shaw should suspect someone had grafted the bill of a duck on to the body of a quadruped. He tried to prise off the bill, and today the marks of his scissors can still be seen on the original skin which is preserved in the British Museum (Natural History).

class	**Mammalia**
order	**Monotremata**
family	**Ornithorhynchidae**
genus & species	***Ornithorhynchus anatinus*** *platypus*

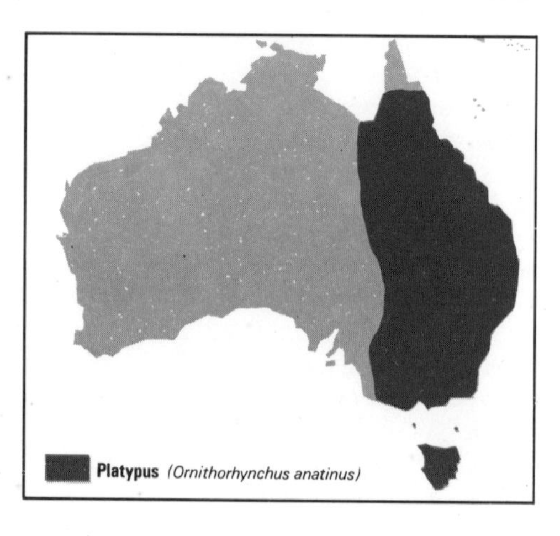

▽ *In the water the platypus uses its strong webbed forefeet for swimming and its hind legs as rudders. On land its forefeet are used for digging and to press the water out of its fur before it enters its burrow.*

Strong foreclaws (left) dig burrows, marked by fan-shaped mounds of earth at entrance (above).

Pocket gopher

The name of these small hamster-like animals is derived from the two external furlined cheek pouches, which run from face to shoulder. They are used for carrying food and are turned inside out when they need cleaning.

There are 30 species of pocket gophers which vary considerably in size, with a body length from 3½ to 13 in. and tails from 1½ to 5½ in. The coat is usually thick and smooth, lighter and thinner on the underside. Pocket gophers living in hot lowland areas have shorter coarser fur. The fur colour varies from almost black through all shades of brown to off-white. Albinos are not uncommon. ***Macrogeomys*** *and* ***Zygogeomys*** *have characteristic white patches.*

The body is adapted for burrowing. The skull is large and angular, the body thickset and tapering towards the tail. Ears and eyes are small, the latter being kept moist with a thick fluid to keep the eyeball free from dirt. The legs are short and powerful, especially the forelegs, which have long digging claws. The tail is almost naked and is very sensitive to touch. By arching the tail to raise the tip just off the ground, a pocket gopher can feel its way as it moves rapidly backwards in its burrow.

All pocket gophers have long curving upper incisors, behind which the lips close to prevent earth entering the mouth while they are burrowing.

Peculiar to North America, pocket gophers are found from western Canada south to Panama. They do not travel far and are found in localised areas, often restricted to valleys by mountain barriers.

The solitary burrower

These solitary rodents spend almost their entire life underground, only very occasionally coming to the surface to collect food. Each gopher lives within its own system of burrows and they only come together at mating times. Young pocket gophers can sometimes be found above ground after leaving their parents, and in drought or after floods adult pocket gophers may be driven out to look for new homes.

The individual burrows are often extensive and are marked by fan shaped mounds of earth around the entrances. These are carefully blocked with earth as protection against predators and to maintain suitable temperature and humidity. There are two types of tunnels: long, shallow ones, used mainly for getting food, and deep ones, used for shelter, with separate chambers for food storage, nesting and latrines. A great assortment of other animals use inhabited and abandoned gopher burrows.

Pocket gophers dig with their strong foreclaws, the curved incisors being used for loosening hard earth and rocks. The incisors are growing continually so replacing the worn surfaces. They may grow as much as 20 in. in one year.

Pocket gophers do not hibernate but may remain relatively inactive for long periods. Evidence of their winter activity is strikingly apparent when the snows melt in high pastures, revealing gopher burrows within the snow lined with earth, and left behind as a mass of criss-crossing 'cores' showing that the animals have burrowed at several levels. The tunnels are up to 40 ft long and 2–3 in. diameter.

Underground vegetarian

The staple diet of pocket gophers is tubers, bulbs and roots of plants which can be found and eaten in the security of the burrow. Occasionally at night, or on overcast days, pocket gophers surface in search of food, but more often plants are seen disappearing in jerks as, from the safety of its burrow, a pocket gopher pulls them down. In agricultural regions, crops often suffer badly as gophers favour sweet-potatoes, sugar cane, peas and fruit. Gophers never seem to drink, probably getting enough moisture from their plant food.

Little is known about the breeding and courtship of gophers, but if the soil and food conditions are suitable their numbers increase rapidly. A female may have one or more litters a year and the number of offspring varies considerably from 2 to 11. The newborn young, weighing about $\frac{1}{12}$ oz, are blind and almost hairless. In most species they leave their mother at about 2 months to make their own burrow network. They are sexually mature within 3 months.

Safe underground

The best protection a pocket gopher has against its enemies is its underground way of life, but in fights with its own kind it is protected by the loose skin and thick hair around the head. Predators, such as coyotes, badgers and skunks, sometimes succeed in digging pocket gophers out and if they have to come above ground for any length of time owls and hawks make short work of them. Athough snakes and weasels hunt pocket gophers down their burrows, man is by far their worst enemy.

Pocket gophers can be very destructive, eating crops, burrowing through dykes and contributing to soil erosion. But on the credit side they do sterling work improving the soil by loosening, aerating and mixing it with organic matter. They also conserve water since, after a heavy snowfall, the melted water sinks deep into the earth through the maze of gopher tunnels instead of flowing straight into the nearest stream.

All manner of snares, traps, sling shots and spears are used to eliminate the pocket gophers and in Mexico the *tucero*, whose official job is to hunt pocket gophers, is a respected member of the village society.

Slim, tapering bodies

To allow manoeuvrability in tunnels a tapering body is essential, and burrowing animals generally have narrow hips. On the other hand, broad hips are an asset in giving birth. In virgin pocket gophers the pelvis is very narrow, with fused pelvic bones as in other mammals, leaving too small a gap for the passage of offspring during birth. Examination after birth shows, however, an enlarged opening which just allows an easy birth. Much of the bone of the pelvic region had been dissolved away through the action of hormones secreted during pregnancy.

class	**Mammalia**
order	**Rodentia**
family	**Geomyidae**
genera	***Geomys, Thomomys Pappogeomys, Cratogeomys Orthogeomys, Heterogeomys Macrogeomys, Zygogeomys***

Polar bear

The polar bear is one of the largest and the most carnivorous of all bears. The males average 7–8 ft long and may reach 9 ft, with a height of 5 ft at the shoulder. The average weight is 900 lb and the maximum over half a ton. The females are smaller, with an average weight of 700 lb. Polar bears have a long head with small ears, and a 'Roman' nose, a long neck, powerful limbs, broad feet with hairy soles, and a mere stump of a tail. Their coat, white with a yellowish tinge, is made up of long guard hairs and a dense underfur.

Their home is along the southern edge of the Arctic pack ice. Although they can swim strongly, they avoid stretches of open water and fast ice. They are carried southwards by the ice in spring and summer and return northwards when the ice breaks up.

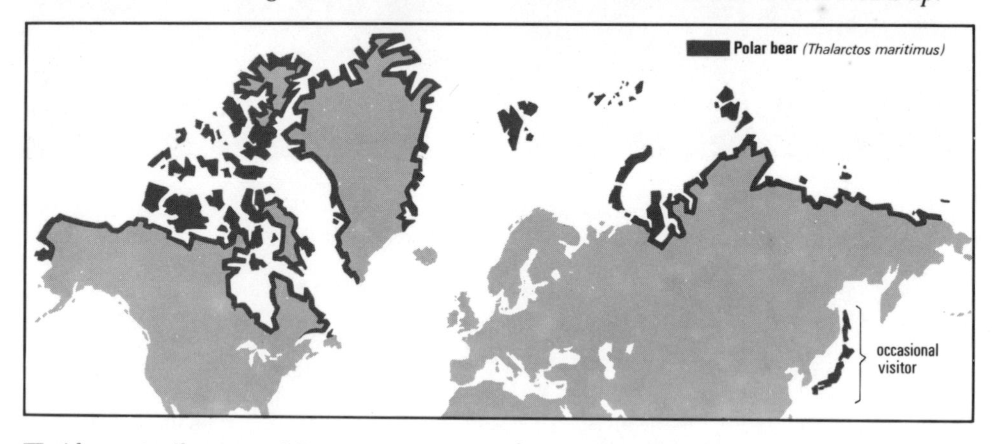

▽ *A happy family—two cuddly cubs satisfy their hunger. Polar bear cubs stay with their mother for at least 10 months. When she is ready to mate again, she drives away any remaining cubs.*

▷ *A solitary polar bear steps carefully over the rocks of the Arctic. Polar bears are nomadic and roam for miles in search of food. Seals, especially the ringed seal, provide them with their favourite food.*

A home of ice and water

Polar bears are expert divers and swim strongly, at about 6 mph, using only the front legs, and trailing the hindlegs. A thick layer of fat under the skin, 3 in. thick on the haunches, helps to keep them buoyant. Polar bears have been seen swimming strongly 200 miles from land. They usually swim with the head stretched forward, but when the sea is rough they put their heads underwater, lifting them periodically to breathe. The fat layer keeps out the cold. When it comes onto land the polar bear shakes itself like a dog.

Polar bears can walk easily over ice because of their hairy soles, and when pursued can run at speeds of up to 18 mph. They swing their heads from side to side as they walk, as if searching or smelling out prey. Essentially nomadic, they wander for miles in search of food.

Preying on seals

Their favourite food is seals, especially the ringed seal, which the bears stalk by taking advantage of snow hummocks. A seal asleep on the edge of the ice falls easy prey. The polar bear swims underwater to the spot, comes up beneath the seal and crushes its skull with one blow of the powerful forepaw. When a ringed seal is about to pup she digs an igloo in a hummock of snow over her breathing hole. Polar bears sniff out the igloos and take the pups. They are said also to hold down a pup with one paw so that its struggles attract the mother within reach of the bear. They will kill young walruses, but in a fight with a grown walrus the bear is likely to come off second best. The bears also eat fish, seabirds—at times their eggs, too—and carrion. A stranded whale will draw bears from a large area. Such a carcase once attracted 24 bears. At certain times of the year, usually in late spring or early summer, polar bears will eat large quantities of grass, lichens, seaweed, moss and other plants, and they are fond of crowberries, bilberries and cranberries.

Cubs driven away

Mating is in April or May and after a gestation of 240 days usually 2 cubs are born in December and January. At birth each cub is a foot long and weighs $1\frac{1}{2}$ lb. It has a coat of short sparse hair. The eyes open at 33 days, the ears at 26 days, although hearing is imperfect until the cub is 69 days old. It starts to walk at about 47 days and is not weaned for 3 months. The cubs stay with the mother until they are at least 10 months old, often longer. When she is ready to mate again, she drives away any remaining cubs, which are by then 200–400 lb weight. Polar bears are sexually mature at $2\frac{1}{2}$–5 years, the females maturing later than the males. Most of this detailed information, which would be difficult to collect in the wild, has come from records of successful breedings in zoos where some polar bears have lived well. Since polar bears are now being rapidly exterminated to satisfy the taste of the affluent for luxurious rugs it is as well that zoos have mastered the art of breeding these fierce but beautiful beasts.

No hibernation

It is often said that cubs are born while the mother is hibernating. Polar bears, like

other bears, do not hibernate in the strict sense and it is now usual to speak of their sleep as winter dormancy. The pregnant she-bear seeks out a bank of snow in the lee of a hill and digs into it. There is difference of opinion as to whether the males 'hibernate'. It seems likely that some do, but for shorter periods than the females.

Man the enemy

Apart from man the polar bear has no enemies, if we except the walrus that may gore it in self-defence. Young bears die of accidents, by being drowned or crushed by ice during storms. Old males will kill and eat the cubs, given the opportunity. Polar bears have long been hunted by the Eskimos for their meat and their pelts. Their long canine teeth, too, are used for ornaments. Bed covers, sledge robes and trousers are made from the pelt. Although Eskimos eat the flesh the liver is not used, not even to feed the dogs. Its poisonous quality is said to be due to the high concentrations of vitamin A, which causes headaches and nausea, and sometimes a form of dermatitis.

Conserving the bear

CR Harrington, writing in *Canadian Audubon,* tells us that the earliest known record of polar bears taken into captivity was for 880 AD, when two cubs were taken from Iceland to Norway. 'At that time the animals were commonly offered to European rulers, who rewarded the donors on various occasions with ships carrying cargoes of timber or even bishoprics.' Intensive hunting of the bear began in the 17th century when whalers reached the pack ice. Two centuries later polar bears were decreasing in many parts of the Arctic. Then came the turn of the sealers, and in 1942 alone Norwegian sealers killed 714. More recently the Eskimos have been hunting the bears not for their own use but to trade the pelts. In parts of arctic North America it is easy to land by aeroplane and shoot the bears for sport, but legislation controlling this is under fairly constant review. In 1965 Harrington estimated the world population at over 10 000 with a total annual kill of about 1 300. In 1968, after several meetings concerning the future of the polar bear, research scientists studying the bear from five Arctic nations—the USSR, the USA, Canada, Norway and Denmark—came together and established the Polar Bear Group. Under the auspices of the IUCN the scientists cooperated and together began the study of the polar bear's life history, ecology, seasonal movements and population.

class	**Mammalia**
order	**Carnivora**
family	**Ursidae**
genus & species	***Thalarctos maritimus*** *polar bear*

Taking the plunge. Protected by its thick layer of fat, a polar bear braves the cold waters of the Arctic where it may swim for miles out to sea.

Porcupine

The tree porcupines of North and South America are very different from the porcupines of the Old World. To begin with, they live mainly in trees and their hindfeet are adapted for climbing. Some species also have a prehensile tail. The best known, the Canadian or North American porcupine, is up to $3\frac{1}{2}$ ft long, of which 1 ft is tail, and has an average weight of 15 lb although large males may weigh up to 40 lb. It is heavy and clumsily built with a small head, short legs and a short, stout, spiny tail. The hindfoot has a well-developed great toe and very long, powerful claws to help the animal climb. The long fur on the upper parts is brownish-black, sprinkled with long white hairs that conceal the short, barbed spines, which are yellowish-white tipped with black.

The South American tree porcupines, of which the Brazilian tree porcupine is typical, differ from the North American species in having a long, prehensile tail, the tip of which is hairless and by having only four toes on the hindfeet with a broad fleshy pad, opposable to the toes, used rather like a thumb in gripping branches when the animal is climbing. It is of lighter build with short, closely set spines, sometimes concealed by long hairs.

The Canadian porcupine inhabits most of the timbered areas of Alaska, Canada and the United States (except the south-eastern quarter), south to the extreme north of Mexico. South American porcupines extend from Mexico through Central America to Colombia, Venezuela, Brazil, Bolivia, Peru and Ecuador in South America.

△ Long fur conceals the porcupine's spines.
▽ North American porcupine revealing its arsenal of over 20 000 sharp-tipped quills.

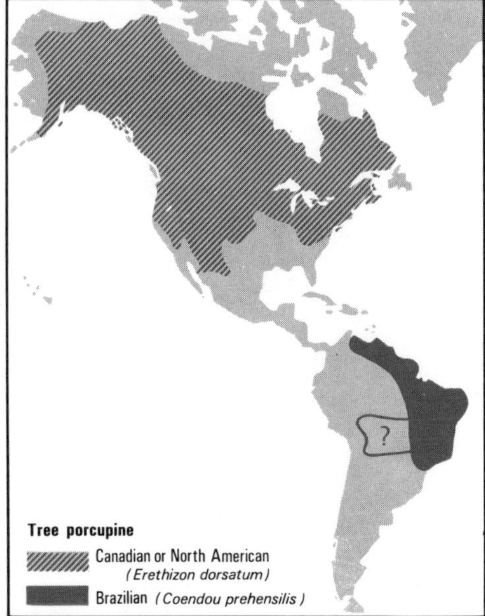

◁ *Picking a precarious path, a South American porcupine* **Coendou** *sp. seeks its typically rodent diet of bark, stems and leaves.*

No hibernation

All the tree porcupines live in wooded areas, the North American species preferring woods of conifers, junipers and poplars. Although clumsily built they can climb well and they will also swim. They lie up during the day among rocks or in hollow trees and feed mainly at dusk and at dawn. They are usually solitary but occasionally several Canadian tree porcupines may shelter together in the same den, especially in winter. They do not hibernate but they take to dens during bad weather.

Salt addicts

The Canadian tree porcupine varies its food with the seasons. In spring it eats the flowers and catkins of the willow, maple and poplar. Later it turns to the new leaves of aspen and larch. In summer it feeds more on herbaceous plants and in winter on evergreens like the hemlock and pine. Its principal food in winter, however, is bark and the porcupines do much damage by ring-barking trees. The young red firs of the Sierra Nevada in California are occasionally destroyed by tree porcupines. When the weather is bad and the snow deep an animal may live in one tree and not leave it until all the bark above the snow-line has been stripped. Tree porcupines also have a strong liking for sweet corn and a few of these animals can completely ravage a field of it.

A more peculiar taste is the porcupine's craving for salt. Handles of farm implements which have been touched by hands moistened with sweat, leaving a trace of salt, will be gnawed. So will gloves, boots, and saddles; even the steering wheel of a car has been gnawed away. The porcupine will also gnaw bones and antlers dropped by deer. But its crowning achievement is to gnaw glass bottles thrown away by campers, presumably for the salt in the glass.

The South American tree porcupines also eat the bark and leaves of trees and tender stems but in addition they eat fruit such as bananas, and occasionally corn.

Well-developed babies

The Canadian tree porcupines mate in the fall or early winter. During courtship the male rubs noses with the female and often urinates over her. Generally a single young is born after a gestation period of 210–217 days. The young are very well-developed at birth; their eyes are open and they are born with long black hair and short soft quills. They weigh about 20 oz and can climb trees when 2 days old. They are weaned in 10 days, and become sexually mature in their second year.

Little is known of the breeding habits of the South American tree porcupines. There is usually a single young at a birth, born from February to May. The young of the Brazilian tree porcupine are comparatively large at birth and are covered with long, reddish hair. Their backs are covered with short spines, which are flexible at birth.

Few natural enemies

Few animals prey on the porcupine because of its spines, but the wolverine, puma and fisher marten will attack the North American species. A tree porcupine is said never to attack an enemy. If cornered, however, it will erect its quills and turn its back on its adversary, striking out repeatedly with its tail. A porcupine does not shoot its quills but they are so lightly attached that when they enter the skin of the enemy they become detached from the porcupine.

Skulls identify species

The crested porcupine is the best known species in the Old World family Hystricidae. Not all the porcupines in that family have such prominent quills as the crested porcupine. One *Trichys lipura* living in Borneo, for example, lacks true quills. It has only short, flat, weak spines and its long tail has a brush of bristles on the end. At first sight it appears not to be a porcupine at all. The same thing can be said of some of the family Erethizontidae. Since the crested and the Canadian porcupines look so alike, the question arises: What is the essential difference between the Old World porcupines and the New World porcupines? The fact that they are widely separated geographically is not important. Both families agree in having species that show a varying tendency to grow quills among the bristly coat, and both families contain a diversity of species. Therefore, those who classify these rodents have to look for something more stable upon which to separate them. They find this in the skull. Any Old World porcupine, whatever it may have in the way of quills, has a very rounded skull which has quite obviously a different shape from that of the New World porcupines.

class	**Mammalia**
order	**Rodentia**
family	**Erethizontidae**
genera & species	***Erethizon dorsatum*** *Canadian or North American porcupine* ***Coendou prehensilis*** *Brazilian tree porcupine, others*

Prairie dog

Prairie dogs are hamster-like, short-tailed ground squirrels that are so named because of their barking calls. The length of the head and body is about 12 in. and the tail about $3\frac{1}{2}$ in. Apart from being slightly flattened, the tail bears little resemblance to the tail of tree squirrels. The head is more squirrel-like except that the ears are very small. The fur is yellowish-grey or brown with lighter underparts and the tip of the tail is black. The five species are all very similar in appearance.

They inhabit the plains and plateaus of North America, from the Dakotas to Texas, and from Utah and Arizona in the west to Kansas and Oklahoma in the east, and they are also found in northern Mexico.

Prairie citizens

Prairie dogs live in vast 'towns', although these are smaller than they used to be because such large concentrations of animals inevitably came into conflict with man. In 1901, one town was estimated to cover an area 100 by 240 miles and to contain 400 million prairie dogs. As with other colonial or social animals prairie dogs have a social organisation, which, as must be expected with such vast colonies, is very complex. A single town is divided into a number of wards, whose boundaries depend largely on the geography of the area. The wards are divided into a number of coteries, each covering less than one acre. The coterie is the base unit on which the prairie dog's life is founded. It is the family unit that defends its territory and individuals rarely venture from it. If they do, they are likely to be chased back by members of neighbouring coteries. A typical coterie consists of an adult male, three adult females and a variable number of young prairie dogs. The members of a coterie recognise each other and are on friendly terms, and except for the very young ones, they jealously guard the coterie's boundaries. Apart from squabbles along the borders, members of the coterie, including the youngsters, advertise their territories with a display. Rearing up on their hindlegs with nose pointing to the sky the prairie dogs deliver a series of two-syllable calls.

Each coterie has a network of burrows with a large number of entrances. From the entrance the burrow descends steeply for 3–4 yd before meeting radial tunnels with nests at the end. From a distance a prairie

▽ *Passion on the prairies? The greeting kiss inhibits aggression between members of one family, at the same time enabling a trespassing stranger to be detected.*

▷ *A couple of prairie dogs survey the scene.*

dog town appears pockmarked with craters because each burrow entrance is surrounded by a volcano-like cone. This is more than the casual accumulation of excavated soil; it is a carefully built rampart of soil 1–2 ft high and up to 6 ft across. The soil is gathered from the surface, brought to the entrance and patted into place where it serves as a lookout post and a protection against floods after heavy rain.

Changing the scenery

Prairie dogs are vegetarian, feeding on grasses and other plants that grow on the prairies. Not surprisingly, the crowds of prairie dogs have a profound effect on the vegetation inside the town limits. The taller plants are eliminated. They are cut down and left to wither if they are not eaten, and the continual cropping of the grasses and herbs encourages fast growing plants with abundant seeds so the optimum vegetation, from the prairie dog point of view, is produced. There is a second advantage to this unintentional agriculture. The removal of tall plants deprives predators of cover and allows the prairie dogs a clear view from their mounds.

Perhaps the main advantage of the coterie system is that each group of prairie dogs has sufficient area for feeding, and over-grazing is prevented by not allowing other prairie dogs onto the pasture. When the population gets too big, some members emigrate to form 'overspill' towns.

Keeping the balance

The rate of reproduction in prairie dogs is slow compared with many other rodents. Each female produces only one litter a year, usually of around four pups, in March, April or May. The pups' eyes open at 33 days and they are weaned in 7 weeks.

Although comparatively few pups are born each year, the population can still increase rapidly, such as from 4 to 15 prairie dogs an acre in 3 months. This would threaten the food supply if it were not for emigrations to 'overspill' towns or suburbs. When the population rises the behaviour of the prairie dogs changes. Usually any member of a coterie can enter any burrow and any female will suckle any pup, but now the females defend their nests while the others dig burrows and feed at the edge of the town, commuting home at night. As the young prairie dogs appear the travelling adults move permanently into their new homes. The population is thus redistributed without disturbing the boundaries.

Early-warning system

Prairie dogs fall prey to many predators, particularly coyotes and birds of prey, but it is usually only the slow and sick or incautious individuals that stray too far from a burrow that are caught.

Prairie dogs have an alarm call that sends them all bolting for cover. This is the bark that is responsible for their name. It is a short nasal yip with several shades of meaning. When high-pitched it is the signal for immediate flight. The territorial call, however, is used as an all-clear signal.

The depredations of so many predators does not affect prairie dog numbers but the prairie dog's use of the grassy plains has led to it being nearly wiped out by man. Man wanted the grasses for his livestock, who were also in danger of breaking their legs in prairie dog burrows. Poisoning was so successful that the towns were wiped out and prairie dogs now survive mainly in national parks.

Kisses seal friendship

The efficient use of pasturage, and the harmonious life within the coteries are essential for the existence of prairie dog towns, and the whole of the elaborate social system is dependent on communication between prairie dogs. Members of neighbouring coteries keep apart by an aggressive ritual but members of a single coterie are drawn together. The basis of the ritual is the exchange of kisses. When two prairie dogs meet near a boundary they drop to their bellies and crawl slowly towards each other. On meeting they bare their teeth and kiss. If strangers, one retreats or a squabble breaks out. If friends, one nibbles the other who rolls over and allows itself to be groomed, so cementing the friendship. Within a coterie all members groom each other and at times the grooming is prolonged and the pair end up side by side. They then go off to feed, with their bodies still pressed together. At other times, two prairie dogs may merely exchange quick 'pecks' but this will confirm identification and keep the coterie harmonious.

class	**Mammalia**
order	**Rodentia**
family	**Sciuridae**
genus & species	***Cynomys ludovicianus*** *others*

Reflecting on more youthful days? An adult prairie dog sits in the sun and relaxes.

Puma

This animal is also known in the United States and Canada as the cougar. However Stanley P Young and Edward A Goldman, American authors, have titled their book on it 'The Puma', and they claim this is the correct name for it. Probably no other animal has received so many common names, another of which is the mountain lion, which gives a clue to the appearance of the puma. It looks like a lioness, its coat is of short close fur and its colour yellowish-brown, although this varies from yellow to red. The maximum size recorded, for a male, was 8 ft long, of which 3 ft was tail, and 260 lb weight, but there is much variation from as little as 4 ft total length and 46 lb weight. The females are generally smaller than males. The puma ranges from western Canada to Patagonia in the southern half of South America, on mountains, plains, deserts and in forests. Although there is only one species, 30 subspecies have been named, based on differences in size and colour, a clear indication how variable these two features are.

Powerful killer

The puma is known for its remarkable strength and stamina. It will cover up to 20 ft in one bound and a leap of 40 ft has been recorded. It can leap upwards to a height of 15 ft and has been known to drop to the ground from a height of 60 ft. Like many other members of the cat family, the puma leads a solitary life keeping very much out of sight. A puma will often kill and, holding its prey, toss it over its shoulder and walk away with it; one has been known to drag a carcase three times its own weight over the snow. It will travel 30–50 miles when hunting. Its trail is marked by the remains of prey lightly buried in the ground, and by the scratchings where it has scraped earth over its urine or dung.

Some of its common names, such as catamount, refer to the puma's voice, but there are differences of opinion on how much this is used. Pumas are said to emit bloodcurdling screams at times and there have been many vivid descriptions of this. It has been described as a weird caterwauling yell or scream, hence the saying 'to yell like a painter'—another name for the puma. By contrast, game wardens working in puma country for years have remarked on the fact that they have never heard a puma scream. It has even been suggested that people have seen a puma in a tree, for it is a good climber, and at that moment a great horned owl, hidden in the tree, has called and this has been credited to the puma. The evidence suggests that both male and female scream but not frequently. Normally they purr when contented.

A unique and magnificent action picture of a puma making a flying leap to the ground. This cat is renowned for its strength and stamina.

Another aspect of the puma's way of life which is much debated has to do with its attitude towards man. It has been represented as highly dangerous yet sober surveys show that attacks on human beings are so rare as to be negligible. It seems likely that the screaming may have contributed to an undeserved reputation for ferocity, together with its habit of stalking people. This seems to arise from an overwhelming curiosity and many naturalists have told how they have been followed by a puma which melts into the undergrowth every time they look round.

Controlling the deer

Its main prey is deer which may make up 50–75% of its food, and wherever the puma is killed off, deer multiply rapidly. It also takes a variety of small mammals including porcupines. It sometimes attacks domestic stock, such as sheep and goats, as well as horses and cattle, and it will take carrion. Ponies are said to be especially attractive to pumas. They stalk the larger prey, suddenly pouncing on their backs with a powerful leap, often with such violence that the prey is carried up to 20 ft along the ground. The typical method of killing is by a bite into the throat.

Spotted kittens

Pumas breed the year round. After a gestation of 90–93 days a litter of 1–4 cubs is born; at birth they are blind, and have spotted fur and a ringed tail. The eyes open at 10–14 days and the cubs are weaned at 1–3 months. The period seems to vary, as does the time they stay with the mother, which is up to 2 years. As they mature the cubs lose their spots and the rings on the tail. The life-span is up to 18 years.

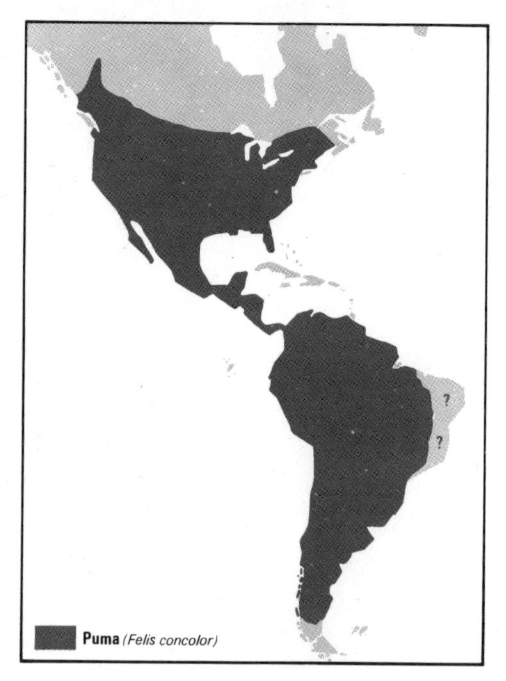

Man the merciless enemy

A puma has few enemies apart from man but where their ranges coincide jaguars and pumas often fight. Wolverines will attack pumas and grizzly bears are also credited with doing so. These hazards are probably very trivial compared with man's activities which have wiped out the puma or seriously reduced its numbers in parts of its range. Pumas are killed with traps and hunted especially with dogs which will usually tree the puma putting it at the mercy of the marksman. In some places and at various times bounties have been paid, the main complaint against the puma being its attacks on domestic stock.

Puma: the friend of man

In contrast with the way pumas are persecuted in North America, these animals have been called 'the friend of man'. The origins of this title go back to 1536 when the Spaniards in Buenos Aires were at war with the Indians. A young woman, Senorita Maldonado, was wrongly accused of treachery when, in fact, she had been carried off by the Indians. She was condemned to be tied to a tree to be eaten by wild beasts. After two nights and a day the soldiers went to collect her bones for burial and found her unhurt. She explained that a puma had stayed by her and driven off the jaguars and other beasts that came to destroy her. The Indians of California had even more cause to regard the puma with friendly feelings. The Jesuit priests who went there to preach Christianity to the natives found that for centuries they had largely fed on the remains of the pumas' prey.

class	**Mammalia**
order	**Carnivora**
family	**Felidae**
genus & species	***Felis concolor*** *puma*

Truce at mealtime. Fastidious as most cats are, the puma cleans a paw while the grizzly bear continues to eat the elk. Pumas have few enemies although grizzly bears are said to attack them. In fact the puma is a cowardly cat and a shy animal leading a solitary life hunting deer and small mammals and birds.

Drinks all round! Rabbits gather round a water hole in Australia. With no natural enemies here their numbers have increased to such an extent that widespread destruction of the grasslands has occurred.

Rabbit

The European rabbit was originally a native of southwest Europe and northwest Africa, but has spread northwards largely with human help. First the Romans, then the Normans were blamed for introducing the rabbit to Britain. Now there is evidence that it was not in Britain until the 12th century and was then regarded as a desirable domesticated animal, farmed in warrens and zealously guarded. It used to be classed as a rodent but is now placed in the order Lagomorpha along with hares and pikas.

The rabbit has long ears and large prominent eyes placed well to the sides of the head. The strong hindlegs are longer than the forelegs and provide the main force in running. Instead of having pads on the soles protecting the feet there is a thick coating of hair which gives a firm grip either on hard rock or slippery snow. The tail, which is white below, is very short and turned up at the end. The coat is a mixture of dusky and buff on the upper parts with grizzled white underparts. The fur is of triple formation: there is a dense, soft woolly underfur, through which project the longer and stronger hairs which give the coat its colour, and among these are still longer but more sparsely scattered hairs. The coat becomes thicker in winter. The bucks are slightly larger than the females and may be up to 16 in. long, and weigh up to $4\frac{1}{2}$ lb. The females are distinguished by the form of the head which is longer and more delicately moulded than that of the male. Under a rabbit's chin are glands, larger in the buck than the doe, which produce a secretion used for marking territory. Rabbits will also rub their chins on each other, especially on a mate or on the young, presumably to provide a ready means of recognition.

Rabbits have been extensively bred for food and cs pets and for laboratory experiments. Black rabbits and, more rarely, other colour mutants, turn up occasionally in wild populations and these have been used to produce domesticated breeds some of which bear little resemblance to the original wild stock. There is, in fact, a wide range of colour, and the fur of some breeds is very long and silky. Those bred for their meat may reach a weight of 15 lb. The name 'rabbit' appears to be of northern French derivation, and originally indicated the suckling young, the adults being known as conies.

The rabbit has now been introduced into many other parts of the world including the Ukraine, the Azores, Madeira, Australia, New Zealand, South America and several parts of the United States. In some places where they have been introduced they have no natural enemies, and consequently have caused widespread devastation of the grasslands and robbed the native animals of their food, particularly in Australia.

Underground community

The rabbit lives mainly in grasslands or open woodlands where it digs extensive burrows. They are gregarious and have their burrows close together so the warrens may cover a wide area. Although so famous as a digger the rabbit is not specially built for burrowing, yet where the soil is light the combined efforts of many generations have resulted in extensive and complicated systems of burrows with bolt-runs as emergency exits and stop-runs for nursery use. Although rabbits use the light sand of dunes, or a sandy heath overgrown with furze and heather, they will also drive tunnels into a firm loam or dry clay. They have also been known to burrow deeply into a surface of coal. The forepaws are principally used to loosen the earth which is then kicked back with the hindfeet. If they meet stones that cannot be loosened by the paws, rabbits have been known to remove these with their teeth. Typical tunnels are about 6 in. in diameter, increased at points along their length to 1 ft to provide passing places. The living quarters are always blind chambers leading from the main passages. Adult rabbits use no bedding materials but rest on the bare soil. The rabbit is mainly nocturnal, coming out of its burrow in the evening and returning in the early morning.

Devastation of crops

Almost exclusively vegetarian, the rabbit's chief food is grass and the tender shoots of furze. In winter when herbs are not available it will eat bark. It will devastate crops of vegetables and before the disease myxomatosis drastically cut down its numbers it inflicted heavy losses on the farmer. Occasionally a rabbit will eat snails or earthworms. A rabbit voids two types of droppings, one kind is eaten again, a process known as refection or 'chewing the cud', and the other is discarded at a special latrine outside the burrow.

△ *A rabbit finds a sunny secluded spot among viper's bugloss* ***Echium vulgare*** *on a cliff.*
◁ *A warm fluffy bundle of white angora wool crouches in the grass. There is a wide range of colour and fur among domesticated breeds.*

Prolific breeders

Rabbits are not promiscuous breeders as has long been thought but are polygamous, one buck mating with several does, each doe keeping to her own territory within the warren. Litters of rabbits succeed one another rapidly at intervals of about a month from January to June and there is some sporadic breeding in other months. Mating is preceded by a courtship in which the buck chases the doe. The gestation period is 28 days.

The pregnant doe makes a bed for her young, of hay and straw lined with fur stripped from her underparts. The nest is usually in the blind end of a 'stab' or 'stop', a short burrow about 2 ft long, just under the surface and well away from the main burrow. This is a necessary precaution against the young being killed by the buck. The doe visits the nest to suckle the young for about 3 minutes once in every 24 hours. The entrance to a stop is closed with earth after each visit, and some does will spread dried grass over this to hide it. The litters vary from two or three to eight, the higher numbers being those of the warmer months. The newly-born rabbits are blind, deaf and almost naked. Their ears are closed and have no power of movement until about the 10th day, the eyes opening a day later. In a few more days the young rabbits can run and make short excursions from the nest. They start taking solid food at 16 days and are weaned after 30 days, when they are capable of an independent existence. Until then the mother will defend them from all dangers, including the weasel and ermine (stoat), using her hindfeet against her adversary. The weight at birth is $1-1\frac{1}{4}$ oz which increases to $9\frac{1}{2}$ oz in 3—6 weeks. Sexual maturity is reached in 3–4 months.

The number of young born does not correspond with the number of eggs fertilised. In recent years it has been established that a percentage of the embryos are resorbed, which means there is a degeneration of the embryonic tissues, the substance of which is taken back into the mother's body through the wall of the uterus. Sometimes whole litters are resorbed. Another piece of recent information is that 36 does, under close observation, weaned 280 young, of which 252 disappeared during the season, presumably dying from natural causes.

Surface nesters

There have been reports in the British Isles and Australia of rabbits' nests, with young, on the surface. Surface nesting became pronounced soon after myxomatosis struck the rabbit, and one view put forward was that non-burrowing individuals may have been less prone to the disease and by their multiplication a population was produced with a high ratio of surface nesters.

Thumping on the ground

The rabbit's chief enemies, in addition to man, are the members of the weasel family, rats, owls, buzzards, ravens, crows, black-backed gulls, and a variety of hawks. Badgers dig out the young and foxes also take a large toll, as do cats and dogs. When disturbed an alarm signal is made, usually by an old buck thumping the ground with both hindfeet together, to which all rabbits within earshot respond by dashing towards their burrows. A rabbit, terror-stricken by the imminent attack of an ermine, will utter a loud scream. Apart from this rabbits are normally silent, except for occasional low growls and grunts, although a doe has also been heard to utter low notes when nursing her young.

Results of myxomatosis

In 1954 and 1955 an introduced disease known as myxomatosis, made the wild rabbit a rare animal in western Europe. The disease has now become attenuated, and there are signs of the rabbit recovering, but usually, as soon as numbers begin to build up in any locality, the disease seems to take its toll once again.

The effects of rabbits' feeding on the countryside in the British Isles were highlighted by the changes following the first wave of myxomatosis. The first results noticed, following the drastic reduction in numbers of rabbits, was that in a year or two green lanes became choked with long grass, seedling trees and brambles, while the margins of cultivated fields were not being eaten bare. On downlands not being cropped by sheep the grass grew taller. At the same time many wild flowers became more plentiful, and wild orchids in particular became more abundant.

A report published in 1960 told what happened in the next stage. Unrestricted growth of the grass led to turf becoming less dense and as a result there was a deterioration in the wild flowers that had so suddenly burst into prominence as the numbers of rabbits fell. There was an increase in brambles, gorse and heather. The most surprising feature of the report was the statement that the vegetation and scenery had reverted to its condition before 1840. It seems that until then there were relatively few rabbits outside the carefully guarded warrens where they had been preserved since the twelfth century in a state of semi-domestication.

class	**Mammalia**
order	**Lagomorpha**
family	**Leporidae**
genus & species	***Oryctolagus cuniculus*** *rabbit*

Raccoon

Commonly known as 'coons', raccoons are one of the most familiar North American animals, if only in folklore and stories. Their adaptability has allowed them to withstand drastic changes in the countryside while their intelligence, cleanliness and appealing looks have combined to make them popular. Their head and body length is 16–24 in. with a tail of 8–16 in. and they weigh up to 45 lb. Their fur is grey to black with black rings on the tail and a distinctive black 'burglar mask' over their eyes. Their feet have long toes and the front paws are almost hand-like and very dexterous.

Raccoons are relatives of pandas, kinkajous and coatis. There are seven species, the best known ranges from Canada to Central America. The crab-eating raccoon lives in southern Costa Rica, Panama and the northern regions of South America. The other species are found on islands.

Adaptable coons

Raccoons originally lived in woods and brushy country, usually near water, but as the woods have been cut down they have adapted to life in open country. They are solitary, each one living in a home range of about 4 acres, with a den in a hollow tree or in a rock crevice. They come out more at night, and are good climbers and swimmers. In the northern part of their range raccoons grow a thick coat and sleep through cold spells. The raccoons of southern USA and southwards, are active throughout the year. Where trees have been cut down raccoons move into fox burrows or barns and they have been known to spread into towns, even to the middle of cities where they live in attics and sheds and raid garbage bins for food.

Raiding garbage bins is one of the raccoon's less popular traits. Apart from the mess, the bins are sometimes carried away bodily. There are stories of ropes securing the bins being untied, rather than bitten through. This is evidence of the raccoon's extreme dexterity. They use their hands almost as skilfully as monkeys; experiments have shown that their sense of touch is very well developed.

Varied diet

Raccoons eat a very wide variety of both plant and animal food. It is the ability to take so many kinds of food that is probably the secret of the raccoon's success and of its ability to survive changes in the countryside. Raccoons are primarily carnivores; earthworms, insects, frogs and other small creatures are included in their diet, and raccoons also search in swamps and streams for crayfish and along the shore for shellfish. The eggs and chicks of birds, both ground and tree nesters, are eaten and raccoons are sometimes pests on poultry

Appealing look from large bundle of fur—a racoon up a tree.

farms and in waterfowl breeding grounds. They are also pests on agricultural land because they invade fields of corn, ripping off the ears and scattering them half-eaten. Fruits, berries and nuts are also eaten.

Irresponsible fathers

Raccoons mate in January or February, each male mating with several females then leaving them to raise the family. The young, usually 3 or 4 in a litter, are born from April to June, after 60–70 days gestation. They weigh 2½ oz at birth and are clad in a coat of fuzzy fur, already bearing the characteristic black mask. Their eyes open in 18 days and at about 10 weeks they emerge from the nest for short trips with their mother. The trips get longer as the young learn to forage for themselves but they stay with their mother until about one year old. Raccoons live as long as 13 years.

Coon currency

Raccoons are a match for most predators and when hunted with dogs the raccoon may come off best, especially if it can lure the hound into water and drown it. Raccoons have always been trapped and hunted in large numbers by Indians and Europeans, both for their hard-wearing fur and because of their attacks on crops. Their fur was the main cause for killing them and even in the 17th century efforts were made by imposing taxes and bans to prevent too many raccoon pelts from being exported. At one time the skins were used as currency and when the frontiersmen of Tennessee set up the State of Franklin, the secretary to the governor received 500 coonskins a year while each member of the assembly drew three a day. Nowadays coonskin is not valuable unless there is a sudden fashion as there was for coonskin hats following the film on Davy Crockett, King of the Wild Frontier.

Why so fastidious?

In the *Systema Naturae* Linnaeus called the raccoon *Ursus* (later *Procyon*) *lotor,* or the 'washing bear'. In other languages the raccoon is similarly named *ratons laveur, ositos lavadores* and *Waschbaren.* Their names testify to the strange habit raccoons have of appearing to wash their food before eating it. This apparently hygienic behaviour has become part of the raccoon folklore and only recently have proper attempts been made to explain it.

Some books state that raccoons always wash their food, others say that the habit may be more common in captive animals, yet there are no authentic reports of food washing in the wild. Naturalists who have studied raccoons deny ever seeing this take place. Food washing must, therefore, be an unnatural habit of captive raccoons.

The first scientific study of food washing was made by Malcolm Lyall-Watson at London Zoo. First, he showed that raccoons do not really wash their food but immerse it, manipulate it, then retrieve it. He suggested that the habit should, therefore, be called dousing.

Lyall-Watson gave a large variety of foods to a number of raccoons. Animal food was doused more often than plant, yet earthworms, the only food that needed cleaning, were doused least of all. In another series of experiments it was shown that the shape, smell and size of food objects governed dousing to some extent, but most important was the distance of the food from water. The nearer the water, the more likely is food to be doused.

The conclusion drawn by Lyall-Watson explains why dousing only occurs in captivity. In the wild, raccoons feed on food found on land and food found in water. In captivity all their food is on land, so they 'go through the motions' of foraging in water by taking their food to water, dropping it then searching for it in an action that has for so long been described as washing. Similar behaviour is seen in captive cats. When presented with dead animals, they often throw them about and pounce on them, 'pretending' to hunt them.

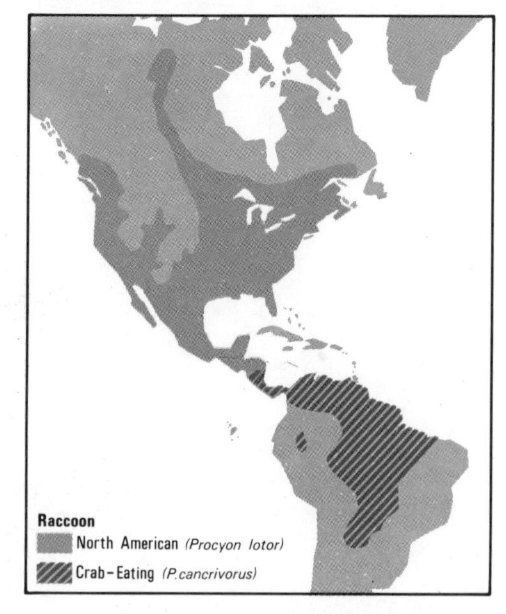

class	**Mammalia**
order	**Carnivora**
family	**Procyonidae**
genus & species	***Procyon lotor*** *North American raccoon* ***P. cancrivorus*** *crab-eating raccoon* *others*

▽ *An unsuspecting racoon enjoys the shallow waters, unaware of the threatened danger of a puma waiting for the moment to pounce.*

Returning to their earth after a hunt: North American red fox and two cubs. Night hunting is a way of teaching the cubs how to fend for themselves.

Red fox

It is usually assumed that, but for its careful preservation by the various 'hunts', the red fox would have become extinct long ago in the British Isles except in the wildest and most remote corners. For centuries it has been persecuted outside the hunt areas because of its alleged poultry-killing habits and even today the killing of a fox is still looked on with approval. Yet, in spite of all this, the fox has survived and at times is unusually numerous.

The head and body of the red fox measure just over 2 ft with a 16in. tail, but there are records which greatly exceed these measurements, especially in Scotland. A well grown fox stands only about 14 in. at the shoulder. The dog-fox and vixen are alike except that the vixen is slightly smaller and has a narrower face as she lacks the cheek ruffs of the male. The fur is sandy, russet or red-brown above and white on the underparts. The backs of the ears are black, as are the fronts of the legs, but these may be brown, and can change from one colour to the other with the moult. The colours may vary, however, not only between one individual and another, but in the same individual from season to season. The foxes (Tods) of Scotland, although of the same species, usually have greyer fur than the English fox. When fully haired the tail is known as a brush. The tip (or tag) is white but may be black. Weights vary considerably but on average a dog-fox weighs 15 lb, a vixen 12 lb.

The sharp-pointed muzzle, the erect ears and quick movements of the eye with its elliptical pupil combine to give the fox an alert, cunning appearance, so many stories of its astuteness have been invented in the past. At the moult, in July and August, foxes lose their characteristic appearance and look thin-bodied, long-legged and slender of tail.

The red fox ranges over Europe and over Asia as far south as central India, as well as northwest Africa. It is found throughout the British Isles, except for Orkney, Shetland and all Scottish islands, but not Skye. In central Asia it lives up to 14 000 ft above sea-level. The North American red fox ***Vulpes fulva*** *is very like the Old World red fox in build and habits. There are several mutants, the cross fox is red with a black band across the shoulders, and the silver fox has a lustrous black coat with white tips to the guard hairs.*

Tree-climbing foxes

The red fox's traditional cunning is a reflection of its adaptability. It prefers wooded or bushy areas but is found in a variety of habitats. Many foxes today are even found living in urban areas or even near large towns, where they probably live off rats and mice and scavenge in dustbins. Although the fox lives mainly on the ground there are many instances of it climbing trees. Usually this occurs when a tree is leaning or when there is a trailing bough that has broken and is hanging down to the ground, up which the fox can clamber. There is one recorded instance, however, of a fox having its sleeping nest at the top of a bole of an elm, 14 ft from the ground, with no branches between it and the ground. Foxes are largely nocturnal, but they can often be seen during the day. Except at the breeding season the dog-fox and vixen lead solitary lives. Most of the day is spent in an 'earth' which is more of a cavity in the ground than a burrow. They may make this themselves or use a badger's set or rabbit burrow.

Foxes use a great variety of calls, the most familiar being the barking of both the dog-fox and the vixen in winter and the screaming of the vixen, generally during the breeding season. It has now been established that, contrary to common opinion, the dog-fox may also scream sometimes.

Poultry killer?

A great deal has been written about the fox prowling round farms looking for an opportunity to kill an unguarded fowl. Certainly foxes will take poultry and they will take lambs, but these habits tend to be local. A vixen that has taken to killing poultry will teach her cubs to do the same. But not all foxes are habitual poultry stealers and there have been instances of foxes repeatedly visiting poultry farms or private gardens containing a few poultry and never molesting them.

More solid information about their food comes from a Ministry of Agriculture investigation of the stomach contents of dead foxes. This showed that, now that rabbits are scarce, the chief items of food are rats, mice and bank voles. Hedgehogs, squirrels, voles, frogs, even snails and beetles are, however, also eaten, as well as a great deal of vegetable matter. Birds such as partridges and pheasants will also be taken. A fox will soon discover offal or carrion, even if buried 2 ft in the earth. Foxes also visit dustbins and a feature of the many foxes now living in towns is that they have turned scavenger. Railway marshalling yards also have their foxes, probably feeding on food thrown out from restaurant cars or on rats living on this food.

Teaching the cubs

Mating takes place from late December to February. The gestation period is 51–52 days. About April the vixen produces her single litter for the year, usually of four cubs. They are blind until 10 days old, and remain in the earth until nearly a month old, the vixen staying close beside them, while the dog plays a large part in supplying the food. When about a month old the cubs come out in the evening and can be seen playing as a group with the parents outside the earth. This continues for several weeks.

After the cubs are weaned it has been noted, in semi-captivity, that the dog-fox continues to bring food for them and the cubs will take the food from his mouth themselves, or the vixen may take it and the cubs take it from her mouth. The cubs have to jump up to reach the parent's mouth and all the time the parent is moving its head, from side to side or up and down. In this way the cubs are being exercised so developing their limbs, and also learning to coordinate movements and senses. During this time the dog plays a great deal with them, more so than the vixen.

Later the vixen takes them hunting at night, so they learn from her example how to fend for themselves. The cubs leave their parents when about 2 months old, reach adult size 6 months after birth, and become sexually mature in their first winter.

'Charming'

Foxes are credited with resorting to a particular stratagem, called 'charming', to attain their end. A story is usually told of a fox which, seeing a party of rabbits feeding and knowing that they will bolt to their holes on its approach, starts rolling about at a safe distance to attract their attention. Then like a kitten it begins chasing its tail, while the rabbits gaze, apparently spellbound, at the performance. The fox continues without a pause, as though oblivious to the presence of spectators, but all the time it is contriving to get nearer, until a sudden straightening of the body enables it to grab the nearest rabbit in its jaws.

There are too many authentic accounts of foxes charming to leave much doubt about the matter. From these, a more likely explanation evolves: foxes are naturally playful. Like some other mammals they will, without obvious cause, suddenly behave as if they have taken leave of their senses, bounding about, bucking, somersaulting, and so on. Rabbits and birds on seeing these antics are drawn to watch out of curiosity. If the fox is hungry then the spectators suffer. It is possible that a fox playing in this way and finding birds and rabbits attracted to it, might use this tactic again, deliberately. Such learning by experience would not be beyond a fox's intelligence, but there is much to be said for the view that charming, as such, is not primarily a deliberate stratagem.

class	**Mammalia**
order	**Carnivora**
family	**Canidae**
genus & species	***Vulpes vulpes*** *European red fox*

▷ With ears pricked, wary eyes glinting from its mask, the red fox with its magnificent brush is a very wily, sometimes vicious, and yet most handsome animal.

Reindeer

Some scientists regard the reindeer of Northern Europe and the caribou of North America as varieties of one species. The reindeer, of arctic Europe and Asia, are now domesticated or at least semi-domesticated. They are up to 43 in. at the shoulder, and a good bull may weigh 224 lb. Reindeer are perhaps the tamest of all domesticated animals, and it is said that even a child can manage a herd of them. The domestication of the reindeer is thought to have begun in the 5th century when they were used as decoys for hunting wild reindeer. The hunter, with four or five tame beasts on ropes, and himself in the middle, would approach a wild herd without alarming them, and then loose his arrows at short range. One tribe in Siberia is known to have used tame hinds in the rutting season to attract wild stags, which were then shot. In time, the tame hinds produced enough fawns to build up herds, and the domestication of the reindeer gradually came about in this way.

Reindeer are all profit

Reindeer are to the Lapps and the northern tribes of the USSR what cattle were to early man farther south. They provide everything he needs. The hide makes a soft leather used for clothing and a variety of other purposes, such as cushions and curtains. Reindeer give milk, cheese and flesh. Their sinews can be used for sewing boots or for covering a canoe, and their bones for needles. The stretched bowel makes a window covering or bag for minced meat.

Reindeer are used as pack animals or for drawing sleighs, and are most economical animals to keep as they can withstand exposure and so need no stabling, and they can forage for themselves, digging down through the snow for the spongy lichen known as reindeer moss. Today, however, the use of reindeer products is declining in favour of manufactured goods or preserved foods especially where towns are accessible.

Man the parasite

The reindeer is a nomad like the people who domesticated it. One valuable quality of the reindeer is its ability to find its way in a snowstorm, and this and its nomadism, together with its many other uses, has made some scientists speak of man's social parasitism on the reindeer. In the association between the animal and the human, the animal gains little besides protection from enemies—an advantage largely offset by man killing some of them for food. Otherwise the advantages are all with the human.

The reindeer's achievements as a draught animal are greater than those of horse or dog over uneven or frozen ground. A reindeer will pull 300 lb at an average of 8 mph, and while a daily journey averages 25–35 miles, up to 75–100 miles a day have been recorded. In an annual race in Sweden

▷ *Reindeer, the cattle of the Scandinavian Arctic, provide all the Lapps' requirements.*

two reindeer pulled a sledge and driver 5 miles in 14½ minutes. At a reindeer festival held on Christmas day on the Kola peninsula a mile course has been covered in 2½ minutes. At Nome in Alaska in an annual race a 10-mile course was invariably covered in under half an hour.

Why reindeer cows have antlers

Because reindeer are domesticated it has been possible to experiment on some aspects of their behaviour. For example, why do female reindeer have antlers, the only deer apart from caribou to do so? A possible answer lies in an experiment carried out in 1962-3 by Yngve Espmark of the University of Stockholm. A group of 16 reindeer were marked and kept in an enclosure, and at the same time observations were made on free-ranging reindeer. The social hierarchy—usually called the peck-order—changes during the course of a season, but generally speaking the larger the antlers the higher the male is in the social order. During the rut the mature bulls, with their large antlers, are the bosses. After the rut the males shed their antlers before the females, and then the females become boss. Moreover, each calf shares its mother's rank. The importance of the antlers was tested by cutting off the antlers of some of the males, who then dropped in the social ranking. One male was castrated and from what is known in other species he should have fallen to a very subordinate rank. He did not and all the circumstances suggest that an old male, experienced in fighting and in holding his rank, has learned to 'know his own strength', which may explain how stags without antlers hold their own.

During the winter the calf stays with its mother and feeds from the 'crater' she makes in the snow. If she had no antlers there is a fair chance both might be driven from the crater by other members of the herd looking for an easy meal, and the calf would starve.

Epic arctic journey

Reindeer once took part in an historic drive across Arctic Canada. In the 1890's a herd of 171 was taken into Alaska from Siberia to save the Eskimos from starvation. These

flourished and increased in numbers, but many mistakes were made, such as poor herding, and by the mid 1940's the number had dropped to 120 000, from a peak of nearly a million 10 years previously. In 1929 the Canadian government introduced reindeer into its Northwestern Territory to build up an industry for the Eskimos there. A herd of 3 400 left Kotzebue Sound in Alaska under the supervision of a Lapp, Andrew Bahr, who delivered 2 370 animals at the Mackenzie River delta 5 years later. The journey was made across unknown territory and included crossing a mountain range, fording one river after another and by-passing lakes. Wolves harried the flanks of the herd, taking a continual toll. Plagues of mosquitoes in summer sometimes held the herd up as did the blizzards in winter, for the journey was made well within the Arctic Circle. On arrival less than 20% of the original herd remained, the rest having been born on the journey.

◁ *Man and his beasts; or beasts and their man? Many think the reindeer domesticated the Lapps.*
◁▽ *Reindeer keep their heads above water.*
▽ *Reindeer crocodile. A Lapp leads the way.*
▽▽ *Reindeer camp. They do not need stabling.*

class	**Mammalia**
order	**Artiodactyla**
family	**Cervidae**
genus & species	***Rangifer tarandus*** *reindeer*

Rhinoceros

Today, rhinos are a vanishing breed, not because they are evolutionary failures—a common fallacy—but because of extermination by man. Like its relatives the horse and the tapir, the rhinoceros bears the weight of each leg on a single central toe, but it has two subsidiary toes on each foot. It differs from them in its bulky build, its thick skin which is naked or sparsely haired, and the one or two horns on its nose.

The five species form four genera in two subfamilies. They differ more from one another than, say, horses from zebras. The Asiatic rhinos, subfamily Rhinocerotinae, all tend to be hairy even in the adult. The horn or horns are short, the skin is thrown into folds superficially like armour-plating, and there are long tusk-like lower canines. The Indian rhinoceros, up to 6 ft 4 in. high and $2\frac{1}{2}$ tons weight, has a deeply folded skin studded with raised knobs, not much body-hair except a stiff brush on the tail-tip, and a single horn. The Javan rhino, $4\frac{1}{2}-5\frac{1}{2}$ ft high and up to 2 tons, has a less heavily folded skin with differently arranged folds in the shoulder region, and the skin is broken up by a network of cracks into mosaic-like polygons, with a short hair in the centre of each, at least in the young. It also has only one horn, which is very short and may be lacking in the female. The Sumatran rhino, up to $4\frac{1}{2}$ ft high, weighing only 750—2 000 lb, has marked skin-folds on the front half of the body only, is distinctly hairy throughout life and has two very short horns. Both Javan and Sumatran rhinos used to be found throughout much of southeast Asia. Today the Sumatran rhino is found in scattered pockets over most of its former range; while the Javan is not known outside a small reserve in Java itself.

The African rhinos, subfamily

Dicerotinae, are quite hairless except for the tail-tip, ear-rims and eyelashes, although white rhino young are quite hairy up to the age of 4 months. They also differ from the Asiatic rhinos in having two long horns, quite close one behind the other, in lacking front teeth and in having skin much less deeply folded. The two species are both plains dwellers. The black rhino is up to 5½ ft high. Its skin is rough with grooves over the ribs, and the lip is pointed and prehensile. The white rhino is up to 6 ft 4 in. high, has a smoother skin with no rib grooves, a hump on the shoulder, and a long head with a wide, square mouth.

Scent signals

All rhinos are short-sighted but have acute senses of smell and hearing. They tend to deposit their dung in communal heaps, each individual that passes adds to the pile until heaps 4 ft high and 20 ft across are formed.

Black rhinos are solitary: 80% of adults and 50% of weaned calves are seen alone, although the latter not uncommonly seek the company of one or two other immature animals. The size of an individual's home range varies according to the environment. In the fertile Ngorongoro crater the average is 6 sq miles (although immature rhinos have bigger ranges) but in the drier habitat of Olduvai gorge it is 8½ – 14 sq miles, while in the subdesert climate of Tsavo Park East,

△ *Big game. Black rhinos are normally solitary and are rarely seen together. They live on the African plains, each keeping to its home range where it browses on low foliage which it collects with its mobile prehensile upper lip.*

Kenya, rhinos are purely nomadic and no proper home range can be made out.

Rhinos are active mainly in the morning and evening. They wallow in mud in the evening to cool off. In Ngorongoro their movements are very regular, and they can often be seen at the same time in the same place, every day. But in Tsavo the black rhino wanders along well-worn paths over a much wider range, and goes to water only every 4 – 6 days.

Social niceties

The home ranges overlap very widely. When two males or two females meet they are wary of each other but may form temporary associations. A male approaches a female with short, cautious steps, often swinging his head from side to side. The female may approach him or even charge him, and he wheels round, gallops off and circles to approach again. This may continue for several hours. When a completely strange rhino enters an area where several residents have their home ranges, the residents threaten the newcomer with lowered heads, rolling eyes, flattened ears and raised tails. Now and again one of them curls its lip and gives vent to a shrill groaning sound, the stranger remaining silent. One of the residents will charge him, stopping short in front of him, and the two will tussle, using their horns as clubs or pikes. If the stranger retreats he will be pursued, perhaps for as much as a mile, but he may eventually establish himself in the area.

A rhino, on coming up to a dung-pile, sniffs it, may push his horn into it, and then snuffles through it with his legs held stiff. After defaecation, the dung is scattered with kicks from the hindlegs. This smears its feet with its own scent and that of its neighbours. Goddard found that he could get rhinos to follow the trail of dung if he dragged it behind a Land-Rover. The exact trail was always followed, even zigzagging where the vehicle had done so. Most of the rhinos followed their own dung or that of animals sharing part of their home range.

△ *Sumatran rhinos, which numbered about 170 in 1966, have been ruthlessly hunted for medicinal, magical and religious purposes.*
◁ *Advancing rhinos. The square-lipped white rhino has a long head with a hump on the shoulder and a relatively smooth skin.*
▷ *Rough-skinned black rhinos with their pointed lips. They are hairless except for the tips of their ears and tails and their eyelashes.*
▽ *In 1966 the total population of Indian rhinos numbered 740, a slight increase in recent years, due to stringent protection.*

Black and white rhinos

The black rhino eats the shoots and twigs of low-growing bushes, such as mimosa and buffalo thorn: dry, thorny woody food, which it gathers with its prehensile upper lip. It often feeds at night. It is a nervous animal, easily startled, but what is taken for aggression is as often as not the result of a stampeded animal running in the wrong direction. There is often, however, a real charge when the animal gallops, head up, towards the intruder, lowering its head at the last moment to strike with its horns. When pressed it can run at 35 mph, but it normally charges at 20 mph.

By contrast, male white rhinos are much more gregarious, and far less truculent. The females are less gregarious, half of them being found accompanied only by their calves; but two or more females with their young may move around together. The white rhinos live in low-lying plains. They feed entirely on grass, are very unwary, and so were quickly exterminated over most of their range. When alarmed, they curl their tails in a loop over the rump and lumber away, reaching a trotting speed of 18 mph. They can also canter at 25 mph.

Asiatic rhinos

The Indian rhino lives in the 12ft high elephant grass, through which it makes well-worn tunnels. The waterlogged meadows and swamps may be divided into territories or used as common land. A common bathing pool may be used at the same time by up to nine rhinos of all ages and both

◁▷ *Armour plating? The back and front of an Indian rhino display its great folds of skin. It can be distinguished by its skin which is studded with raised knobs and is quite hairless.*

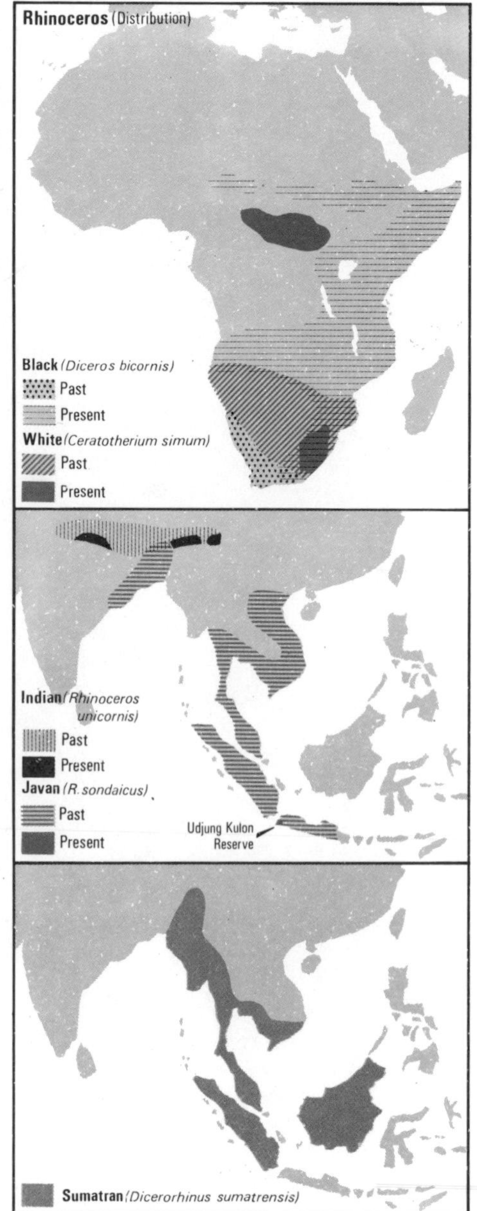

sexes, most of them lying jammed up together in it. The dung-heaps are near wallows or pools, or mark the entrances of tunnels. Unlike the African species, Indian rhinos have glands in the forefeet which leave scent trails in addition to those of the dung. Indian rhinos eat young grass, twigs and water-plants (especially water-hyacinth, which often forms a carpet on the bathing pools). They run at 25 mph, with the tail held down, are active during the day, sleep, hidden away in the high grass, from midnight till dawn and again at midday.

Javan rhinos live in deep forest, preferring areas of secondary growth. They feed mainly on young saplings, pushing against them till they break and eating only a little from the crown. The Javan rhino, especially the male, has a home range which it leaves every day on wide, regular wanderings. During the rains they bathe and wallow in streams, but in the dry season they use moist low-lying places or larger rivers, estuaries and the sea. Dung is deposited in large heaps as in the Indian rhino, often more scattered, and sometimes in streams. Bulls have another method of scent-marking, using highly pungent orange urine spattered over the bushes.

The habits of the Sumatran rhino seem to be similar to those of the Javan. It eats branches and shoots, leaves, fruits, lichens and fungi. It has a reputation as a mountaineer, having been found as high as 8 580 ft in Java. Where the Sumatran and Javan species were found together, the Javan was much more of a lowlander, while the Sumatran tended to visit the lowlands only towards the end of the rains.

Nursed for a year

Before mating, the males may fight over the female. In the black rhino one will charge, halting suddenly 5 yd away, and the two joust or try to club each other with their horns. In the white rhino, generally less aggressive, these fights may be lethal.

The gestation period averages 455 days for the black rhino, 486 for the Indian, 547 for the white. Birth takes about 15 minutes, and the infant stands up after about an hour. Only in the white rhino have twins been recorded. The calf may begin to graze at one week, but suckles for at least a year. Young African rhinos are 20 in. high and weigh 44–55 lb at birth, young Asiatic rhinos stand 2 ft at the shoulder and weigh up to 50 lb. A small place for the front horn is visible at birth, the horn beginning to grow in about 5 weeks and reaching $1\frac{1}{2}$ in. long in 5 months. Females are sexually mature at 3 years, males at 7. Rhinos have lived 45 years in captivity.

Attacks with teeth and horn

A rhinoceros's main enemy is man—and this is enough! In Africa, lions take young rhinos when they can, even occasional adults, and adults caught in snares or stuck in the mud have been mauled by hyaenas. There is supposed to be a traditional enmity between rhinos and elephants and the famous hunter FC Selous recorded the exceptional sight of a rhinoceros being pulled into the water by a crocodile.

African rhinos attack with their long horns, tossing the victim several times. Asiatic rhinos fight more with their tusks, slashing with effect. The Sumatran rhino uses its teeth but is said occasionally to toss and trample its victims. The black rhino in Africa and the Sumatran rhino in Asia have received a not altogether fair reputation for being very aggressive.

Three-horned rhinos

Rhinoceros horn is made of tubular horny fibres secreted from the skin of the nose and firmly cemented together. Some black rhinos grow three horns, one behind the other and there used to be a high proportion of three-horned rhinos around Lake Young in Zambia. The record length for a black rhino's horn is 54 in., for a white rhino's 65 in. When the front horn becomes very long, it often grows as much forward as upward. Two famous black rhinos with such forward-pointing horns were Gertie and Gladys, both of Amboseli National Park. In 1959, both of them broke off most of their horns, but they began to grow back very quickly. In January 1962, Gladys was killed by poachers, and one of the most photogenic animals in Africa was gone.

class	**Mammalia**
order	**Perissodactyla**
family	**Rhinocerotidae**
genera & species	***Ceratotherium simum*** *white rhinoceros* ***Dicerorhinus sumatrensis*** *Sumatran rhinoceros* ***Diceros bicornis*** *black rhinoceros* ***Rhinoceros sondaicus*** *Javan rhinoceros* ***R. unicornis*** *Indian rhinoceros*

Saki

Sakis are South American monkeys with unusual hair styles. They have bushy tails and thick dark fur as well as the elaborate hair-patterns on the head. Like all New World monkeys, they have broad noses with nostrils pointing sideways and separated by a wide septum. Like many of their relatives they are unusual in grasping objects between the second and third fingers, the thumb being 'just a finger', not opposable as it is in Old World monkeys and apes. Sakis have nails on their fingers and toes, although these are convex and rather pointed, almost claw-like. Their lower incisors are long and lean forward, a characteristic shared only with the related uakaris (page 226). The white-faced saki has a black coat, and a hood of hair originating in a whorl on the nape of the neck hangs over the forehead. The female's coat is more brindled. The sexes are very distinctive: the male's face is white, the female's dark, with white streaks from the sides of the nose to the corners of the mouth. Both sexes of the related monk saki are similar to the female white-faced saki.

Another group of sakis with less distinctive faces have really extravagant hair-styles. The white-nosed saki is black with a pinkish-white nose and upper lip. Its hair lies forward in a fringe over the forehead. The black saki has a parting on the crown, on either side of which the hair rises in a 'bouffant' style, and it has a bushy beard. These two species are larger than the other two, weighing about 6–7 lb, whereas the white-faced and monk sakis weigh only 3–3½ lb. Sakis grow up to 40 in. long including the tail.

Sakis occur from the Orinoco river north of the Amazon to the Guianas, and around the upper Amazon except for the white-nosed saki which is restricted to an area south of the Amazon between the rivers Madeira and Xingu.

Hairy hand as a sponge

Sakis are active tree-dwelling monkeys. They move around by day, feeding mainly on fruit. Although spending most of their time in the high canopy, they descend at times to the shrub layer to feed. They drink by soaking the fur on the backs of their hands in water and licking it. In captivity, however, they will learn to drink with their lips. The white-faced and monk sakis are said to live more on the forest borders, but the other two reportedly live deeper in the forest, especially along rivers where the forest grows very thick, due to the greater amount of daylight reaching the ground-vegetation.

Little is known about the general behaviour of sakis. They probably live in pairs and defend their territories like titis

▷ *Up in the clouds: monk saki holds on tight.*

and marmosets. They jump and run through the trees, sometimes hanging by their arms. The young at first cling to the mother's belly—unusual, for most New World monkeys sit on their mothers' backs.

Devoted pet

The black saki was well known to Humboldt, the 18th-century German naturalist, and his contemporaries, who were struck with its manlike face and hair-style. Broderip, who refers to it as the 'Capuchin of the Orinoco', says it is fierce, with a mixture of melancholy and ferocity in its eyes; when angry it rises on its hindlegs and grinds its teeth. He also says its habit of drinking from its hands is supposed to derive from an anxiety not to wet its beard: 'If any malicious person wishes to see this Homunculus in a most devouring rage, let him wet the Capuchin's beard, and he will find that such an act is an unforgivable sin.' On the other hand, early explorers found the monk saki easy to tame. One was said by Bates to follow its master around like a dog, and to sit on his shoulder during the day. It protected its master against everyone else—even other members of the household—and was most deeply attached to him. When one morning its owner left the house without it, the little fellow set out in search of him. They always used to pay a call on Bates during the morning, and the saki headed straight for Bates's house looking for its master. A neighbour observed it taking a short cut over gardens and through thickets, a route it had never taken before, having always gone the long way through the streets with its master. Finally it got to Bates's house; its master was not there, so it climbed onto a table and sat 'with an air of quiet resignation' waiting for him.

class	**Mammalia**
order	**Primates**
family	**Cebidae**
genera & species	***Chiropotes albinasus*** *white-nosed saki* ***C. satanas*** *black saki* ***Pithecia monachus*** *monk saki* ***P. pithecia*** *white-faced saki*

▽ *White-faced saki with a dour expression.*

Sea lion

The sea lion familiar in circus performances is the Californian sea lion, the smallest of the five species, all of which look very like the fur seals.

Both sea lions and fur seals are eared seals with small external ears, and both turn the hind flippers forward to move by bounding on their flippers. Sea lions, however, have broader muzzles than fur seals. The Californian sea lion is the smallest sea lion, the males measuring 7 ft and females about 6 ft. The males lack the lion-like mane characteristic of other sea lions. When wet the fur appears black but it dries to a chocolate brown. Californian sea lions are found on the coasts of California and northern Mexico, especially on offshore islands, on the Galapagos Islands and islands off the Japanese coast.

The other northern sea lion is Steller's sea lion that lives in the North Pacific, from Japan around to California. Adult males measure 10½ ft and weigh up to 1 ton. Females are very much smaller, measuring about 7 ft 6 in. and weighing only 600 lb. Adult males develop very thick necks and shaggy manes. The southern sea lions are similar to the Steller's sea lion but are smaller, except for the Australian sea lion. The Australian sea lion is found on the coasts of south-west Australia and Hooker's sea lion is restricted to Auckland, Campbell and Snares Islands, to the south of New Zealand. The southern sea lion is found on the coasts of South America, from northern Peru, around Cape Horn to southern Brazil and the Falkland Islands.

Outside the breeding season sea lions live in large mixed herds on rocky shores, but some migrate. Californian sea lions and northern sea lions often move northwards in winter, some of the latter reaching the Bering Strait, but they return south when the sea freezes.

▷ *The return to the sea. A few mammals that once lived on land have returned to the sea. Among these is the sea lion, its name belying its link with the land. The sea lions seen here along the turbulent Oregon coast are Steller's sea lions. They are gregarious animals, inquisitive and intelligent. Although they are easy to tame they can be dangerous in their 'rookeries'.*

Rubber ribs

Except when defending their pups or territories, sea lions are usually quite tame and it is possible to walk close to them. In fact, they are more wary than aggressive and sometimes a whole herd will panic and rush into the sea. On reasonably smooth ground they can outpace a man but are less manoeuvrable. In the Falkland Islands sea lions trample paths through the tall, dense tussac grass and, when disturbed, they rush down these paths to safety in the sea. A man walking along such a path must keep a watch behind to avoid being run over by a sea lion in its headlong rush. Sea lions display remarkable agility in leaping over rocks and down steep slopes and they may even jump over a low cliff. The shock of the fall is absorbed by the front flippers, the blubber and the soft, cartilaginous ribs. The last are presumably an advantage when swimming around rocks in heavy seas.

Sea lions' profit and loss

Although their eyesight is bad sea lions are expert fishermen. In murky water they probably use their long whiskers to detect their prey and other objects. Recently it has been found that sea lions, and probably other seals, use echo-location or sonar, but this is not as efficient even as the seals' eyesight. Steller's sea lion, which has been studied in detail, eats a wide variety of food including squid, herring, pollack, halibut, sculpin and salmon. Wherever sea lions live near a commercial fishery they are blamed for damage both to fish and equipment, but Steller's sea lion sometimes offsets this damage by feeding on lampreys, a pest of salmon. Californian sea lions, also unpopular with fishermen, appear to prefer squid to salmon. Around the Falkland Islands crustaceans and squid are the main food but fish are also eaten.

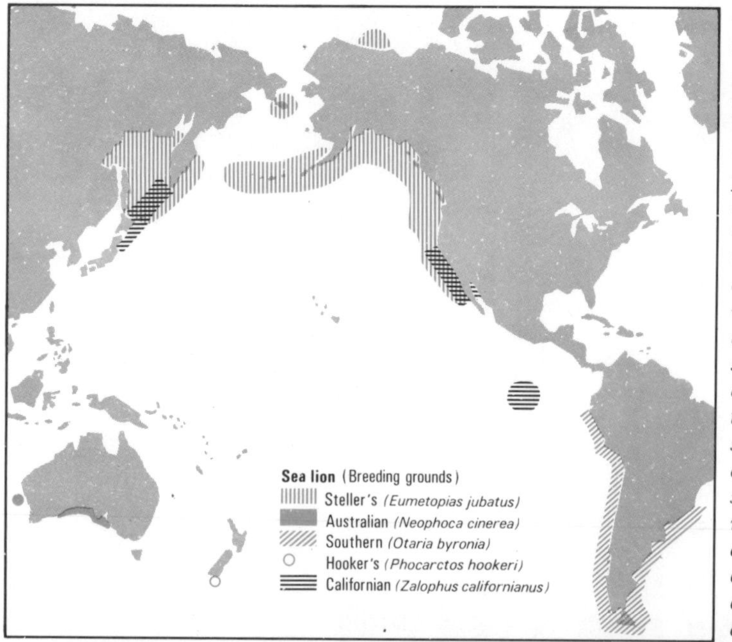

◁ Map showing the scattered distribution of the five species of sea lion.

▽ Australian sea lion, mother and young. Breeding takes place from October to December when the whole community is ashore, usually on a sandy beach. There is a single young which is carefully guarded by the mother. This species does not fast during the breeding season but feeds mainly on penguins caught as they come out of their shelters or as they waddle across the beach.

Breeding habits

At the start of the breeding season, the mixed herds split up as each mature bull attempts to stake out a territory. As with elephant seals and fur seals, only the old bulls can form a territory; the younger bulls are driven off the beach and spend their time just offshore or on common ground where all the sea lions mix. Each bull gathers a harem of 10–20 cows in his territory. The cows come ashore 2–3 weeks after the bulls, when the territory boundaries have been decided and most of the serious fighting has finished. Within a day or two the single pups are born. A few days later the cows mate again with the bull in whose territory they have pupped and then return to the sea to feed, returning every so often to suckle their pups. The pups can swim at an early age and are suckled for up to a year. Many die in storms, although their mothers help them, carrying them by their scruffs away from the waves. Depending on the species, females are sexually mature at 3–4 years and males at 4–5 years, although the males do not mate until large enough to hold a territory.

Performing sea lions

Californian sea lions are familiar to everyone as the 'performing seals' at circuses. Many seals are naturally playful but the sea lions are more fun-loving than most. They have even been seen chasing their own air bubbles as they float to the surface. The idea that they balance stones on their noses in the wild is, however, very far-fetched. The familiar act of balancing a ball has to be taught by repeatedly throwing a ball at a sea lion until it balances, when a reward of fish is given. Eventually the sea lion understands what it is supposed to do.

A new use for performing seals is running errands between underwater laboratories and the surface. So far wild sea lions have been taught to come for food at a signal but carrying messages is more difficult.

class	**Mammalia**
order	**Pinnipedia**
family	**Otariidae**
genera & species	***Eumetopias jubatus*** *Steller's sea lion* ***Neophoca cinerea*** *Australian sea lion* ***Otaria byronia*** *southern sea lion* ***Phocarctos hookeri*** *Hooker's sea lion* ***Zalophus californianus*** *Californian sea lion*

Sheep droving in Australia. Commercial sheep flocks of many thousands, such as this one, are kept on the open ranges in Australia and America.

Sheep

The sheep, next to the goat, was the earliest ruminant to be controlled by man, probably in southwest Asia as early as 12 000 years ago. Very little is known about the wild ancestor of the domesticated sheep but it is thought that the urial, the wild sheep of southern Asia, was the first to be domesticated. There are, however, other theories, one being that domestic sheep are descended from species of wild sheep that have since died out, but it is more likely that the urial and the mouflon and possibly the argali gave us the sheep we know today, but in what proportions these contributed is far from certain.

There is a fairly constant ratio of one sheep to every three human beings; the world sheep population being approximately 1 000 million.

Today there are about 450 breeds of domestic sheep. With normal variations and selective breeding, with crossings of different breeds, mutations, and with the production of many strains to suit particular geographical conditions, it is not surprising that the shape, size and colour of sheep should present such a bewildering variety. There are breeds of domestic sheep for every kind of pasture, for most climates and most altitudes. There are tall, short, fat, lean, grey, or tawny sheep; some are hornless, some have two horns, others four or even eight; some are white-faced, others black-faced. Some sheep are lop-eared, others fat-tailed, some have short fleeces, some have dense, woolly fleeces, while yet others are without wool.

Many variations occur in the fleece. The double coat, similar to that of wild sheep, consists of hair-like fibres which form the outer covering of the coat, providing mechanical protection. The shorter, finer fibres, lying among the coarse outer kemps, provide insulation from temperature changes.

Breeds for all purposes and places

For the finest wool we are indebted to the Merino, a Spanish breed with white face and legs, extensively farmed in Australia, the United States, USSR, South Africa, the Argentine, France and Germany and also the Rambouillet, a French breed derived from it. For the dual purposes of meat and wool, there are a large number of breeds originating in the British Isles. The oldest British breed is the Leicester, developed by Robert Bakewell early in the 18th century. The Southdown, medium wool dark face, without horns was originated on the Sussex Downs as an early maturing breed, towards the end of the 18th century. Its face varies in colour from dark-brown to mouse-coloured. It is a good breed for mutton and its fleece, though of good quality, is light. The Welsh mountain sheep, a medium wool white face, which thrives best on mountains and is bred mainly for its mutton, is interesting because it has been found there are some two dozen strains, each confined to a particular locality, each differing from the others and known by the name of the district in which it is found—for example, the Glamorgan strain.

Many British breeds have been exported for farming and ranging in distant parts of the world and other countries have produced their own breeds and exported them. In 1880, for example, New Zealand produced the Corriedale, by crossing Lincoln rams with Merino ewes, and the breed was adopted in Australia and the United States. In 1912 the United States produced the Columbia, a large rugged white-face sheep giving good wool and good marketable lambs, by crossing the Lincoln rams with Rambouillet ewes. The Netherlands was responsible for the Texel, a medium wool white face. It is good for growing on large ranges, bears a heavy fleece and produces many lambs which mature rapidly. This sheep is favoured in France, Belgium, Denmark, Spain, South Africa, Indonesia, South America and Mexico.

In central Asia where the Karakul sheep is raised we find an entirely different trade. Its lambs are black with tight-curled coats and when these are skinned at 1–3 days old they provide the wellknown commercial astrakhan fur. The breed, although native to central Asia, has been taken to other parts of the world.

Some sheep are bred, rather surprisingly, especially for their milk, examples of which are the East Friesian and the La Razza Sarda of Sardinia.

Following sheep-like

Commercial sheep flocks are maintained in a variety of conditions, depending on the nature of the available grassland. They vary in size from a few sheep kept on a small pasture to large flocks of many thousands kept on the open ranges in Australia, South America and the western part of the United States.

An interesting feature of sheep behaviour is that in most breeds almost all activity is carried out within a flock and is highly co-ordinated. When there is some outside disturbance, such as a stray dog or strange person coming into their field, the sheep at first all put up their heads to watch and then, if further alarmed, they run towards each other forming a solid mass. Then they run away from the disturbance, usually led by one of the older animals. This pattern of behaviour has undoubtedly made possible the herding of sheep and the ease with which a dog and shepherd can control them. In fact the dog was domesticated before the sheep and it was probable that this made the domestication of sheep possible.

Grazing ruminants

In general a sheep's day is broken up into periods of intensive grazing, resting and ruminating. The longest periods of grazing occur in the early morning and between late afternoon and dusk. During extremely hot weather sheep may graze more frequently at night than in the daytime. Large flocks do not generally all graze together but break up into subgroups feeding in separate areas of the pasture. Sheep have a cleft upper lip which allows for very close grazing. The lips, the lower incisor teeth and the dental pad combine for grasping the food. There are no upper incisors so leaves and stems have to be severed by the lower incisors against the dental pad. Sheep have definite preferences of herbage, although if food is scarce they become less discriminating. Left to themselves on a sufficient area they will range widely and graze selectively.

A characteristic of sheep which is of great service to the farmer is its power of improving and restoring fertility to land upon which it is fed. On sandy soils and poor, light land the value of sheep for this purpose cannot be overestimated.

Spring lambs

Mating takes place in the temperate zones during the autumn months although not all breeds of sheep are restricted to an autumn breeding season. In Britain the Dorset Horn breed can lamb at most times of the year. The Merino has an extended breeding season as have many other breeds especially those originating in more equatorial zones. Mating is preceded by a certain amount of courtship between the sexes and fighting amongst the rams. The gestation period ranges from 144 to 153 days so that the one or occasionally two lambs are born in late winter or spring. In most commercial flocks the breeding season is artificially controlled so that the lambs may be born as early as February or as late as May. The ewes suckle the lambs for about six weeks, the duration of the suckling periods becoming shorter as the lambs get older. The lambs follow the ewes about soon after birth giving voice to high-pitched baas if they become separated from their mothers. When they are only two days old they will suck a blade of grass and let it go but after two weeks they will eat the grass blades.

Unusual diet

You might be excused for supposing that the best-flavoured mutton comes from sheep that have fed on the best pastures. In England, at least, it is believed that a snail contributes most to the superior flavour of Southdown and Dartmoor mutton. Sheep are basically herbivores, yet it is claimed that if, in a pasture, there is one area richer in snails than elsewhere the sheep will automatically make for it. In some places these snails are very numerous, so much so that after rain they are seen in such vast numbers, drawn from their hiding places by the moisture, that there have been sensational reports of showers of snails! Whether the sheep eat the snails as such or take them into their mouths accidentally with the grass is really beside the point. The sheep do consume the snails, and this it is believed imparts the special flavour to the mutton that they yield.

class	**Mammalia**
order	**Artiodactyla**
family	**Bovidae**
genus & species	***Ovis aries*** *sheep*

A sign of spring, ewe with her lambs. Usually born any time between February and May, the lambs are suckled by their mother for about 6 weeks.

Shrew

Shrews are the smallest mammals, and for their size probably the most belligerent. Their ancestors are probably among the first true mammals on the earth.

The 170 species are distributed over the world apart from Australasia and the polar regions. They are divided into red-toothed and white-toothed shrews. The red-toothed shrews include the common shrew of Europe and Asia, 4 in. long including a $1\frac{1}{2}$in. tail, and the water shrew, nearly 7 in. long with a 3in. tail. The 34 North American shrews are also red-toothed and they include the common or masked shrew, the pygmy shrew, and the water shrew, all three species being much the same size as their European counterparts.

Of the remaining North American shrews mention can be made of the short-tailed shrew, reputed to have a poisonous bite, the smoky shrew whose tail swells in the breeding season, and the least, little, or lesser short-tailed shrew, 'the furry mite with a mighty fury' (E Laurence Palmer).

Red-toothed shrews are similar in appearance and in habits. They are mouse-like but with small ears and eyes and a tapering snout with many long bristles. The fur is greyish to dark brown on the back and dirty white on the belly.

Nose to the ground: the common shrew is always on the move, searching for insects with its long snout.

Three-hourly rhythm

Shrews live solitary lives in tunnels in grass, among leaf litter or in surface tunnels, where they are seldom seen and are only revealed by their high-pitched squeaks. There is reason to believe that shrews also use ultrasonics for echo-location but not to the high degree found in bats. The common and pygmy shrews, and probably all the other species, have a 3-hourly rhythm of alternately feeding and resting, but are active for a longer proportion of each of the 3 hours during the night. Shrews are also relatively short-lived. In the common shrew and most red-toothed shrews, 15 months represents extreme old age.

There are many legends about shrews. One English legend is that a shrew cannot cross a human path and live, and many people have reported seeing a shrew tottering towards them and dropping at their feet. Such a shrew may have been extremely old, so that it dropped dead in its tracks. Another possibility is that the shrew died of 'cold starvation'.

Vicious food circle

The smaller the animal the greater its surface area in proportion to its bulk, and the more readily it loses heat by radiation. This can only be replaced by food, so a very small animal must move restlessly, searching for food to make good the loss of heat—and the loss of energy in searching for food. One result is that it cannot go without food for long periods, hence the 3-hourly rhythm. A shrew deprived of food for 2 or 3 hours will die, and the lower the temperature the shorter the period of fasting it can endure. Most reports of shrews seen dropping dead relate to the early morning. This is the time, especially in autumn, when cold starvation is most likely to occur.

Death from shock?

The susceptibility to cold starvation led to another misunderstanding, that shrews were especially prone to death from shock. Naturalists of the late 19th and early 20th centuries reported shrews dying if a gun were fired near them, and even if a blown-up paper bag were burst near them. In fact, shrews are tough and will survive all manner of misadventures so long as they are well fed. When suffering from cold starvation, however, to the point where they are moribund, almost anything will kill them.

Another wrong idea

Shrews are basically insect-eaters but they will eat any animal small enough for them to overpower, such as snails and worms. They will also eat carrion. But they do need some cereal, and this was discovered by Australian zoologist Peter Crowcroft, when keeping them in captivity. He found they remained healthy if given a little cereal or seed each day. It is often said shrews eat more than their own weight of food a day. A more exact estimate puts it at three-quarters their own weight.

High death rate

From spring to autumn each mature female has at least two litters of 4–8 young after a gestation of uncertain length which may be between 13 and 21 days. A newborn shrew weighs $\frac{1}{100}$ oz. Its eyes open at 18–21 days and it is weaned 2 days later. There is probably a heavy infant mortality, in spite of the musk gland in each flank of most shrews which emits a foul odour. Domestic cats, for example, will kill shrews but do not eat them, presumably finding them unpalatable. Birds of prey will eat them, however, especially various owl species, and so will foxes and weasels.

Singing contests

There has long been the idea that shrews are not only savage, ready to use their teeth when handled, but also are extremely quarrelsome among themselves. Naturalists have reported seeing two shrews apparently locked in mortal combat and squeaking furiously. Crowcroft, who made a special study of European shrews, and careful observation of them in captivity, has put a different light on this. He found that if two shrews meet they go towards each other until their whiskers touch, then they squeak. As a rule one of them, usually the intruder will retreat. If it does not do so both rear up onto their haunches, still squeaking. If at this stage neither gives way, they throw themselves onto their backs squeaking even more and wriggling about. Then it usually happens that the muzzle of one comes in contact with the tail of the other and it seizes it with its teeth. As the two shrews continue to wriggle, almost inevitably the second shrew finds the other's tail, seizes it, and the two continue wriggling and squirming apparently in close embrace. Seldom do they hurt each other, or if they do the injuries are not severe. Because food is so vital to them they may become overcrowded and these 'singing contests' are the best way to keep the population evenly spaced over the ground to ensure a maximum food supply for all.

class	**Mammalia**
order	**Insectivora**
family	**Soricidae**
genera & species	***Blarina brevicauda*** *short-tailed shrew* ***Cryptotis parva*** *least shrew* ***Microsorex hoyi*** *N. American pygmy shrew* ***Sorex araneus*** *European common shrew* ***S. fumeus*** *smoky shrew* ***S. minutus*** *European pygmy shrew* *others*

Skunk

The skunk belongs to the family Mustelidae along with the badger, weasel, mink and otter. All mustelids have musk or stink glands at the base of the tail but the skunk is the best-known owner as it can squirt a nauseating fluid at its enemies.

The bold black-and-white colour pattern of the skunk's fur makes it highly conspicuous and acts as a warning to any would-be predators. All species of skunk have long fur and long, bushy or plumed black-and-white tails. The legs are short and the soles of the feet are nearly naked and have strong, curved claws.

The commonest species is the striped skunk, ranging from southern Canada through most of the United States to northern Mexico. It grows up to about 30 in. long including a tail which is up to about 8 in. long. The females are usually smaller than the males. The fur, which is long and harsh with soft underfur, is black with white on the face and neck, dividing into two white stripes diagonally

along the sides of the body. The hooded skunk of southwestern United States and Central America is similar to the striped skunk but the tail is longer than the head and body and the white stripes more widely separated. The spotted skunk ranges from British Columbia through most of the United States and Mexico into Central America. It is distinguished by its small size, being only 22 in. maximum length of which 9 in. is tail. It also has a white spot on its forehead and a pattern of white stripes and spots.

The hog-nosed skunk is not so well-known as the other skunks. It ranges in the southwestern United States and is the only genus of skunk with representatives in South America. It is much the same size as the striped skunk but with shorter, coarse hair and usually with the top of the head, back and tail white. It has a long, naked, pig-like snout used for rooting up insects.

△ *Dissatisfied with its offspring's slow progress, a striped skunk employs its own strength in a rough but certainly effective method of speeding it to their destination.*

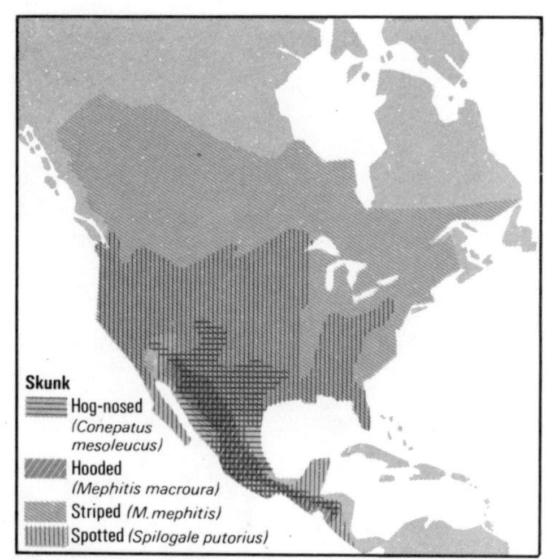

Fluid squirter

Skunks are found in a variety of habitats including woods, plains and desert areas. They live in burrows that they dig for themselves or in abandoned burrows of badgers, foxes or woodchucks, or under buildings, denning up by day and only coming out at night to forage. Although not true hibernators, most skunks, especially those in the northern parts of their range, settle down in their dens and sleep for long periods during the cold weather. The dens are lined with dry leaves and grass. Occasionally several skunks den together. Spotted skunks, however, are active throughout the year.

When disturbed or attacked a skunk lowers its head, erects its tail, stamps a warning with its front paws and, if this does not deter the enemy, turns its back and squirts an amber-coloured foul-smelling fluid from its anal glands for as far as 12 ft with unerring accuracy. This pungent spray can cause temporary blindness if it touches the eyes and its odour can be detected half-a-mile away. A striped skunk will turn its back to a predator and eject its fluid at the enemy. A spotted skunk may discharge from a 'hand-stand' position.

Carnivorous feeder

Skunks feed mainly on insects such as beetles, crickets, grasshoppers and caterpillars. They will also take mice, frogs, eggs, small birds and crayfish and occasionally they cause damage by entering poultry runs, killing the birds and taking the eggs. As the winter approaches the striped skunk becomes fat, adding leaves, grain, nuts and carrion to its diet. Some skunks will feed on snakes and in the Andes the hog-nosed skunks are immune to the venom of pit vipers. It may be that the spotted skunks are resistant to rattlesnake venom.

Blind and hairless kits

In the striped skunk breeding takes place in late winter or early spring preceded by boisterous but relatively harmless fighting among the males. After a gestation of 42–63 days 4 or 5, occasionally up to 10, young are born in the den. The young skunk or kit weighs about an ounce at birth, is blind, hairless and toothless but the black-and-white pattern shows plainly on its skin. Its eyes open in 21 days and it is weaned in 6–7 weeks but towards the end of this period it is taken on hunting trips by its mother. By the autumn the family breaks up and each youngster goes its own way to fend for itself. Skunks become mature in about a year. They have lived for 10–12 years in captivity.

Spotted skunks' young are usually born in late May or June but they may be born at any time in the southern parts of their range where two litters may be raised in a year. As in the striped skunk the usual litter size is 4 or 5 but the young weigh only $\frac{1}{3}$ oz at birth and are only 4 in. long. They are not weaned until 8 weeks old when the youngsters begin to take insects.

Left alone by most predators

Most predators give skunks a wide berth but pumas and bobcats occasionally kill and eat them when their usual prey is scarce. Only the great horned owl preys regularly on the skunk and it usually has a noticeable odour of skunk about it. Many skunks are killed every year on the road by cars, particularly at dusk. They have never learned to run away from a car but stand their ground and eject their fluid in futile defiance, as if at a predator.

Some skunks are still trapped for their fur and in South America the hog-nosed skunk is hunted by the local people as its meat is said to have curative properties. The skin is used for capes and blankets.

Routed by the smell

'There is no quadruped on the continent of North America, the approach of which is more generally detested than that of the skunk. Even the bravest of our boasting race is, by this little animal, compelled to break off his train of thought, hold his nose, and run–as if a lion were at his heels.'

These words were written by Audubon. He also tells the story of how an entire church congregation was routed by skunk odours. At the request of an asthmatic clergyman Audubon obtained the musk glands of a striped skunk. The clergyman kept them in a tightly corked smelling bottle and whenever he felt an asthmatic attack coming on he uncorked the bottle and inhaled deeply to have immediate relief. One Sunday morning half way through the service, the clergyman, on the verge of an attack, quickly took out his smelling bottle and vigorously waved it to-and-fro under his nose. The congregation, according to Audubon, 'finding the smell too powerful for their olfactories made a hasty retreat', as many in less confined quarters have done since.

class	**Mammalia**
order	**Carnivora**
family	**Mustelidae**
genera & species	***Conepatus mesoleucus*** *hog-nosed skunk* ***Mephitis macroura*** *hooded skunk* ***M. mephitis*** *striped skunk* ***Spilogale putorius*** *spotted skunk*

▽ *In acrobatic pose, a spotted skunk prepares to discharge its nauseating deterrent fluid at a pair of lynx. The stance is typical of spotted skunks; striped skunks just turn their backs, with all feet firmly on the ground.*

Sloth

Sloths are bizarre mammals that spend nearly all their lives hanging upside down. With the anteaters and armadillos, these three South American groups belong to the order Edentata which means without teeth. The anteaters are the only edentates wholly without teeth but sloths have teeth in the cheeks. There are nine on each side and they grow throughout life. Their bodies show some remarkable adaptations for an upside down life in the trees. Sloths hang by means of long curved claws like meat hooks and their hands and feet have lost all other functions, the fingers and toes being united in a common fold of skin. The arms are longer than the legs and the pelvis is small. The back muscles which are well developed in other animals are weak in sloths. The head can turn through 270° so that it can be held almost the right way up while the rest of the body is upside down. The hair lies in the opposite direction to that of other mammals, from belly to back, so rain water still runs off it. The individual hairs are grooved and are usually infested by single-celled algae which make sloths look green.

The seven species of sloth are divided into two-toed and three-toed sloths. The three-toed sloths are about 2 ft long with a short stump-like tail. The two-toed sloths are a little larger but lack a tail. Sloths live in forests, the three-toed sloths from Honduras to northern Argentina and two-toed sloths from Venezuela to Brazil.

Slow-motion animals

As is usual for nocturnal animals living in the forests of South America, very little is known about their habits. Indeed sloths appear to have very few habits for they live in slow motion. Their movements along the branches are so slow that it is often said that a sloth may spend all its life on one tree. They eat, sleep, mate, give birth and nurse their young upside down, although they do not hang from branches all the time, as they will sit in the fork of a tree. They sleep with the head on the chest between the arms, looking very inconspicuous.

Sloths occasionally come to the ground, presumably to reach another tree when they cannot travel overhead by branches and creepers. They are just able to stand on their feet but they cannot walk on them. They move by sprawling on their bellies and dragging themselves forward with their hands. They can, however, swim well.

Despite their sluggish habits sloths can defend themselves well by slashing with their claws and by biting. It is often suggested that this is sufficient defence against their main enemies, jaguars and ocelots, but it is difficult to imagine a nimble cat being unable to outmanoeuvre a sloth. The sloths may benefit more as regards their enemies, from their camouflage of green algae and their sluggish habits which often make them look like a mass of dead leaves.

△ *Just hanging around: a two-toed sloth picks its leisurely path through the tree canopy. All but a small fraction of a sloth's life is spent upside down; it eats, sleeps, mates and gives birth in this position, relying on formidable claws and a close resemblance to a bundle of leaves for defence.*
▽ *Earthbound ignominy: ill-adapted for anything but hanging, a three-toed sloth heaves and grovels its way to the next tree.*

Plant eaters

Sloths eat mainly leaves and shoots with some fruit, which they may hook towards their mouths with their claws. Their stomachs are complex like those of ruminants such as cattle and sheep.

One baby

A single baby is born at the beginning of the dry season after a gestation of 17–26 weeks. It immediately hooks itself into the fur of its mother's breast and stays there until old enough to leave. Sloths have lived for 11 years in captivity.

Strange lodger

If a crop of green algae is not enough, sloth's fur harbours another guest—moths rather like a clothes moth. Three species of pyralid moth have been found on the two species of sloth. They are about $\frac{1}{3}$ in. long with flattened bodies and can run agilely through the dense mat of hair. This makes them difficult to collect, especially as the collector has to avoid the sloth's attempts to defend itself. No one has been able to find out why the moths live in sloths' hair. They do not feed there, nor have their eggs or caterpillars been found in the fur. The caterpillars may live a normal life on plants where the eggs were laid by the adults on forays from their host.

class	**Mammalia**
order	**Edentata**
family	**Bradypodidae**
genus & species	***Choloepus didactylus*** *two-toed sloth* ***Bradypus tridactylus*** *three-toed sloth, others*

▽ *Look, no hands: a young two-toed sloth* ***Choloepus hoffmanni*** *giving a clear illustration of the 'backward' growth of hair, necessary for adequate insulation and waterproofing of the animal.*

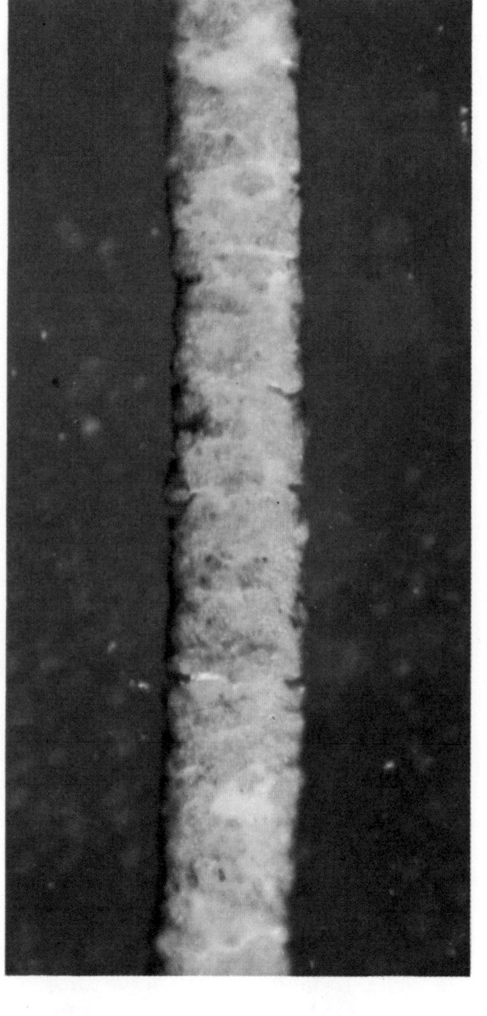

▽ *Individual sloth hair showing the one-celled algae, which make sloths look green.*

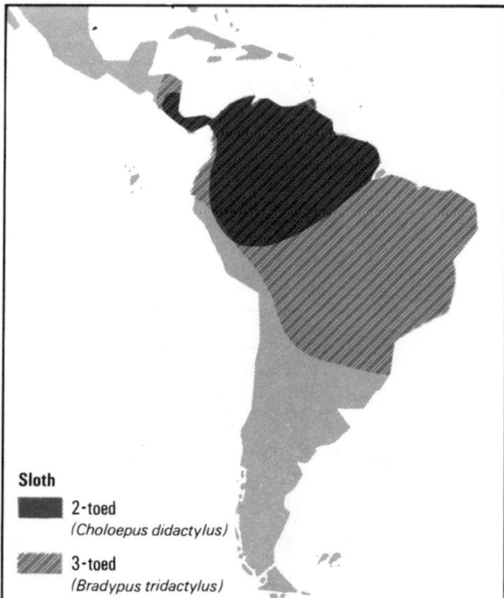

Squirrel

Plenty of squirrels live in trees to a greater or lesser extent. Tree squirrels are the many species belonging to the genus **Sciurus** *that are distributed over Europe, most of Asia and America from southern Canada to the eastern United States. Two of the best known and most widespread are the grey squirrel of North America and the red squirrel of Europe and Asia, and attention will be given to these.*

Tree squirrels differ from ground squirrels (page 108) and flying squirrels in having bushy tails about as long as head and body combined. They range from 16 to 26 in. in total length and weigh up to 3 lb. Their coat is grey to red, sometimes black, with white or cream, sometimes yellow or orange, underparts. The winter coat is usually slightly different to the summer coat and although the body fur is moulted twice a year the tail is moulted only once. The tail is always well furred and in many species it is flattened—feathery rather than bushy. The ears usually have tufts of hair at the tips but these are retained for only part of the year in some species. The four toes on the front feet and the five toes on the hind feet bear sharp claws.

△ *Red squirrel gnawing a titbit.* ▽ *Squirrel ranges before the spread of grey and decline of red.*

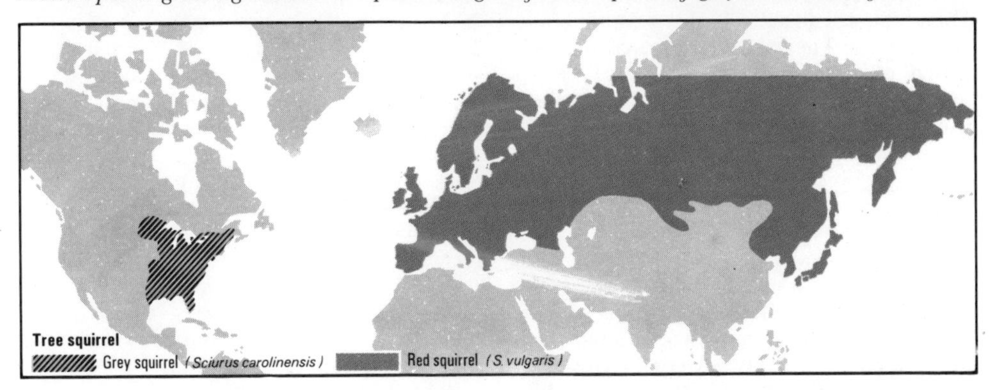

Arboreal acrobats

Tree squirrels forage on the ground but quickly escape to trees when disturbed. They shin rapidly up a trunk, their first leap taking them 3–4 ft up after which their sharp claws are used in a bounding climb. They can run along a branch or hang from it, travelling upside down by using all four feet hand over hand. The grey squirrel, especially, is a skilled acrobat. It will hang by its hindfeet from one branch to reach food on a branch below. It will leap gracefully from the outer branches of one tree to those of another over a distance of 12 ft. As it sails through the air the legs are spreadeagled and the flattened tail acts both as a balancer and a rudder.

A tree squirrel's usual reaction to an intruder is to scold, then to disappear behind a trunk or stout branch, all the time keeping the trunk or branch between itself and its enemy. When forced to do so it will drop to the ground for escape, from heights of 30 ft or more, plunging into undergrowth like an arrow or dropping straight down on all fours onto bare earth or short turf.

Scattered stores

Although primarily vegetarian, eating nuts, berries, soft fruits, buds and some fungi, most tree squirrels take birds' eggs and nestlings, even carrion. They are traditionally hoarders but so far as the grey and the European red squirrels are concerned they seldom cache food in the way traditionally attributed to them. Stores of nuts buried in the ground or in a hollow tree are more likely to be the work of fieldmice. A squirrel buries nuts, acorns or berries singly and well spaced out. It carries the food in its mouth, stops at a chosen spot, digs a hole with a quick action of the forefeet, just deep enough to take the nut or acorn, then pushes the earth back with the forefeet.

Importance of chewing

It has been shown that a young grey squirrel does not open a nut efficiently the first time but learns to gnaw it open. Although all adult grey squirrels open nuts in the same way, showing that the pattern of this behaviour is species specific—all members of the species do it the same way, 'by instinct'—the instinct has to be reinforced by learning.

The front teeth of squirrels grow continuously at the roots and it is usually said that their habit of chewing objects other than food is necessary to keep the teeth worn down to normal length. Possibly some of the damage done to trees comes from this chewing habit, but the way grey squirrels chew the lead labels on ornamental trees in parks and gardens is perhaps the best illustration of the habit. Recently, however, it has been shown that squirrels, and rats also, are constantly grinding their teeth, when not otherwise engaged, and it is this that keeps them the required length.

Nest of twigs and leaves

Tree squirrels build nests, or dreys, in the branches of trees and these are used for several purposes, one being for a nursery. Each squirrel usually builds several in adjacent trees. They are bulky structures composed of twigs, strips of thin bark, leaves and moss. Some are cupshaped, others domed. Sometimes an old bird's nest, such as one made by a crow, is used as a foundation. The breeding or nursery nest is often a huge ball of leaves and sticks in a roomy hollow in a tree trunk. The breeding season of the European red squirrel varies with latitude. In the northern parts of the range the only breeding period is spring. In the southern parts there are two periods, January–April and May–August. Gestation is 46 days, the litters consisting of 1–6 young, usually 3 or 4, born blind and naked. They are weaned at 7–10 weeks. In the grey squirrel mating takes place between early January and August. There are usually two litters a year. Gestation is 44 days, and other details of development are much the same as in the red squirrel.

Declining enemies

Tree squirrels are remarkably skilful not only in moving among trees but in keeping out of sight once they are alerted to possible danger. This is well illustrated by the occasion when a professional pest exterminator was brought in to clear a garden of a dozen squirrels–the estimated number made by the people living there. In a few days he shot more than three times that number. In much of Europe the natural enemies of squirrels have been largely eliminated. What these could be, throughout the range of tree squirrels, can be judged from the natural enemies of grey squirrels in North America. They include the goshawk and red-shouldered hawk, barred owl and horned owl, and tree-climbing snakes, and, on the ground the fox and the bobcat; finally, and probably most important, the pine marten, which can move through trees at least as skilfully as the nimble squirrel.

△ *Already confident, a young grey squirrel climbs among the branches of a horse-chestnut, dangling its tail. This acts as both balance and rudder as the squirrel bounds among foliage.*
▷ *Nursery in the tree tops–the grey squirrel builds several dreys in adjoining trees, one of which acts as a nursery. The nest is built from twigs, bark, leaves and moss and is easily visible among the bare branches in winter.*

Rapid tooth growth

At one time the only way to demonstrate the remarkable speed of growth of the incisors of rodents was to show what happened if two of the incisors were awry and failed to bite on each other. They curved into the mouth eventually locking the jaws so that the animal died of starvation. In the last 30–40 years accurate measurements have been made of their rate of growth. Although the figures are not known for tree squirrels we can be reasonably sure they compare fairly closely with those that are known. The incisors of the common rat grow up to 6 in. in a year, those of the guinea pig up to 10 in. and those of the pocket gopher up to 14 in. in a year. The pocket gopher uses its teeth for digging and needs a rapid growth rate to counteract a high rate of wear. So it would seem that the rate of growth is related to the uses to which the teeth are put.

class	**Mammalia**
order	**Rodentia**
family	**Sciuridae**
genus & species	***Sciurus carolinensis*** *grey squirrel* ***S. vulgaris*** *red squirrel* *others*

Tarsier

The tiny, 3–4oz tarsier is one of the most fascinating of all the primates. It has a short snout and enormous eyes, a hairy nose with the nostrils facing sideways, mobile and membraneous ears, and a long, almost naked tail. Its fur is thick and woolly. Its hindlegs are very long, with a lengthened ankle section or tarsus (hence the name) and its fingers and toes have tiny triangular nails and large, rounded sucker-like disc-pads at the ends. The second and third toes have long toilet-claws instead of nails.

The three species, restricted to certain islands in the East Indian archipelago, differ in size, colour and the underside of the tail. Horsfield's tarsier, 5–6 in. long with a tail of 8–10 in., is dark reddish brown, and the underside of the tail is naked, with alternate ridges and V-shaped grooves. It is found on Borneo, Billiton, Bangka and part of southeastern Sumatra, around the town of Palembang. The Philippine tarsier is the same size but greyish in colour. Its tail is not tufted at the end as in the last, and is smooth underneath. It is found on the islands of Mindanao, Samar, Leyte and Bohol. The Celebes tarsier or spectral tarsier is only 4–5 in. long and is dark greyish in colour although paler below, with more pointed ears, and a heavily tufted tail, which is scaly and has three hairs sticking out from each scalex. It is found on Celebes and a few nearby islands.

Progress by leaping

Tarsiers live in low-lying, usually coastal forests, in thick cover. They frequent areas around plantations where there is abundant food. They live singly or in pairs, each with its own territory which they spend much time marking with urine. This is either squirted directly onto the branches, or it is squirted onto the toepads and then transferred to the branches. Sometimes they mark the territory with certain glands in the anal region and on the lip. They are nocturnal, hence the large eyes and mobile ears. The eyes are relatively immobile, but the head can turn through almost 360°.

Tarsiers are noted for their leaping, using their long hindlegs. When moving about on branches a tarsier is rather awkward, and it spends much time sitting on vertical trunks and stems or asleep, clinging to a stem with its head sunk forward. When it does move, its legs suddenly straighten like a frog's, and it turns to leap through the air, landing with its tail up and feet extended forward. It can turn and change direction in mid-air, and has been known to leap 6 ft. The male and female of a pair do not often groom one another, but they often seek contact, one crawling under the other. Self-grooming is common, by scratching with the long claws on the foot, biting and licking, and rubbing the head against a branch. As a rule, the only sounds tarsiers make are loud squeakings during sexual activity.

△ *All the better to see you with: Horsfield's tarsier is nocturnal, hence the large eyes.*

Wholly flesh-eating

They are almost the only primate which are entirely carnivorous. Their food consists of grubs, insects, lizards, frogs and perhaps small birds. The food is held in one hand and pulled at with the teeth, the hands being little used for manipulation. The palm is very small, the fingers long and spidery except for the discs at the end, and the thumb is not opposable.

Precocious babies

The tarsier breeds throughout the year and has a monthly cycle, like the higher primates. The female even shows a slight degree of menstrual bleeding. The gestation lasts 6 months–a strikingly long time for such a small animal. The single young is born well-furred, with its eyes open, and able to cling to its mother's fur and climb about on its own. When left alone, it sits squeaking quietly until rejoined by its mother. When moving some distance the mother may take her baby with her in her mouth. Tarsiers have been known to live for 12 years.

Alert to danger

The tarsier is eaten by owls and possibly by small cats but it is always on the alert, its head swinging around and its ears constantly moving back and forth. Even when resting it may leave one eye open, or partially so, which looks rather uncanny. When threatened it opens its mouth, showing its teeth which can bite hard. Above all, its jumping is a first-class method of escape.

Safety at night

There is something grotesque and macabre yet irresistibly attractive about the tarsier. Zoologically, too, it is a contradiction and a puzzle. It is perhaps the most primitive living primate. At the same time it is quite close to the 'higher primates'–monkeys, apes and man himself, and has many features which are much closer to them than to the 'lower primates'–lemurs, lorises and bushbabies–among which it is usually classified. Thus, its nose is quite hairy, and the upper lip is free from the gum, as in higher primates, whereas the lemurs and their relatives have a dog-like nose, with a large moist, naked area, and the upper lip is bound down to the gum. Nevertheless, its teeth are extremely simple in structure, and of a type one would expect in a very primitive mammal. Its skeleton, muscular system and, indeed, virtually all its internal organs are very primitive, including the reproductive system, the two sexes being difficult to distinguish externally. Moreover, the tarsier, like the South American tapir, has 80 chromosomes, the highest number known among mammals, and this is generally supposed to be a primitive feature.

More important, the family Tarsiidae, today represented by only these three closely related species, was very numerous during the Eocene period, 40–55 million years ago. Some of these long extinct forms were almost identical with the living tarsiers except that they did not have the large eyes, so presumably were not nocturnal. The only conclusion possible is that the immediate ancestors of present-day tarsiers were able to survive only by becoming nocturnal, so avoiding competition with the monkeys and apes which began to increase in numbers at that time and dominated the scene by day.

class	**Mammalia**
order	**Primates**
family	**Tarsiidae**
genus & species	***Tarsius bancanus*** *Horsfield's tarsier* ***T. spectrum*** *Celebes tarsier* ***T. syrichta*** *Philippine tarsier*

◁ *Savage snarl: the voracious Tasmanian devil will feed on any flesh, dead or alive.*

△ *Now agricultural development has reached Tasmania, the devils may soon be in danger.*

Tasmanian devil

The Tasmanian devil is the second largest of the living carnivorous marsupials, after the thylacine. It is stockily built, rather like a small bear, being 2–3 ft long including a 1ft long tail, and weighing 12–20 lb or more. The female is smaller than the male. The Tasmanian devil has a short, heavy head with a pinkish white snout, strong jaw muscles and powerful teeth. Its eyes are small, its ears large and pink. Its coat is black with white patches on the throat, shoulder, rump and occasionally the tail tip.

The Tasmanian devil, now thought to be confined to Tasmania, was once widespread on the mainland of Australia, judging from the bones found in Aborigines' middens. There are occasional reports even today of its continued existence on the mainland but these reports have not been confirmed. One was killed in 1912 in Victoria but it was believed to have escaped from a zoo. When all is said and done, nobody is very sure when the 'devil' became extinct on the mainland of Australia or what wiped it out. The dingo has been blamed, but that is only guesswork. It could just as easily turn up again, as the noisy scrub bird and other animals have done.

Savage 'devil'

The Tasmanian devil lives in woodlands and rocky, inaccessible places, sheltering in the daytime in caves, hollow logs or among tree roots and coming out only at night to hunt. It takes readily to water, diving in when pursued and swimming underwater to surface some distance away, usually under a patch of water plants. It is not a particularly good climber although it may run up leaning tree trunks and stout branches. It received the name 'devil' because of its reputation for savagery, which can be justified by its behaviour towards animal prey; but there is little to prove that it is savage towards humans. It seems to be indifferent to danger when eating and will allow itself to be caught and handled, in marked contrast to other flesh-eating animals which tend to be more savage when feeding. According to Mrs Mary Roberts, who kept and bred devils in her zoo, they make delightful and affectionate pets. They were very clean animals and loved bathing themselves, washing their faces by licking their cupped fore-paws and wiping them over their heads. They also enjoyed basking in the sun. One Tasmanian farmer had two tame devils which he used to take out walking with him on a lead and even took to Melbourne.

The devil's voice is a whining growl, followed by a snarling cough or low yell.

Flesh, dead or alive

The Tasmanian devil feeds on any flesh, dead or alive, and it will even pull down animals larger than itself. According to Gould, it used to invade sheep-folds and hen-roosts, but small wallabies, rat kangaroos, small mammals, birds and lizards now make up most of its diet. It is thought that frogs and crayfish may also be taken, as devils have often been seen on the banks of streams.

Unusual breeding

From the little known about them Tasmanian devils seem to have unusual breeding habits. In April the male and female come together but they do not mate for nearly 2 weeks. During that time, curiously enough, the male does not let the female out of his den. After they have mated, however, the female turns the tables on him and snarls and bites whenever he approaches her. The young, never more than four, are born at the end of May or beginning of June in a closed pouch which is made of a flap of skin directed backwards and enclosing four teats. The baby devil is less than ½ in. long at birth but grows to 2¾ in. after 7 weeks. It holds on to a teat all the time for 15 weeks, after which time its eyes have opened and its coat has grown. In late September, feet or tail are sometimes seen sticking out of the pouch and soon after this both male and female make a soft bed of grass or straw for the young, in a hollow tree-trunk, under a rock or even in a wombat's burrow. The young are not weaned for 5 months and it is thought they do not become sexually mature until their second year. They may live for 7–8 years, perhaps more.

Few enemies

The Tasmanian devil's disappearance from the mainland of Australia is said by some people to be almost certainly due to the spread of the dingo. Fortunately this wild dog never penetrated to Tasmania. Today, owing to its inaccessible habitat, the devil has few enemies and is not considered to be in danger of extermination.

Tasmanian conservation

Tasmania is a small state of approximately 25 000 square miles. As it was not occupied until 1801 there are as yet no great population pressures and until recently the countryside has remained largely intact, so sheltering Tasmania's interesting and varied wildlife. Agricultural development is, however, beginning to increase, and although ⅛th of the entire country is controlled in the form of reserves and sanctuaries by the Animals and Birds Protection Board, agricultural development is taking place largely at the expense of sclerophyll forest. This type of forest ranges from open savannah woodlands to dense stands of timber and it supports much of the wildlife of Tasmania including the thylacine and Tasmanian devil, the forester kangaroo, Bennett's wallaby, the wombat, opossums and wild cats. In addition many of the small ground-dwelling species, such as bandicoots and dasyures, are numerous here, mainly because of the absence of dingoes and foxes.

Destruction of its habitat has undoubtedly contributed to the decline of the thylacine, although the 'devil' has not yet been too badly affected. Pressure on it will no doubt increase in the future but it is hoped that conservationists will be successful in making more sanctuaries in the forested areas to keep the Tasmanian fauna intact.

class	**Mammalia**
order	**Marsupialia**
family	**Dasyuridae**
genus & species	***Sarcophilus harrisii*** *Tasmanian devil*

Tiger

One of the largest of the 'big cats', the tiger's sinuous grace, splendid carriage and distinctive colouring make it one of the most magnificent of all animals. A large male averages 9 ft – 9 ft 3 in. in length including a 3ft tail. It stands 3 ft or more at the shoulder and weighs 400 – 500 lb. Females are a foot or so less in length and weigh about 100 lb less. The various races of tigers vary considerably in size from the small Bali Island tiger to the outsized tiger found in Manchuria which may reach 12 ft in total length. The ground colour of the coat is fawn to rufous red, becoming progressively darker southwards through the animal's range, the Balinese tiger being the darkest. The underparts are white. There have been rare cases of white tigers in India. The coat is overlaid with black to blackish-brown transverse stripes, and these contrasting colours provide an excellent camouflage in forest regions.

In cold climates such as Siberia and Manchuria, tigers have thick, shaggy coats which become shorter and denser in the warmer climates. The hair round the face is longer than on the rest of the body, forming a distinct ruff in adult males.

From its original home in Siberia, the tiger spread across almost the whole of Eurasia during the Ice Ages. Today it is found only in Asia where a number of geographical races are recognised, including those of Siberia, Manchuria, Persia, India, China, Sumatra, Java and Bali. The races differ only in size, colour and markings.

▷ *Solitary splendour: tiger caught by flash.*
▷▷ *Transport solution: a helpless tiger cub is carried in the same way as a domestic kitten.*

△ *Reflected glory: unlike many of the cat family, tigers often take to the water and are strong swimmers. In times of flood they have been known to feed on fish and turtles and to swim in search of stranded prey. They cannot bear excessive heat and will sometimes sit in shallow water in an attempt to keep cool.*

Solitary prowler

Although its original home was in the snowy wastes of Siberia the tiger's natural preference is for thick cover. It has, however, become adapted to life in rocky mountainous regions, the reed beds of the Caspian, and the dense steaming jungles of Malaya and islands such as Java and Bali. It cannot, however, tolerate excessive heat and during the heat of the day it will lie up in long grass, caves, ruined buildings, or even in swamps or shallow water.

The tiger is an excellent swimmer and in times of flood has been known to swim from one island to another in search of food. Unlike most members of the cat family it is not a good climber and seldom takes to the trees, but there is a record of a tiger taking a single leap of 18 ft from the ground to pull a man off a tree. Its hearing is very good and is the sense most used in stalking prey. It does not appear able to see unmoving animals, even at a short distance.

The tiger has a variety of calls ranging from a loud 'whoof' of surprise or resentment to a full-throated roar when disturbed or about to launch an attack.

Strength widens choice of prey

A tiger preys on deer, antelope, wild pig and smaller animals such as monkeys and porcupines. It will take fish and turtles in times of flood and locusts in a swarm. It occasionally attacks larger animals such as wild bull buffaloes, springing on their backs and breaking their necks. When food is short it may steal cattle, and an old or injured tiger too weak to hunt may attack humans. Game is, however, its natural food and it is interesting that tigers have completely deserted some forested areas of India where game animals have disappeared even though there were still plenty of wandering cattle about.

A tiger stalks using stealth for the first part of its hunt, finally attacking with a rush at its victim, grasping a shoulder with one paw and then seizing the throat. It then presses upwards, often breaking the neck in the process. After a kill it withdraws to a secluded spot, preferably under cover, taking its prey with it. If it cannot do this, or hide its kill near its lying-up place, it is forced to have a hurried meal and leave the rest of the carcase to the hyaenas and vultures and other carrion eaters.

Small striped cubs

Only while the tigress is in season do male and female tigers come together; according to some authorities, this could be for less than two weeks. During this time a tiger will not allow another male near him and will fight, sometimes to the death, over possession of the female. In India the mating season is variable, but in Malaya it is from November to March and in Manchuria it is during December. A female starts to breed at about 3 years of age and then has a litter every third year, or sometimes sooner. After a gestation of 105–113 days, 3–4 cubs are born, occasionally as many as 6. The mortality among cubs is high and usually no more than two survive to adulthood. They are born blind and helpless, weighing only 2–3 lb, but they have their parents' distinctive striped pattern from the beginning. The cubs grow rapidly; their eyes open after 14 days and they are weaned at 6 weeks. At 7 months they can kill for themselves, but stay with their mother until 2 years old, during which time she trains them in hunting. They are fully grown at 3 years.

Man the hunter

Although the tiger has few natural enemies it has been hunted by man from very early times, at first by the local people and later for sport. In India especially, the coming of the British and the introduction of firearms was disastrous to the tiger and it is estimated that in 1877 alone 1 579 tigers were shot in British India. Today the reduction of game animals and the reduction of its natural habitat is further diminishing its numbers, and as a result six of the eight races of tiger are listed as being in danger of extinction.

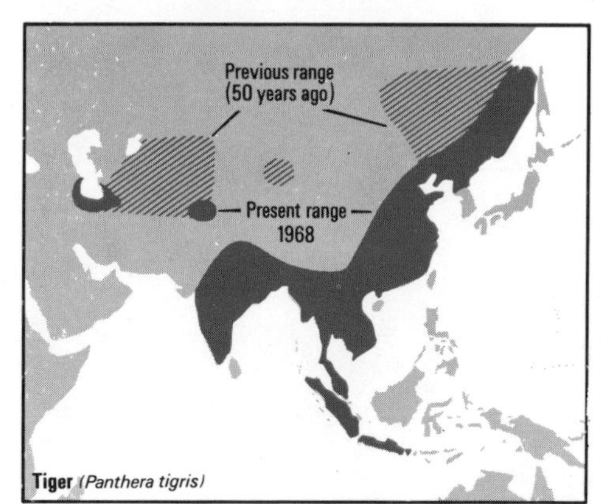

Tiger *(Panthera tigris)*

Not normally dangerous

Tigers have a respect and fear for man which is difficult to explain. Even if harassed by curious humans or sportsmen a tiger will not normally react until its patience is well-nigh exhausted. Normally a man can walk in a tiger's habitat without fear or hindrance and there have been several instances of a tiger approaching a man while sitting quietly near his camp and passing by, doing no harm even though it was obvious that it had seen him. Men have been followed for many miles by tigers and have come to no harm; they were probably being escorted off the territory. It is only when its normal hunting routine is disturbed that it becomes really dangerous. It may then become a man-eater, especially when shot at indiscriminately, incapacitating it rather than killing it. A wounded tiger left to its fate, without the strength to hunt, will resort to man-eating or cattle killing, out of necessity, as it will when injured by natural mishap. One of the commonest causes of injury is damage by porcupine quills. If the quills enter the paws or lower limbs the tiger cannot pull down and kill natural prey or cattle. Occasionally the quills may even penetrate the tiger's jaw and the animal starves to death. Old age may also cause a tiger to attack cattle or humans. Once a tiger has turned man-eater or cattle-killer, for whatever reason, every man's hand is against it. Whole villages will turn out and not rest until it is killed, even in areas where the tiger is protected by law.

class	**Mammalia**
order	**Carnivora**
family	**Felidae**
genus & species	***Panthera tigris*** *tiger*

Titi

To most people who know it well the titi hardly looks like a monkey at all, with its short legs and bushy tail. It is, in fact, one of the most primitive of the South American monkeys, and it shares with the rest of them the wide-spaced, sideways-looking nostrils and the short and very monkeyish face. It is not very large, normally being 12–16 in. long in head and body with a tail of 13–20 in. It weighs about $1\frac{1}{2}$ lb. A titi has long shiny fur which varies between grey and reddish brown from race to race.

As many as seven species and 33 subspecies have been named but today only two species are recognised. The dusky titi, the more widespread, ranges over most of the forest south of the Amazon. It is grey to dark brown, often with orange or reddish underparts. Its hands and feet are black or they may be the same colour as the rest of the limbs, and the tail is reddish or greyish. There may be a white tail-tip and possibly a grey or black band on the forehead. The second species, the widow monkey, is found around the tributaries of the upper Amazon, and is reddish-black with a white to orange patch on the throat, a black tail, black forearms and hindlimbs but usually whitish hands.

△ *The short, monkey-like face of a titi.*

▽ *Keeping in touch. A resting pair of dusky titis sit touching each other, their backs arched, hands and feet grasping the branch and their tails half-twined together helping them to balance.*

Daily hymn of hate

Titis usually live low in the trees, in pairs, each with one or two youngsters. Each pair holds a territory, in which there are favourite trees for feeding and sleeping. Almost every day there are conflicts at the borders of the territories. The occupants sit facing one another, calling at each other and take up a variety of aggressive postures, their hair standing on end, their backs arched and their tails lashing. The calls are mournful, wailing sounds. These conflicts, which help to maintain the territorial boundaries, are usually no more than a show of force, but they sometimes erupt into actual violence, one male chasing the other into his territory. While he is chasing the other male he may suddenly stop and mate with the opposing female! As well as defining its territory by a show of force the titi marks the boundary by rubbing the secretions of his chest gland along the branches. This scent-making usually takes place after the daily ritual of hurling insults and making faces at each other.

Carnivorous vegetarian

Titis have certain favourite feeding trees within their territories. They make their way to these after leaving the 'sleeping' trees, and again after the morning wailing match. Their food consists of fruit and some buds, as well as insects, birds' eggs, and even it is said some small birds which they creep up on unawares.

No guarantee of paternity

Many aspects of the titi's breeding cycle resemble those of marmosets (page 146). The fact that titis live in mated pairs which seem to be long-lasting—perhaps even permanent—restricts each male to one female, and ensures that a male looks after his own offspring—unless by chance his female has been fertilised by a more dominant male during a territorial chase. Just like marmosets, male titis carry the young, handing them to the female only for suckling. There seems to be no specific breeding season, although in eastern Colombia, William Mason, who studied the behaviour of titis in the wild, found a birth peak from December to March. Unlike marmosets, titis bear only one young at a time. It clings to its father with its hands, feet, and its tail, which is not, however, prehensile.

Unusual monkey

The titi is a New World monkey of isolated and primitive status. It has some anatomical similarities to marmosets, especially in its skull, teeth and skeleton. This may, however, be no more than two animals sharing certain primitive features. Its locomotion is also primitive, for although its legs are short, the hindlegs are relatively long and used for jumping. A titi does not, however, cling to vertical branches like other primitive primates such as bushbabies, tarsiers and some marmosets. Instead it sits on horizontal perches with the body hunched and all four feet together clutching the branch beneath, its tail hanging down behind. The male and female or an adult and young often sit side by side, huddled close with their tails intertwined. The titi is a nervous animal, with rapid, jerky movements and is of a restless temperament, continually jumping and running around. It does, however, have long periods of rest when the pair sit in contact, with their tails twined. They also occasionally groom one another but not as much as some primates.

Although classified in the same family, the titi is a very different type of monkey from a species such as the spider monkey, with its prehensile tail and long arms, or the saki (page 203) with its mobile fingers and long, comb-like incisors. It has been compared most closely to the douroucouli, but the characteristics it shares with this monkey are, like those it shares with marmosets, due to primitive characters surviving in both of them. So we are left with a 'mixture-monkey' that howls like a howler monkey, jumps like a squirrel monkey, and is in its social structure most like the smallest of the apes, the gibbon!

class	**Mammalia**
order	**Primates**
family	**Cebidae**
genus & species	***Callicebus moloch*** *dusky titi* ***C. torquatus*** *widow monkey*

Uakari

The uakari (pronounced wakari) is a little-known monkey of South America, closely related to the saki (page 203) but differing from it in a number of features, especially the short tail. It is the only South American monkey whose tail is actually shorter than the head and body, which measure 16–18 in., the tail being only 6–7 in. The body is covered with shaggy hair, variable in colour, but underneath it the uakari is a skinny, spidery animal. The head is naked, with only very short, sparse hair or none at all. This utter baldness is accentuated by an almost complete lack of fat under the skin, making the face incredibly lean and bony. In the adult male the jaw muscles become big and bulky, and can be seen clearly beneath the bare skin, bulging out on the top of the skull. The bare face pokes out from a mass of shaggy hair which begins behind the ears and on the neck and the back of the head, so the uakari looks rather like a bald monk with a cape.

There are two species of uakari, found around the Amazon on both sides of its upper course. On the north side, between the rivers Branco, and its tributary the Rio Negro, and Japurá, lives the black-headed uakari, which is chestnut-brown in colour with black hands and feet, and a naked black face. To the southwest, on the other side of the Rio Japurá, lives the bald uakari, which differs in its skull characteristics, and has a longer coat and a pink or red face. The face turns pale if the animal is kept from sunlight and becomes bright crimson if it is allowed to live in the full sun. The bald uakari is divided into two very distinct races: the white one, in which the coat is white or silvery, and the red one, with a coat that is red like the face. The first ranges from the Rio Japurá to the Rio Içá. The second extends south from the Içá to about 7° S, its range bounded on the west and east by the rivers Ucayali and Juruá.

Active in the treetops

Uakaris have rarely been observed in the wild. Their superficial thickness of body has led people to suspect they are clumsy and lethargic, but on seeing how thin a shaved uakari really is, it is not at all surprising to learn that they are in fact agile and active. In the wild they have been seen making leaps of 20 ft or so, launching themselves into the air with arms stretched forwards. On the ground they are somewhat ill at ease, walking with the hands partly flexed, and turned out sideways, but in captivity they invent games for themselves, sliding along the cage floor or turning back somersaults. When feeding they are very dexterous. They hold the food in the whole flexed hand, the thumb not being divergent, or between the index and middle fingers, or even

▷ *Sweet dreams for a relaxed bald uakari.*

between the hand and wrist. They have projecting lower incisors like their relatives the sakis, and these are probably used for spearing the fruits which form part of their diet along with buds, leaves and seeds.

Uakaris have been seen both in small troops and in large gatherings of about 100. They may go right up to the treetops but they come down to the lower branches when travelling through the forest.

Breeding is a closed book

It is a pity that these remarkable animals seem neither to have been bred in captivity, nor to have been observed in the wild to any great extent. So nothing is known of their breeding habits. Even the length of the gestation period is unknown, which is the one thing that is usually known about a mammal even when other details of its breeding remain obscure.

Sedation an old trick

The only thing we possess that approaches an adequate record of uakaris in their natural habitat was written over a century ago. Bates, in 1855, wrote that the bald uakari is captured alive by shooting it with arrows and blowpipes. The curare poison always used with these arrows was, however, diluted in order to capture the monkeys alive. The uakari, when shot, would run quite a long way, and it took a really expert hunter to track one and be underneath it when, weak with the poison, it fell from the branches. As soon as the animal fell a pinch of salt would be put into its mouth, which, so the story ran, acted as an antidote to the poison and revived the monkey. Animals caught in this way were kept as pets, often being traded far from their native haunts. They seem to have developed a great devotion to their owners and they were fairly easy to keep, although the initial death rate was high. Nowadays they are not too uncommon in zoos, but have not yet bred there. Some have lived as much as eight years in captivity. Their weird appearance, especially in the case of the bald species, and their off-beat antics, performed in silence, have made them popular zoo inmates from Bates's time to the present day.

Perhaps the most interesting feature in this story of how the South American Indians caught the uakari alive is that the method anticipated modern usage. Today we are familiar with the way animals that are marked for study purposes, or for transport to wildlife parks, are first immobilized with a drug in a dart shot from a crossbow. There is no substantial difference between this and the blow-pipe darts used long ago by the South American Indians.

class	**Mammalia**
order	**Primates**
family	**Cebidae**
genus & species	***Cacajao calvus*** *bald uakari* ***C. melanocephalus*** *black-headed uakari*

◁ *'Old monk' of the forest: the bald uakari seems to be scarlet with rage.*

△ *Cutting closeup: the bloodstained mouth and shear-like teeth of a common vampire.*

Vampire bat

The vampire bat of fact is totally unlike the vampire of fiction except that it feeds on blood. In one way it is worse than the fictional vampire—it is a carrier of rabies, a disease feared the world over. True vampires, of tropical and subtropical America, feed only on the fresh blood of mammals and birds. Unlike the man-sized vampires of fable, they are only 2½—3½ in. long with a forearm of 2—2½ in. The weight of an adult varies in the different species from ½ to 1¾ oz. Their fur is various shades of brown. They have no tail. The ears are small and the muzzle short and conical without a true noseleaf. Instead there are naked pads on the snout with U-shaped grooves at the tip, which may be sensory. The upper incisor teeth are large and razor-edged, well adapted for gently opening a small wound to take blood. The grooved, muscular tongue fits over a V-shaped notch in the lower lip, so forming a tube through which the blood is sucked. The stomach is also adapted for liquid feeding, the forward end being drawn out into a long tube. The saliva contains substances that prevent the blood from clotting.

There are three genera, each with a single species. The common vampire bat, the most numerous and widespread of the three, is distinguished by its pointed ears, longer thumb with a basal pad and its naked interfemoral membrane. It has only 20 teeth. It ranges from northern Mexico southward to central Chile, central Argentina and Uruguay. It is now one of the most common and widespread mammals in eastern Mexico.

The second species, the white-winged vampire, is much less numerous. The edges of its wings and part of the wing membrane are white. It has a peculiar short thumb about ⅛th as long as the third finger, and has a single pad underneath. It is the only bat known to have 22 permanent teeth. The white-winged vampire is mainly confined to the tropical regions of South America from Venezuela and the Guianas to Peru and Brazil, but it has also been found on Trinidad and in Mexico.

The hairy-legged vampire, smaller than the common species, is not well known. It has shorter, rounded ears, a short thumb without a basal pad and softer fur. Its interfemoral membrane is well-furred. It has 26 teeth and is unique among bats in having a fan-shaped, seven-lobed outer lower incisor tooth which resembles the lower incisor in the order Dermoptera, the gliding lemurs. This species is found in eastern and southern Mexico, Central America, and southwards to Brazil.

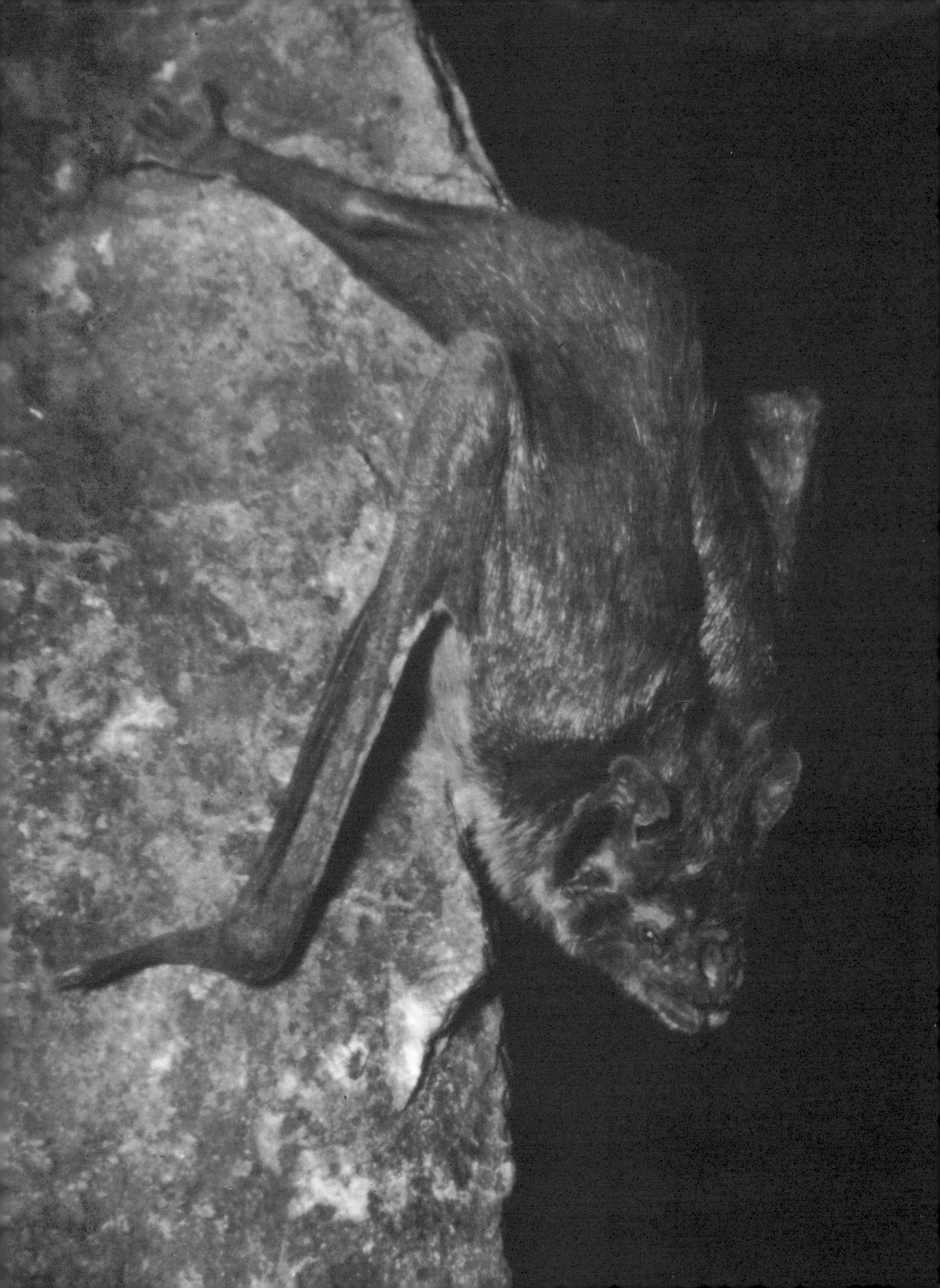

Victims attacked while asleep

During the day vampire bats roost in caves, old mines, hollow trees, crevices in rocks and in old buildings. Colonies of the common vampire may number as many as 2 000 but the average is about 100. The sexes roost together and they may share the caves with other species of bats. They are very agile and can walk rapidly on their feet and thumbs either on the ground or up the vertical sides of caves. Shortly after dark the bats leave their roosts with a slow noiseless flight, usually only 3 ft above the ground. The bats attack their victims while they sleep, sometimes alighting near them, crawling up to them, looking rather like large spiders. They make a quick shallow bite with their sharp teeth in a place where there is no hair or feathers. They cut away only a very small piece of skin, making a shallow wound from which they suck the blood without a sound, so the victim does not wake. Unlike other bats they do not cling with their claws but rest lightly on their thumbs and small foot pads, so lightly that even a man is unlikely to be wakened by the visit of a vampire. The common vampire bat in particular can drink such large quantities of blood that it is barely able to fly for some time afterwards.

The common vampire attacks only large mammals such as horses, cattle and occasionally man. Cattle are generally bitten on the neck or leg and a human on the big toe. The white-winged vampire, so far as is known, attacks only birds, biting the neck or ankle, and the hairy-legged vampire appears to prey mainly on birds such as chickens, but it is possible it may also attack some mammals.

In captivity vampire bats have been kept alive on blood defibrinated to prevent clotting. One survived for 13 years in a laboratory in Panama.

Echolocation in vampires

Like all bats, vampires find their way about and detect their prey by echolocation. Since their source of food is large and relatively stationary they do not have the same difficulty in finding their prey as bats that feed on fast-moving insects, or even those that catch fish. Like the fruit-eating bats, which also feed on stationary food, their echolocation is by pulses having only $\frac{1}{1000}$th of the sound energy of those used by bats feeding on insects or fish. It is noteworthy that vampires very seldom attack dogs, presumably because they have more sensitive hearing than larger mammals such as cattle and are able to detect the bat's higher sound frequencies.

Babies left at home

Little or nothing is known of the breeding habits of the white-winged and hairy-legged vampire bats. The common vampire gives birth to a single young after a gestation of 90–120 days. They breed throughout the year and it is possible there is more than one birth a year. The young are not carried about by the mother, as in most other bats, but are left in the roost while she is out foraging.

◁ *Disturbed while sleeping: common vampires prefer retreats of almost complete darkness.*

Desperate measures

The real danger of vampire bats lies not so much in their feeding on the blood of domestic animals and man, although this is bad enough, but in the transmission of disease resulting from the bites and risk of secondary infections. Vampires can transmit rabies which may be fatal in cattle or even in man. They may also transmit the disease to other species of bats and they may die of it themselves. In Mexico alone it is necessary to inoculate thousands of head of cattle each year against the disease. The disease is always fatal to uninoculated cattle.

Various control methods have been tried in the past, including dynamiting the caves where the bats roost and the use of flame-throwers and poison gas. These have been found to be largely ineffective and also highly destructive to other harmless species of bats. The only solution to the problem seems to lie in biological control, including sterilisation, habitat management and the use of selective chemical attractants and repellents. A research centre has now been set up in Mexico City for the ecological study of vampire bats and for research into biological methods of control.

class	**Mammalia**
order	**Chiroptera**
family	**Desmodontidae**
genera & species	***Desmodus rotundus*** *common vampire bat* ***Diaemus youngi*** *white-winged vampire bat* ***Diphylla ecaudata*** *hairy-legged vampire bat*

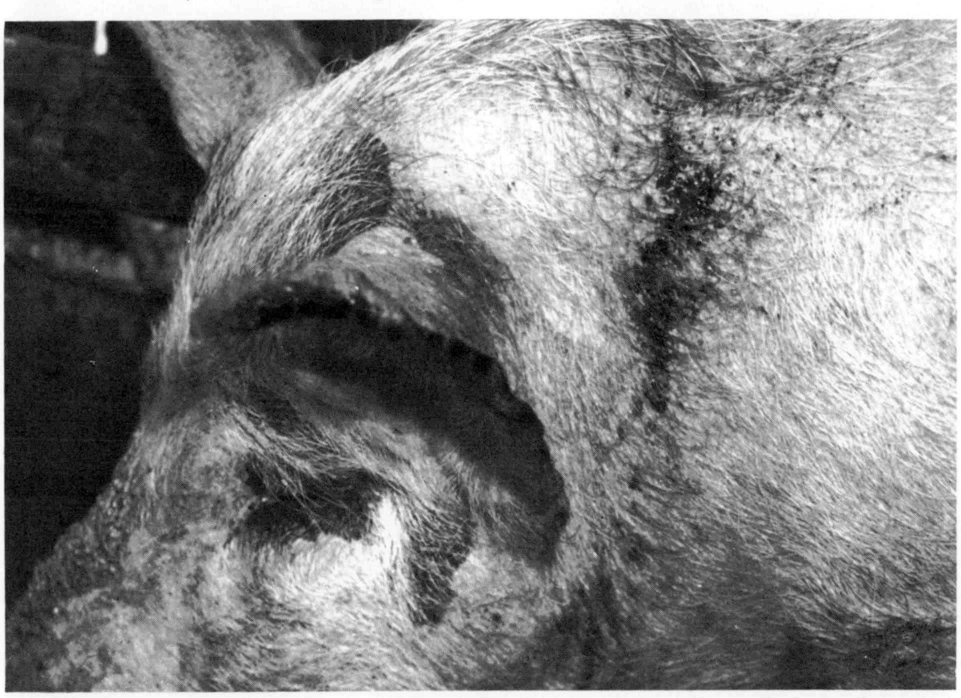

△ *Fresh bites on a pig's neck are grisly evidence of a common vampire's feeding methods. It only feeds on large animals, and because horses and cattle provide such a large supply of easily accessible blood, stockrearing in many tropical areas has proved uneconomical.* ▽ *The culprit on all fours.*

Wallaby

There are more than a score of wallabies, such as the pademelon, quokka and the rock wallaby. The way of life of the wallabies described here brings out some interesting aspects. In Australia itself the wallabies' story is largely one of persecution or extinction. There are also wallabies in New Guinea and one species of wallaby has become acclimatized to living wild in parts of Europe.

Most wallabies are the size of a hare or slightly larger, but the brush or scrub wallabies are up to 3 ft long in head and body, with a tail 2½ ft long and a weight of up to 50 lb. There are three species of hare wallaby, which are greyish brown with some red in places. The single species of banded hare wallaby is greyish with many dark bands across the back, from the nape to the base of the tail. The three species of nail-tailed wallabies are mainly grey but the eleven species of scrub wallaby are sandy to reddish brown. The five New Guinea forest wallabies are greyish brown to blackish brown.

The distinction between kangaroos and wallabies is hard to fix satisfactorily. In general, wallabies are small, looking like miniature kangaroos, but some are large and these are sometimes called 'kangaroos' locally.

◁ *Relief! A contented wallaby closes its eyes as it sits down to scratch itself.*

Hares and organ grinders

Hare wallabies are so named not only for their size but for their habit of lying in a 'form', a shallow trench scratched in the ground, under a bush or grass clump. They also run fast and will double back on their tracks like hares. They are solitary, nocturnal and make a whistling call when pursued. Nail-tailed wallabies have similar habits except that when bounding along they hold their small forelegs out to the sides and swing them with a rotary movement, which has earned them the nickname 'organ grinders'. The purpose of the spur or nail at the tip of the tail is not clear. The scrub wallabies spend much of the day in cover but often come out to feed—although they never stray far from cover. They may be solitary or live in pairs, or in large groups. The New Guinea forest wallabies live in the rain forest from sea level to 10 000 ft, but apart from this there is little information on them, and it is assumed that they live much as other wallabies do.

Generalised diets

All wallabies are vegetarian, most of them eating grass, although the nail-tails eat mainly the roots of coarse grasses and the scrub wallabies take succulent roots and eat leaves as well. Probably, as with other herbivores, the diets as given here are oversimplified. Some wallabies, normally grass eaters, have been seen to eat bark and fruits, as well as leaves. There is also a lack of detail on the life history except that they probably carry one young in the pouch at a time.

Unfortunate wallabies

Piecing together such information as is commonly available, it seems the wallabies dealt with here are among the 'unfortunates'. They are small enough to be at the mercy of introduced foxes and dogs; the fur of some of them is valued for export; their speed makes them targets for sport; and their flesh is agreeable as human food. The aborigines of Australia and New Guinea have hunted them for food, some species, such as hare wallabies, being persecuted more than others. The banded hare wallaby, now confined to a few islands in Sharks Bay, southwest Australia, was once numerous on the mainland. The aborigines used to burn the undergrowth to drive out the game, and this particular hare wallaby suffered badly as a result. The larger scrub wallabies have suffered from the general accusation that they are outgrazing the sheep and cattle. Recent research in Australia suggests this may have been exaggerated for all wallabies and kangaroos. Already it had been noted that the reverse was true for the banded hare wallaby—it was displaced by grazing sheep.

Some species of wallabies have become extinct, others are much reduced in numbers and range, and the only hope of ultimate survival for all seems to lie in creating suitable reserves and sanctuaries for them, and generally improving the climate of opinion in Australia.

◁ *Meal time. Incredibly this large young wallaby fits comfortably into mother's pouch.*

Sunning itself: a swamp wallaby ***Wallabia bicolor****. This position is adopted when the ground is cool.*

Fate of immigrants

Australia is not the only part of the world where wallabies have had their ups and downs. Bernard Grzimek has summarised, in his *Four-legged Australians,* what has happened to Bennett's wallabies brought over to Europe. In 1887 Baron Philipp von Böselager released two males and three females in 250 acres of forest near Heimerzheim, in western Germany. In spite of one hard winter they multiplied to 35–40 in six years, but were later exterminated by poachers. At the beginning of this century Count Witzleben successfully bred Bennett's wallaby on his estate near Frankfurt-on-Oder—then shot them all because, he said, they frightened the deer. Prince Gerhard Blücher von Wahlstatt released wallabies on the island of Herm, Channel Islands, where they did well until British troops occupied the island and 'the entire herd found its way into their cookhouse'. Today feral wallabies live on the moors in Derbyshire, England.

One danger peculiar to temperature climates comes from ice on lakes, which tends to break, with fatal consequences, under the strain of a wallaby's heavy, rhythmical leaps.

The wallabies imported into New Zealand have flourished exceedingly. Two does and a buck of the red-necked wallaby found the climate most favourable and their progeny became a nuisance. Between 1948 and 1965 control measures by government resulted in 70 000 being shot, in addition to many thousands shot or killed by poisoning.

class	**Mammalia**
order	**Marsupialia**
family	**Macropodidae**
genera & species	***Dorcopsis veterum*** *New Guinea forest wallaby* ***Lagorchestes leporoides*** *hare wallaby* ***Lagostrophus fasciatus*** *banded hare wallaby* ***Onychogalea frenata*** *nail-tail* ***Wallabia rufogrisea frutica*** *Bennett's wallaby* *others*

Walrus

Although hunted since the time of the Vikings, almost to the point of extinction, the walrus has survived and today, with strict conservation measures, some herds are very slowly recovering their numbers. The two subspecies, the Pacific walrus and the Atlantic walrus, differ in only minor details. The Pacific bulls average 11 – 11½ ft long and weigh a little over 2 000 lb but they can reach 13¾ ft and weigh up to 3 700 lb when carrying maximum blubber. The Atlantic bulls average 10 ft long and up to 1 650 lb in weight but may reach 12 ft and weigh 2 800 lb. The cows of both subspecies are smaller, 8½ – 9½ ft and 1 250 lb, but large Pacific cows may reach almost 12½ ft and a weight of 1 750 lb.

The walrus is heavily built, adult bulls carrying sometimes 900 lb of blubber in winter. The head and muzzle are broad and the neck short, the muzzle being deeper in the Pacific walrus. The cheek teeth are few and of simple structure but the upper canines are elongated to form large ivory tusks, which may reach 3 ft in length and are even longer in the Pacific subspecies. The nostrils in the Pacific subspecies are placed higher on the head. The moustachial bristles are very conspicuous, especially at the corners of the mouth where they may reach a length of 4 or 5 in. The foreflippers are strong and oar-like, being about a quarter the length of the body. The hindflippers are about 6 in. shorter, very broad, but with little real power in them.

The walrus's skin is tough, wrinkled and covered with short hair, reddish-brown or pink in bulls and brown in the cows. The hair becomes scanty after middle age and old males may be almost hairless, with their hide thrown into deep folds.

The Pacific walrus lives mainly in the waters adjacent to Alaska and the Chukchi Sea in the USSR. The Alaskan herds migrate south in the autumn into the Bering Sea and Bristol Bay to escape the encroaching Arctic ice, moving northwards again in spring when it breaks up.

The Atlantic walrus is sparsely distributed from northern Arctic Canada eastward to western Greenland, with small isolated groups on the east Greenland coast, Spitzbergen, Franz Josef Land and the Barents and Kara Seas. They migrate southward for the winter.

Walruses also inhabit the Laptev Sea near Russia and do not migrate in the winter. It is thought that this herd may be a race midway between the Atlantic and Pacific subspecies.

◁ *A long-in-the-tooth bull walrus of the Pacific subspecies. The elongated upper canine teeth are put to a variety of uses, among them defence and digging for clams.*

◁ *Playful pups: young walruses nuzzle each other with their sensitive moustaches.*

Tooth-walking bulls

Walruses associate in family herds of cows, calves and young bulls of up to 100 individuals. Except in the breeding season the adult bulls usually form separate herds. They live mainly in shallow coastal waters, sheltering on isolated rocky coasts and islands or congregating on ice floes. Since their persecution by man, however, walruses have learnt to avoid land as much as possible and to keep to the ice floes, sometimes far out to sea. They are normally timid but are readily aroused to belligerence in the face of danger. There seems to be intense devotion to the young, and the killing of a young one will rouse the mother to a fighting fury, quickly joined by the rest.

Walruses can move overland as fast as a man can run and because of their formidable tusks hunters, having roused a herd, have often been hard put to it to keep them at bay. Walruses have even been known to spear the sides of a boat with their tusks or to hook them over the gunwales.

As well as using them as weapons of offence and defence the walrus makes good use of its large tusks for digging food out of the mud and for keeping breathing holes open in the ice. It also uses them as grapnels for hauling itself out onto the ice, heaving up to bring the foreflippers onto the ice. The horny casing of bare hard skin on the palms of the flippers prevents the walrus from slipping. The walrus also uses its tusks for hauling itself along on the ice—indeed the family name Odobenidae means 'those that walk with their teeth'.

Walruses sunbathe and sleep packed close together on the ice floes with their tusks resting on each other's bodies. If the water is not too rough, adult walruses can also sleep vertically in the water by inflating the airsacs under their throats.

Monstrous swine

The walrus was associated in the Middle Ages with a variety of sea monsters. Named the whale-elephant in the 13th century it also became the model for the original sea-horse and sea-cow. In addition it was described as 'a monstrous swine . . . which by means of its teeth climbs to the top of cliffs as up a ladder and then rolls from the summit down into the sea again.'

Clam grubbers

The walrus's diet consists principally of clams, which it grubs out of the mud with its tusks, and sea snails. It will also take mussels and cockles. The snout bristles help in detecting the shellfish. Clams are swallowed whole and no shells have ever been found in the stomach of a walrus, although it is not known how they are disposed of. A walrus also swallows a quantity of pebbles and stones, possibly for helping to crush the food in its stomach. Walruses usually dive for their food in shallow water of about 180 ft or less but occasionally they go down to 300 ft. It is not known how they deal with pressure problems at this depth but it may be in the same way as seals.

Occasionally a walrus, usually an adult bull, will turn carnivorous and feed on whale carcases or it may kill small ringed or bearded seals. Having sampled flesh it may continue to eat it in preference to shellfish.

Hitch-hiking pup

Most matings take place from late April to early June and after a gestation of just over a year one pup is born, every alternate year. Birth takes place on an ice floe. The new-born pup is 4 ft long with a coat of short silver grey hair and weighs 100–150 lb. It is able to swim immediately, although not very expertly, and follows its mother in the water. After a week or two it can swim and dive well. Even so, it usually rides on its mother's back for some time after birth, gripping with its flippers. After a month or two the silver grey hair is replaced by a sparser dark brown coat of stiff hairs. The cow nurses the pup for 18 months to two years but they remain together for several months after weaning. The pups grow quickly, males becoming sexually mature at about 5–6 years, the females at about 4–5 years.

Killed in the rush

Killer whales and polar bears attack walruses but not often, the polar bear particularly being wary of attacking an adult bull even when he is ashore and therefore more vulnerable. Panic when killer whales are near may, however, cause high mortality. In 1936 a large herd was attacked by killer whales and driven ashore on St Lawrence Island. They hauled out onto the beach in such panic that they piled up on each other and 200 of them are said to have been smothered or crushed to death.

Slaughter by man

Walruses have been hunted by man from early times. The Eskimo and Chukchee have always depended on the annual kill to supply all their major needs, including meat, blubber, oil, clothing, boat coverings and sled harnesses. Even today they are largely dependent on it. The annual killings by the local people, however, had no very marked effect on the numbers of the herds. It was the coming of commercially-minded Europeans to the Arctic that started the real extermination. From the 15th century onwards they used the walrus's habit of hauling out on the beaches in massed herds to massacre large numbers in the space of a few hours. After 1861, when whales had become scarce, whalers from New England started harpooning walruses. Then they started using rifles and the Eskimos followed suit. More walruses could be killed but large numbers of carcases fell into the water and could not be recovered. An even greater wastage has been that caused by ivory hunters, who kill for the tusks and discard the rest of the carcase.

By the 1930's the world population of walruses had been reduced to less than 100 000 and strict conservation measures have now been enforced. The Pacific walrus now seems safe from extinction but the Atlantic walrus is still in danger.

class	**Mammalia**
order	**Pinnipedia**
family	**Odobenidae**
genus	***Odobenus rosmarus divergens*** *Pacific walrus* ***O. r. rosmarus*** *Atlantic walrus*

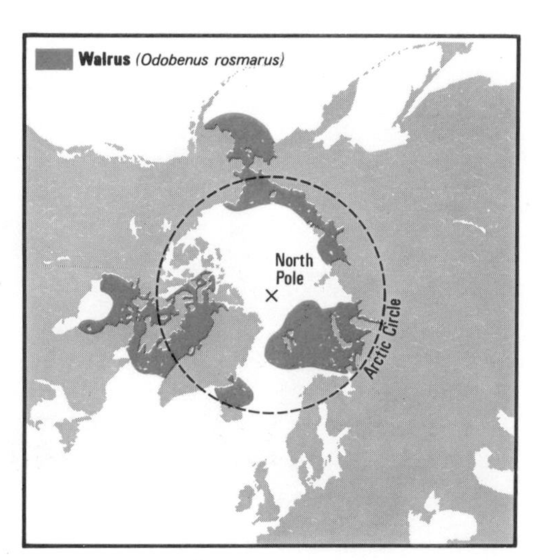

Wapiti

The wapiti of North America is often referred to as cousin to the red deer, although the closeness of the relationship remains open to question. In the 17th century the wapiti—the name given by the Algonquin Indians—was an abundant and widely distributed deer in North America. Its range fell just short of the Pacific coast in the west across almost to the Atlantic coast in the east, and from British Columbia in the north to New Mexico and Arizona in the south. Its numbers then were about 10 million. The total today is half a million.

The wapiti is larger than a red deer, up to 9 ft long in head and body, 5 ft high at the shoulder and weighing up to 1 000 lb, the hinds being smaller than the stags, as in red deer. It resembles the red deer in colour except that it is less reddish in summer and it has a more prominent light rump patch. Wapiti means 'white deer' and probably refers to this light patch. The antlers of red deer go up to 43 in.; those of wapiti may reach 66 in.

There are four forms of wapiti living in widely differing environments. The largest, living towards the Pacific coast, was named after Theodore Roosevelt in 1897. The smallest, the dwarf or Tule wapiti, lives on the hot, dry plains of southern California. Its coat is much paler than in the other three.

Incompatible neighbours

In spite of the tremendous reductions in its numbers, the wapiti gets in man's way. Most of the half-million survivors are in national parks or other wildlife refuges, mainly in the western states. Two small herds were introduced into the Virginian Jefferson National Forest in the eastern United States, and there have been reintroductions elsewhere. Where they invade arable or other settled land they tend to damage crops or compete with domestic livestock for browse and grazing. They also bark trees, especially the aspen, and wapiti introduced into Australia barked the pokaka. They present the same problem as the red deer in wildlife refuges in Britain; their natural ability to build up numbers under protection is apt to lead to the destruction of their habitat and the need to control them. Regrettably, the hand of man must always be to some extent against the deer in his vicinity.

Musical wapiti

The breeding habits of wapiti are very like those of red deer. There is, however, a marked difference in the stag's calls at the beginning of the rut. Instead of a roar, the wapiti stag makes an undulating bugling which starts in a low key, rises to a high pitch in a prolonged note which abruptly drops to a harsh scream, and ends in a few grunts. Otherwise the details are much the same, with the mature bulls rounding up the hinds into harems in September and October, to the accompaniment of a clatter of antlers and a clashing of foreheads from furious fighting, as the subordinate stags challenge the bulls. In May or June of the following year the hinds leave the herds and go into thickets to drop their dappled calves. There is usually one at a birth, rarely twins and exceptionally triplets, after a gestation of 249–262 days. The calf is up to 30 lb at birth, stands within minutes of being dropped and can run after a few hours. It starts to feed itself at three months, and is weaned and loses its spots in September or October.

Lion-hearted deer

The enemies of wapiti, a prey species itself, are much reduced in numbers at the hands of man. They include wolf, coyote, puma and bear, which prey especially on the calves. Adults can use speed to escape or can turn and defend themselves, striking down with the front hoofs. A big wapiti stag is credited with breaking the back of a wolf with one kick. One wapiti stag reduced a bear to a lacerated carcase with the tines of its antlers, then trampled it furiously until only a battered pulp remained. Such behaviour seems to be unusual until we recall that in the 18th century the Duke of Cumberland, so we are told, had a stag and a tiger brought together in an enclosure and that the stag made 'so bold and furious a defence, that the tiger was at last obliged to give up the contest'.

Dance of the deer

Contemporary accounts tell how the North American Indians sometimes hunted the wapiti in parties, forming a wide crescent around a stag, with the horns of the crescent half a mile apart. As they closed in the stag dashed first one way, then the other until, with the circle closed, the exhausted stag could be taken alive. Nevertheless, unless the animal was completely worn out, he would stand at bay, and usually one or more of the hunters was hurt before the stag was secured. Apparently this 'dance of the deer', as it was called, was for sport. The wapiti stag was also killed by Indians for his upper canine teeth, which were worn as charms. Surprisingly, the carcase was not used but left to rot. Later, some Americans of European origin formed a fraternal order or brotherhood named after this deer, not after its Indian name but after the one wrongly given it by early white settlers, which was 'elk'. Members of the order, it seems, used the canine teeth as emblems, for which a stag was killed, its canines extracted—and the carcase left to rot.

class	**Mammalia**
order	**Artiodactyla**
family	**Cervidae**
genus & species	***Cervus canadensis*** *wapiti*

▽ *Wildlife refuges, such as this one beneath the snow-covered Teton Range, harbour most of the American population of wapiti.*

△ *That lean and hungry look: timber wolf in alert pose. The wolf would seem, at first sight, the most probable ancestral dog, but modern opinion favours an extinct wild dog.*

Wolf

Thought by many to be at least one of the possible ancestors of the domestic dog, the wolf was once widespread over Europe, most of Asia and North America and had a range probably greater than any other land mammal. Today, there are two distinct species of wolf, the grey or timber wolf which is still found in the wilder parts of northern Europe, Asia and North America and the red wolf which is restricted to the south-central United States. There are numerous local races, however, differing in size and colour.

The grey wolf is the larger of the two, with a head and body length averaging 42–54 in. and a tail of 11½–22 in. The height at the shoulders is up to 38 in. The weight varies from about 60–150 lb. The grey wolf tends to be heavier in the more northern parts of its range—one shot in east-central Alaska weighed 175 lb. The red wolf is more slender, weighing on average 33 lb, occasionally 70–80 lb.

Both species are dog-like in appearance. They have large heads with erect rounded ears and long muzzles with strong jaws which contain 6–7 cheek teeth on each side, including a well-developed carnassial. The limbs are long and slender with four toes on the hindfoot and five on the forefoot, each bearing non-retractile claws. When angry, wolves erect the long hair on the nape of the neck, so that it looks like a mane. There is a scent gland on the upper side of the tail near its base which is used for recognition.

The colour of the grey wolf varies considerably over its range but it is usually grey, sprinkled with black apart from the legs and underparts which are yellowish-white. Black and light-coloured phases occur quite commonly. On the Arctic coast of Alaska and in western Canada, wolves are sometimes white throughout the year but more usually they are a mixture of white and grey tinged with brown. The red wolf is more tawny and sometimes small ones look like coyotes.

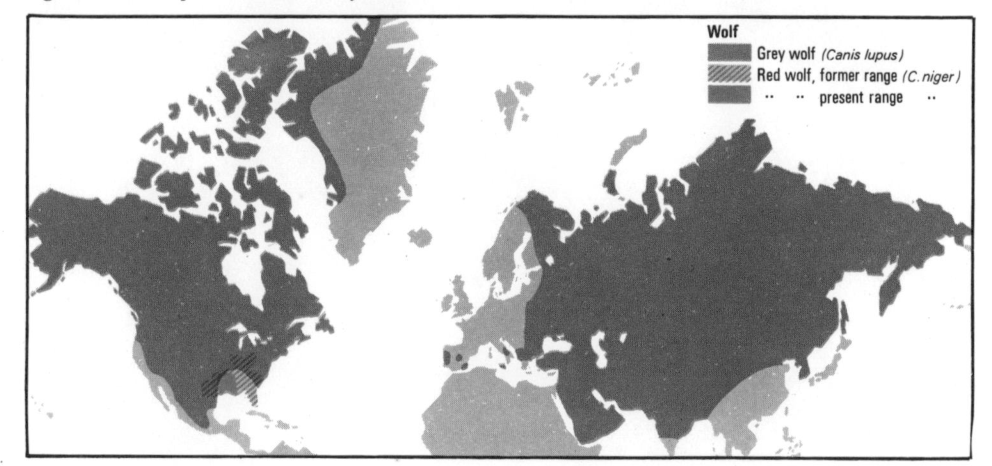

Savage and intelligent

The wolf is a ruthless and ferocious carnivore. It is also very courageous and has great fighting ability, intelligence and endurance. It lives in open country and forests, hunting by day and hiding at night among rocks, under fallen trees or in holes dug in the ground. Wolves sometimes hunt singly or in pairs but more often they move in a family party or pack of from three to two dozen individuals. A single large wolf can bring down and kill a large steer but a pack can tackle much larger animals such as moose or elk. Although a wolf runs at an average speed of only 22–24 mph, reaching a maximum of 28 mph for short distances, it has remarkable powers of endurance and can keep up a loping run for mile after mile, running right through the night if necessary. It has greater powers of endurance than most large game animals and so it can usually outrun its prey. It travels widely and there is a record of a red wolf in Oklahoma which covered 125 miles in two weeks, crossing four mountain ranges. The wolf is a good swimmer when necessary, sometimes pursuing deer into the water.

The wolves make use of pathways through the territory, which is usually in open country, often incorporating game trails and cattle tracks in these hunting routes which sometimes cover more than 100 miles and have numerous 'latrines' which also function as scent posts and vantage points on high ground for observation.

Large feeder

The wolf is an enormous feeder. It can eat ⅕ of its body weight at one meal and then go without food for a considerable time, rather like a snake. Although it will kill large animals such as caribou, musk-oxen, deer, moose and horses, much of its food is made up of small animals such as mice, rabbits and squirrels. Fish and crabs are sometimes taken, as well as carrion. When natural food is short the wolf will take to killing any domestic livestock and poultry within its range and will even resort to man-killing when driven to it by starvation.

A large family

The breeding season is usually from January to March with a gestation of 60–63 days. There are 5–14 cubs in a litter, usually 7, born in a den prepared by the female. At birth the cubs are blind with a sooty brown fur, except in the Arctic where the white colour phase predominates and the cubs' fur is light blue or dull slate. The eyes open 5–9 days after birth and at 18 months the cubs are well-grown. Both parents teach the cubs to hunt and kill prey and the family may keep together for some time. The males are fully mature in 3 years and females in 2. They are thought to mate for life.

Slaughtered by man

Over many centuries the wolf's chief enemy has been man. Constant efforts have been made to exterminate it because of its destructiveness to domestic stock and even to human life. Numerous methods have been used to kill it, including poison, steel traps, shooting and hunting by dogs and it says much for the wolf's cunning and endurance that it still survives over much of its former range. Indeed, it is only the encroachment of human settlement on its habitat that has been most effective in reducing its numbers.

The wolf in legend

From earliest times the wolf has been depicted in literature and legend as a symbol of savagery, courage and endurance. Beowulf, the legendary Teutonic hero, and many Anglo-Saxon kings and nobles incorporated 'wolf' into their names as an indication of their fighting prowess and in North America the Indians used the name for their most powerful warriors. Yet, strangely enough, there have been many stories, from way back in history, of wolves that have raised human children from infancy, lovingly looking after them and protecting them from other predators. The most famous story is that of Romulus, the legendary founder of Rome and his twin brother Remus, but in more recent times there is a story from India, purported to be true, of a child raised by a wolf until she was about 9 years old.

class	**Mammalia**
order	**Carnivora**
family	**Canidae**
genus & species	***Canis lupus*** *grey or timber wolf* ***C. niger*** *red wolf*

▽ *Wolves travel a great deal, and although a grey or timber wolf does not attain great speeds, its stamina enables it to keep a steady pace and journey for miles at a time.*

△ *Casual encounter: powerful enough to fear nothing in its Swedish National Park home, a wolverine pauses to size up the photographer.*

Wolverine

The wolverine, or glutton, is the largest of the weasel family. It has been aptly referred to by one writer as the super-weasel; yet its appearance is not that of a weasel, ermine or stoat, but rather of a bear or badger. A full-grown male may be up to 4 ft long, including nearly a foot of tail, stand 14–17 in. at the shoulder and weigh 30–60 lb. The females are smaller and lighter. The wolverine is powerfully built, thick-bodied with short legs set widely apart and ending in broad powerful paws armed with long sharp claws. It has a shaggy coat of thick dense fur, very dark brown above, with a pale brown band on the sides and dark brown below.

The wolverine ranges across the Arctic and sub-Arctic regions of Europe and Asia and in North America from the Arctic to the northern United States.

Ferocity and courage

Wolverines live in the cold evergreen forests from 800 to 13 000 ft above sea-level. They are solitary except during the breeding season, inhabiting very large but definite territories. They hunt mainly at night although they tend to have a 3- or 4-hourly rhythm of alternate activity and rest. They do not burrow or make any permanent home but use whatever shelter is available in the particular locality they are hunting. Where they use man-made shelters in this way they may tear down the timbers of a cabin to effect an entry and also wreck the contents, consuming what food they need and carrying other articles away. This has given them a false reputation for wanton destructiveness.

The wolverine cannot move with speed and, unlike the smaller members of its family, has little skill in stalking. At most, it may hide behind a rock in a kind of ambush or drop from a branch of a tree onto a victim's back. Mainly, it depends for survival on its unusual courage in driving other predators from their food. The wolverine bares its teeth, raises the hair on its back, erects its bushy tail and emits a low growl. Even bears have been driven from a carcase by such a display. It has remarkably strong teeth and jaws and is said to crack large bones to powder, to snap branches up to 2 in. diameter with ease and to have bitten a lump out of a rifle butt. The more reliable reports suggest, unless disturbed with young, that it is not aggressive towards man.

Trappers' tales

Among the exaggerated stories about the wolverine is its alleged high skill not only in avoiding the traps of the fur trapper but in robbing those in which marten and fox have been caught. When satiated it is said to exude its musk on the remaining carcases to prevent any other beast taking them. It has even been said that parent wolverines teach their young how to spring traps. The truth seems to be that it does sometimes rob traps and there may have been occasions when a whole line of traps has been cleared but, as a rule, the depredations are not as wholesale as most accounts suggest. As for skill in avoiding traps, this seems to be explained by the small size of the traps used. Wolverines are usually caught by the toes in marten traps and often escape leaving a toe or two behind. There seems to be no evidence that they can avoid, or escape, the larger traps set deliberately for them.

A libellous name

Its diet is a wide one. Mice, rats, small mammals of many kinds, eggs, ground-nesting birds, ducks and even snails are included. Above all, carrion, especially the kills of other carnivores, is eaten. It is reputed to be powerful enough to kill a reindeer or even a moose or elk and to drag a carcase three times its own weight for some distance over rough ground. Uneaten food is cached, either covered with soil or snow or wedged in the fork of a tree. A wolverine has a reputation for eating more than any other carnivore – hence its name of glutton – but probably the many stories about its excessive feeding habits are also exaggerations. Indeed, some of them take no account of the size of the animal's stomach.

Possible delayed implantation

The young, usually 2 or 3, occasionally 5, are born from February to May. The gestation period seems to be uncertain, the records varying from 60 – 120, or even 183 days, suggesting that delayed implantation occurs. The young are born in a hollow tree, among rocks or even in a snow drift. They have thick woolly fur at birth and are weaned at 8 – 10 weeks. They stay with their mother for as long as 2 years, then she drives them away to find their own territories and fend for themselves. They are sexually mature at 4 years of age and in captivity have been known to live for 16 years.

Persecuted by man

Being so powerful the wolverine has little to fear from natural enemies. It has, however, been persecuted by man for its destructiveness and also because of its reputation for killing reindeer. For over 100 years attempts have been made in Norway to stamp out the wolverine and premiums have been paid for each one destroyed. The eskimos hunt it for its fur as this does not hold moisture and then freeze, so it is invaluable for trimming the hoods of their parkas. Although its numbers have been reduced everywhere by persecution it is still not uncommon in parts of its range.

Exaggerated beliefs

One of the many stories of the wolverine concerns its stratagem for catching deer, or other large prey. The animal is said to climb into a tree carrying a quantity of moss in its mouth. When a deer approached the wolverine would let the moss fall. Should the deer stop to eat it the wolverine would then drop onto its back, fix itself firmly between the antlers and tear out its victim's eyes. Following this, either from pain or to rid itself of its tormentor, the deer would bang its head against a tree until it fell dead.

As alleged proof of their amazing strength we have the 18th century account from Churchill, on Hudson Bay, of some provisions hidden by several of the Hudson Bay Company's servants in the top of a wood-pile. On their return from Christmas festivities, the wood-pile, over 70 yd round, had been thrown down and scattered about. And this 'notwithstanding some of the trees with which it was constructed were as much as two men could carry.' The large quantity of provisions had been consumed, or carried away, except for the sacks of flour and cereals, which had been ripped to shreds.

class	**Mammalia**
order	**Carnivora**
family	**Mustelidae**
genus & species	***Gulo gulo*** *wolverine*

▽ *Profile of a powerful predator: a wolverine shows its strong head and claws as it looks out from a snowy vantage point.*

The soft-furred or hairy-nosed wombat.

Wombat

Like its nearest relative, the koala, the wombat looks like a bear, but it is more like a badger in its habits so it is often called 'badger' in Australia. Its head and body length varies from 27 to 47 in. and its weight from 33 to 80 lb. It is thickset with little or no tail, its legs are short and strong and its toes are armed with stout claws used in digging. The wombat's teeth are unlike those of other marsupials, being more similar to those of rodents. The 24 teeth are rootless and there are two incisors in both upper and lower jaws, like those of a beaver. There are traces of cheek pouches.

The two genera of wombats each contain a single species. The common or coarse-haired wombat lives in the hilly or mountainous coastal regions of south-eastern Australia and on Tasmania and Flinders Island in the Bass Strait. It is the larger of the two and has a naked muzzle, rounded ears and coarse fur ranging from a yellowish buff to dark brown or black.

The soft-furred or hairy-nosed wombat lives in the sandy or limestone coastal country and drier inland areas in the southern half of South Australia. It was once plentiful in the hilly parts of inland southern Queensland but it is probably nearing extinction in this area. It is distinguished from the common wombat by its smaller size and larger skull bones, and by having a haired nose and relatively pointed ears. The fur is soft and silky, the upperparts a grizzled-grey and the underparts white or grey.

Shy and nocturnal

Both species of wombats are nocturnal and shy and therefore difficult to observe in the wild. They sleep by day in burrows dug out with the powerful claws of their forefeet, the soil being thrust back with the hindfeet. The burrows are large, usually 10–15 ft long or as much as 100 ft at times, and one series of burrows is reported to have been ½ mile long and 60 yd across. There is a

△ *Greedy nibbler: the wombat tears up grass roots with its clawed forefeet to expose the tasty bases of grass stems and other roots.*

sleeping chamber at the end of the burrow which contains a nest of bark. The entrance to the burrow is large and arched and a few yards away there is usually a shallow depression scraped in the surface of the ground, at the base of a tree, where the wombat goes to bask in the morning sun. Wombats are solitary except in the breeding season and quite inoffensive to man unless interfered with, although a female has been reported to attack someone in defence of her young. Although heavily built they are quick in their movements and can run swiftly for short distances. The only sound heard from a wombat is a hoarse grunting cough, rather like that made by a large kangaroo.

Wombats are easily kept in captivity and make affectionate and amusing pets. One is recorded as living for 26 years in captivity.

Diet of grass

There are often well-defined paths leading from the wombats' burrows to feeding areas in open country. Their food is mainly grass and roots and occasionally the inner bark of certain trees and fungi. They tear out and grasp the grass stems with their forefeet and they sometimes damage pasture and crops near settled areas.

One young in the pouch

During May to July the female gives birth to one young which is carried at first in the pouch, in the usual way of marsupials. Later it runs free, but stays with the mother until the end of the year. During this time she feeds it on sword grass, pulling out the stems singly and dropping them on the ground so that the youngster can feed on the tender bases.

Man the chief enemy

The wombat has few natural enemies but it has suffered severely at the hands of man. In many areas it has disappeared entirely. Although its skin is not used commercially the aborigines make string out of the fur of the hairy-nosed wombat, coiling it round their hair. Wombats have been ruthlessly banished from settlements from early days because of their habit of tearing down fences to reach the grass in sheep pastures or cultivated crops of various kinds. Another similarity with badgers is that they have been killed for their hams. Their burrows have been eradicated because of the danger to horse riders and because rabbits sheltering in them could not easily be destroyed.

Resembling a badger

The first account of the common wombat in New South Wales was supplied by a former convict, James Wilson, in the 18th century, while on a journey into the southern highlands across the Nepean River. His account was written down by one of his companions, a young servant to Governor Hunter. When evidently near the present town of Bargo on January 26, 1798: 'We saw several sorts of dung of different animals, one of which Wilson called Whom-batt, which is an animal about 20 inches high, with short legs and a thick body with a large head, round ears, and very small eyes, is very fat, and has much the appearance of a badger.'

As has been said before in previous articles many marsupials resemble closely in appearance and habits well-known animals in other parts of the world. The wombat is no exception with its marked resemblance to the badger. Although the wombat shows some characters of the koala and possums it is so unlike them in other features, particularly in feeding, that it has been classed in a family on its own. In the past wombats have not excited much interest and have therefore received little attention but they are quite as interesting zoologically as the koala and other marsupials, and it would be a pity if their remaining numbers were allowed to diminish any further. In view of the very great similarity to the badgers of the northern hemisphere, it would be of unusual interest for a complete study to be made of the wombat.

class	**Mammalia**
order	**Marsupialia**
family	**Phascolomidae**
genera & species	***Lasiorhinus latifrons*** *soft-furred or hairy-nosed wombat* ***Phascolomis ursinus*** *common or coarse-haired wombat*

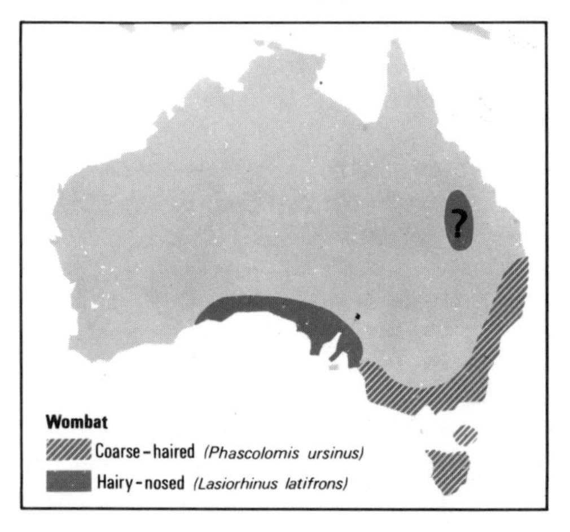

Woodchuck

The woodchuck, also known as a whistle-pig and popularly called a 'chuck', is a common rodent with a reputation for forecasting the weather in most of Canada and the eastern United States. Its body is thickset and 18–27 in. long. It has a 6in. hairy tail. Its weight varies from a little over 5 lb when the animal emerges from hibernation in the spring to 10 lb in September. The largest woodchucks may weigh as much as 14 lb at this time. The head is flattened with small ears, and the four toes have long claws used for digging. The colour of the fur varies from yellow to reddish-brown.

The woodchuck's range extends from Labrador and Nova Scotia, south to Virginia and Alabama, west to Kansas and north through Minnesota and Central Canada to the northern Rocky Mountains.

True hibernation

The woodchuck frequents woods and farmland and digs its burrow on a rocky hillside or in a gully, but preferably in bushy woods at the edge of meadowland. The burrow has several exits and the entrance has a large pile of freshly removed earth around it. The burrows vary in depth according to the soil; in soft earth a tunnel may be as much as 6½ ft below the surface. The burrow consists of several compartments for sleeping, hibernating and for toilets. The toilets are cleaned regularly and the waste taken up and buried in the entrance mound. The woodchuck is solitary and diurnal, feeding in the morning and again in late afternoon or evening. It never wanders far from its burrow, its home range being only about 100 yd. It can swim well and unlike most large rodents will climb for food.

In autumn the woodchuck hibernates, starting earlier in the northern parts of its range. It settles down in one of the chambers in its burrow, sealing it off with earth. It then rolls up in a ball and goes into a deep sleep. Its breathing slows down until it almost stops: 14 breaths per minute against the normal of 262. Its temperature gradually drops to 4–14°C/40–57°F, as against the normal of 37–40°C/94.8–104°F. When the woodchuck emerges in the spring it looks very thin and hungry and has lost as much as half its weight.

A pest to farm crops

The woodchuck is mainly vegetarian, feeding on grass, leaves and flowers, particularly clover, and sometimes acorns. It will eat bark from trees and also takes fruits such as blackberries, raspberries, cherries and windfall apples. It will occasionally eat snails, insects and small birds. In some farmland areas the woodchuck has become a pest, eating all kinds of farm produce and cereal crops. In addition to feeding voraciously it spoils the crops by trampling, and in some parts of the United States the woodchuck's numbers have had to be controlled by gassing it in its burrow.

▷ *Favourite stance for a curious woodchuck.*

Looking for a mate

As soon as the male woodchuck comes out of hibernation in the spring it goes to look for a mate. Fights often break out between two males and considerable damage may be inflicted before the weaker animal retreats, leaving the victor in possession of the female. Mating takes place in March and April and, after a gestation period of 28–32 days, 2–8, on average 4, babies are born in the burrow. The young 'chuck is pink, naked and blind, less than 4 in. long and weighing only 1–1½ oz. The female feeds her babies sitting on her haunches or standing on all fours. At the end of a month the youngsters' eyes open and they make their first trip out of the burrow. They are weaned in 35 days, and by midsummer have been driven out of the home burrow but continue to live in one nearby. The mother still watches over the youngsters and at the first sign of danger warns them with an alarm whistle. The young are sexually mature at one year.

Many enemies

The woodchuck has many natural enemies. It is preyed upon by bears, coyotes, wolves, mountain lions, as well as by eagles and hawks, and is often attacked by farm dogs. In addition, man hunts it for sport and for its flesh which, except when old, makes very good eating. The hunters, however, have to be very quick and sharp-eyed as 'chucks are very wary and bolt for cover at the slightest hint of danger.

Weather forecaster

The woodchuck is a legendary weather forecaster in America. It is said that if it emerges from hibernation on Candlemas Day, February 2nd, and sees its shadow, it returns to sleep for another six weeks and there will be another six weeks of winter. If, however, it does not see its shadow then winter is over and spring has come. According to Will Barker, in *Familiar Animals of America*, this legend dates back to early Colonial times. He states that in European folklore it was the badger that was supposed to look for its shadow on Candlemas Day, and early European settlers in America transferred the myth to the woodchuck. It is, however, hard to find confirmation of this belief regarding the badger.

class	**Mammalia**
order	**Rodentia**
family	**Sciuridae**
genus & species	***Marmota monax*** *woodchuck*

◁ *A young woodchuck on a cautious outing from the safety of mother's burrow. When the 4 young, born in spring, are a month old their eyes open and they take their first tentative steps outside. They are weaned at 35 days and are driven from the burrow by the mother within a month or two. The young woodchuck take up residence nearby where she can keep an eye on them and warn them when danger threatens.*

Zebra

Zebras are distinguished from horses and asses by the stripes on their bodies. Their mane is neat and upright. The tail is tufted as in asses, but the hard wart-like knobs known as 'chestnuts', are found on the forelegs only, and not on the hindlegs as in horses. There are differences from both the horse and the ass in the skull and teeth. Three species of zebra live in Africa today. The commonest and best-known is Burchell's zebra, which extends from Zululand in the southeast, and from Etosha Pan in southwest Africa, north as far as southern Somalia and southern Sudan. In this species the stripes reach under the belly, and on the flanks they broaden and bend backwards towards the rump, forming a Y-shaped 'saddle' pattern. Although the races in the southern and northern parts of the range look quite different, the differences are only clinal. That is, there are gradual changes from south to north, but they all belong to one species. In the southernmost race, the 'true' Burchell's zebra, now extinct but once living in the Orange Free State and neighbouring areas, the ground colour was yellowish rather than white; the legs were white and unstriped; the stripes often did not reach under the belly; and between the broad main stripes of the hindquarters and neck were lighter, smudge-grey alternating stripes commonly known as 'shadow-stripes'.

Further north a race known as Chapman's zebra is still found. It has a lighter ground colour than the true Burchell's, the stripes reach further down the legs—usually to below the knees—and the shadow-stripes are still present. All zebras still living from Zululand north to the Zambezi are referred to as members of this race; but at Etosha Pan there are some zebras that have almost no leg stripes and closely resemble 'true' Burchell's.

North of the Zambezi is the East African race, known as Grant's zebra. Its ground colour is white, the stripes continue all the way down to the hoofs and there are rarely any shadow-stripes. Grant's zebra is smaller than the southern races, about 50 in. high, weighs 500—600 lb, and has a smaller

mane. In the northern districts the mane has disappeared altogether. Maneless zebras occur in southern Sudan, the Karamoja district of Uganda and the Juba valley of Somalia.

South and southwest of the Burchell's zebras' range lives the mountain zebra, about the same size as Burchell's but with a prominent dewlap halfway between the jaw angle and the forelegs. Its stripes always stop short of the white belly. Its ground colour is whitish and, although the stripes on the flanks bend back to the rump, as in Burchell's, the vertical bands continue as well, giving a 'grid iron' effect. The southern race, the stockily built, broad-banded Cape mountain zebra, is nearly extinct, preserved only on a few private properties. The race in southwest Africa, Hartmann's zebra, is still fairly common. It is larger and longer-limbed than the Cape mountain zebra, with narrower stripes and a buff ground colour.

The third species is Grévy's zebra, from Somalia, eastern Ethiopia and northern Kenya, a very striking, tall zebra. The belly is white and unstriped, and there are no stripes on the hindquarters, except the dorsal stripe which bisects it. On the haunches the stripes from the flanks, rump and hindlegs seem to bend towards each other and join up.

▽ At the waterhole: a herd of southeast African Burchell's zebra. This race generally has striped legs and a paler ground colour than the now extinct true Burchell's zebra.

Belligerent stallions

Burchell's zebras are strongly gregarious. Groups of 1–6 mares with their foals keep together under the leadership of a stallion, who protects them and also wards off other stallions. Sometimes, for no apparent reason, the male simply disappears and another one takes his place. The surplus stallions live singly, or in bachelor groups of up to 15 members. Burchell's zebras are rather tame, not showing as much fear of man as the gnu with which they associate. When alarmed they utter their barking alarm call, a hoarse 'kwa-ha, kwa-ha', ending with a whinny. Then the herd wheels off, following the gnu. When cornered, however, the herd stallion puts up a stiff resistance, kicking and biting.

Mountain zebras, said to be more savage than Burchell's, live in herds of up to six, although sometimes they assemble in large numbers where food is plentiful. They seem to have regular paths over the rugged hills and move along them in single file. The call of the mountain zebra has been described as a low, snuffling whinny, quite different from that of the Burchell's.

Although in Grévy's zebra there are family groups as well as bachelor herds, the biggest and strongest stallions, weighing up to 1 000 lb, are solitary, each occupying a territory of about a mile in diameter.

Slow breeding rate

A newborn foal has brown stripes and is short-bodied and high-legged like the foal of a domestic horse. It is born after a gestation of 370 days. It weighs 66–77 lb and stands about 33 in. high. The mares come into season again a few days after foaling, but only 15% are fertilised a second time; usually a mare has one foal every three years. They reach sexual maturity at a little over 1 year, but do not seem to be fertile before about 2 years. Young males leave the herd between 1 and 3 years and join the bachelor herd. At 5 or 6 years many of them attempt to kidnap young females and if successful a new one-male herd is formed. The unsuccessful ones remain in the bachelor herd, or become solitary. Zebras live about as long as horses.

Lions beware

Man still hunts the zebra for meat but in protected areas, at least, very little of this continues. The zebra, with the gnu, is the lion's favourite prey. Because zebras are potentially dangerous, the lion must make a swift kill and young lions have been routed by zebra stallions that turned on them. Astley Maberley, the wildlife artist and writer, tells the story of an African poacher who was killed and fearfully mangled by an irate troop of Burchell's zebras after he had killed a foal.

(1) Linear line-up: a row of Grévy's zebra, large, handsomely-marked animals recognised by the huge ears and narrowly-spaced stripes.
(2) Topi antelope and Grant's zebra — a species having stripes that reach below the knees. The odd-looking animal on the left of the picture is a rare melanistic form of Grant's zebra.
(3) Grant's zebra spar in the dust of Ngorongoro crater. Zebra stallions are aggressive not only to males of the same species but also to predators, including man.

The lost quagga

A fourth species of zebra, the quagga, was extremely common in South Africa 150 years ago. It has since been completely exterminated. Most closely resembling Burchell's zebra, the quagga was distinctly striped brown and off-white on the head and neck only. Along the flanks the stripes gradually faded out to a plain brown, sometimes extending to just behind the shoulders, sometimes reaching the haunches. The legs and belly were white. Its barking, high-pitched cry, after which it was named, was rather like that of the Burchell's zebra.

The early explorers, around 1750–1800, met quaggas as far southwest as the Swellendam and Ceres districts, a short way inland from Cape Town. The Boer farmers did not appreciate quaggas except as food for their Hottentot servants. Their method of hunting was to take a train of wagons out onto the veldt and blaze away at everything within sight. Then large numbers of carcases would be loaded onto the wagons, and the rest of the dead and dying animals were simply left to rot. It is no wonder that today Cape Province is virtually denuded of wild game. When Cape Province was emptied, the trekkers to the Orange Free State repeated the process there. By 1820 the quaggas' range was already severely curtailed; they were almost gone even from the broad plains of the Great Fish River, which had been named 'Quagga's Flats' from the vast numbers of them roaming there. A few lingered for another 20 years or so in the far east of Cape Province and in the Orange Free State, the last wild ones being shot near Aberdeen, CP, in 1858, and near Kingwilliamstown in 1861. Strange to say, no one realised that they were even endangered. Zoos looking for replacements for their quaggas that had died were quite shocked to be told, 'But there aren't any more'.

class	**Mammalia**
order	**Perissodactyla**
family	**Equidae**
genus & species	***Equus burchelli burchelli*** *true Burchell's zebra or bontequagga* ***E. b. antiquorum*** *Chapman's or southeast African Burchell's* ***E. b. boehmi*** *Grant's or East African Burchell's* ***E. b. borensis*** *maneless zebra* ***E. grevyi*** *Grévy's zebra* ***E. quagga*** *quagga* ***E. zebra zebra*** *Cape mountain zebra* ***E. z. hartmannae*** *Hartmann's mountain zebra*

(4) '. . . along a mountain track' – Hartmann's mountain zebra, a race of the Cape mountain zebra described in 1898, has a large dewlap between chin and forelegs and stripes that end short of the belly. The stripes form a 'grid-iron' effect on the rump.
(5) Nearly extinct: less than 200 Cape mountain zebra live in specially protected areas of high tableland in western Cape province.
(6) Extinct: the quagga was hunted in large numbers by early white settlers to South Africa.

Index

Page numbers in *italics* refer to illustrations.

Acknowledgments

This book is adapted from 'Purnell's Encyclopedia of Animal Life', published in the United States under the title of 'International Wild Life'.

Photo sources: AFA (A. Boxall, Geoffrey Kinns); Alan Band Associates; Toni Angermayer; Atlas (Drogesco); Australian News & Information; William W. Bacon III; Albert Barber; Barnaby's Picture Library; Bavaria (Helmut Heinpel, Walter Schmidt, H. W. Silvester); Carlo Bevilacqua; Ron Boardman; K. Boldt, J. Breeds; British Antarctic Survey; A. Brown; Fred Bruemmer; Kent Burgess; Robert Burton; Carolina Biological Supply Co.; J. Allan Cash; Lynwood M. Chace; Chicago Zoological Society; Arthur Christiansen; F. Collet; Gerald Cubitt; Mary Evans; M. Fogden; Forestry Commission, London; Forestry Fish & Game Commission, USA; Harry & Claudy Frauca; W. Goodpaster; E. O. Happ; Bruce Hayward; Tierpark Hellabrunn; Robert C. Hermes; Peter Hill; Eric Hosking; Jacana (A. R. Devez, A. R. Dover, Gerard, A. Kerneis, B. Lomstrand, Jean Philippe Varin, P. & C. Vasselet, Jacques Vielliard, Albert Visage); Roy Jarris; Peter Johnson; Keystone; Jurg Klages; H. Klinget; Yves Lanceau; Leonard Lee Rue III; HAE Lucas; Wolfgang Lummer; Michael Lyster; Malcolm McGregor; Steve McGutcheon; Mansell Collection; Marineland, Florida, USA; John Markham; Walter Marsden; Meston; Walter Miles; Wilfrid Miller; Mondadori Press; M. Morcombe; P. Morris; G. Mundey; NHPA (Andrew M. Anderson, Anthony Bannister, Stephen Dalton, Graham Pizzey); Les Noke; Okapia (A. Root); Klaus Paysan; Photographic Library of Australia; Photo Library Inc; Photo Res (Australian News & Information Bureau, Des and Jen Bartlett, Jane Burton, Rod Allen, C. Guggisberg, Sven Gillsater, Peter Jackson, Russ Kinne, L. Lee Rue III, Tom McHugh, N. Myers, C. J. Ott, Ed Park, D. C. Pike, Graham Pizzey, Masood Quraishy, Vincent Serventy, H. W. Silvester, J. Simon, Dick Robinson, Simon Trevor, H. E. Uible, Joe Van Wormer); Graham Pizzey; Popperfoto; Michael Price; Masood Quraishy; Cecil E. Rhode; Roebild; G. Ruppell; San Diego Zoo; Phillipa Scott; Fritz Siedel; E. Slater; South African Tourist Corporation; K. H. Stanley; A. J. Sutcliffe; John Tashjian; Edwin W. Teale; Ronald Thompson; Sally Anne Thompson; G. Thomson; William Vandvert; Jean-Philippe Varin; R. W. Vaughan; R. S. Virdee; Walt Disney Productions Ltd; John Warham; A. N. Warren; We-Ha; Birgit Webb; H. Weghofer; WLF (N. Myers); Joseph Van Wormer; Margaret Wunsch; WWF (C. Baucher, Fraser-Brunner, E. P. Gee, W. Myers, Peking Zoo, K. Stennuher); Zoological Society, London.

PDO 78/381 1:4